Roadside Geology of NEW MEXICO

Second Edition

Magdalena Sandoval Donahue
and Lucy Chronic

Illustrated by Chelsea M. Feeney

2024
Mountain Press Publishing Company
Missoula, Montana

ROADSIDE GEOLOGY

Roadside Geology is a registered trademark
of Mountain Press Publishing Company.

© 2024 by Magdalena Donahue and Lucy Chronic
First Printing, April 2024
All rights reserved

Geologic maps and figures were created by Chelsea Feeney (www.cmcfeeney.com).
Photographs by authors unless otherwise credited.

BACK COVER IMAGE: *Chimney Rock, composed of Entrada Sandstone, towers to the northwest of the road into Ghost Ranch.*

Library of Congress Cataloging-in-Publication Data

Names: Donahue, Magdalena Sandoval, 1982- author. | Chronic, Lucy M., author. | Feeney, Chelsea McRaven, 1980- illustrator.
Title: Roadside geology of New Mexico / Magdalena Sandoval Donahue and Lucy
   Chronic ; illustrated by Chelsea M. Feeney.
Other titles: Roadside geology series.
Description: Missoula, Montana : Mountain Press Publishing Company, 2024. |
   Series: Roadside geology series | Includes bibliographical references
   and index.
Identifiers: LCCN 2024001950 | ISBN 9780878427178 (paperback)
Subjects: LCSH: Geology—New Mexico—Guidebooks. | Roads—New Mexico. | New
   Mexico—Guidebooks. | LCGFT: Guidebooks.
Classification: LCC QE143 .D66 2024 | DDC 557.89—dc23/eng/20240124
LC record available at https://lccn.loc.gov/2024001950

PRINTED IN THE USA

P.O. Box 2399 • Missoula, MT 59806 • 406-728-1900
800-234-5308 • info@mtnpress.com
www.mountain-press.com

This book is written for the people of New Mexico. They are as beautiful and enduring as the state's landscapes. New Mexico has nurtured me as a scientist and as an author from childhood. I hope this book finds the curious minds of residents and bright eyes of travelers passing through, and that it offers insight into the countryside around you and sparks future geologic inquiry.

Thank you to my husband. This book is a chronicle of our adventures together. John, you have been my partner in exploration and geology since we were children. Thank you for listening to my geologic musings and appreciating my hand-waving at far-off mountain ranges. Also, thank you to my children, whose playful exploration and constant query for information have given me new and exciting perspectives into the world around me. You three, my little companions in this world, have taught me how to explain and appreciate geology far better than I could before you walked alongside me.

—MAGDALENA SANDOVAL DONAHUE

To my wonderfully creative daughters, Betsy and Haley; my sweet husband, Chris; and eager dog, Rosco.

To friends Robin and Barbara, who kept me laughing on trips around New Mexico.

And to my amazing sister, Felicie, and mother, Halka, two fantastic geologists whose words and ideas grace these pages. I miss them every single day.

—LUCY CHRONIC

Rosco, Chris, and Haley at White Sands National Park

CENOZOIC
QUATERNARY

Qa	alluvial deposits
Ql	landslide and rockfall deposits
Qpl	playa lake deposits
Qp	piedmont alluvial deposits (early Pleistocene to Holocene)
Qw	windblown deposits; includes gypsiferous deposits (middle Pleistocene to Holocene)
Qwp	windblown and piedmont deposits (middle Pleistocene to Holocene)
Qd	glacial deposits (Pleistocene)
Qoa	older alluvial deposits of upland plains and piedmont areas (latest Pliocene to late Quaternary)

QUATERNARY–PALEOGENE

QTt	travertine (Pliocene to Holocene)
QTp	older piedmont alluvial deposits and shallow basin fill (late Pliocene to middle Pleistocene)
QTsf	Santa Fe Group, undivided (latest Oligocene to middle Pleistocene)
QTg	Gila Group (latest Oligocene to middle Pleistocene)

NEOGENE and PALEOGENE

Tus	younger sedimentary units (late Miocene to Pliocene)
To	Ogallala Formation (middle Miocene to early Pliocene)
Tfl	Fence Lake Formation (Miocene)
Tlp	Los Pinos Formation (Miocene and late Oligocene)
Tc	Chuska Sandstone (middle to late Oligocene)
Tps	older sedimentary units (Paleogene)
Tsj	San Jose Formation (Eocene)
Tn	Nacimiento Formation (Paleocene)
Toa	Ojo Alamo Formation (Paleocene)

PALEOCENE–LATE CRETACEOUS

- **TKpc** Poison Canyon Formation
- **TKr** Raton Formation
- **TKpr** Poison Canyon and Raton Formations, undivided
- **TKa** Animas Formation

MESOZOIC
CRETACEOUS

- **K** sedimentary rocks, undivided
- **Kmc** McRae Formation
- **Kvt** Vermejo Formation and Trinidad Sandstone
- **Kkf** Kirtland and Fruitland Formations
- **Kpcl** Pictured Cliffs Sandstone and Lewis Shale
- **Kpn** Pierre Shale and Niobrara Formation
- **Kmv** Mesaverde Group
- **Ktma** Tres Hermanos Formation, Morena Hill Formation, and Atarque Sandstone
- **Kcc** Crevasse Canyon Formation
- **Kg** Gallup Sandstone
- **Km** Mancos Shale
- **Kd** Dakota Sandstone
- **Kdg** Dakota Group
- **Kgc** Graneros Shale, Greenhorn Formation, and Carlile Shale
- **Kmb** Mancos Shale and Beartooth Quartzite

JURASSIC

- **J** sedimentary rocks, undivided
- **Jm** Morrison Formation
- **Jze** Entrada and Zuni Sandstones
- **Jsr** San Rafael Group

TRIASSIC

- **Tr** sedimentary rocks, undivided
- **Trc** sedimentary rocks; includes Chinle Group, Garita Creek, and Moenkopi Formation
- **Trr** Redonda Formation
- **Trb** Bull Canyon Formation
- **Trt** Trujillo Formation
- **Trs** Santa Rosa Formation; includes Anton Chico Formation

VOLCANIC AND IGNEOUS ROCKS

- **Qb** basalt (Quaternary)
- **Qv** basalt and lavas near vents (Pleistocene)
- **Qbo** basalt (Pleistocene)
- **Qr** ring fracture rhyolite lava domes of the Valles Caldera (early to latest Pleistocene)
- **Qbt** Bandelier Tuff (early Pleistocene)
- **QTb** basaltic to andesitic lava flows (late Pliocene to early Pleistocene)
- **Tb** basalt (Pliocene and Miocene)
- **Tnv** volcanic rocks (Neogene)
- **Thb** Hinsdale Basalt (Miocene and late Oligocene)
- **Tua** andesites of the Mogollon Group (late Oligocene)

PALEOZOIC

- **Pz** sedimentary rocks, undivided (Paleozoic)

PERMIAN

- **P** sedimentary rocks, undivided
- **Pqr** Quartermaster and Rustler Formations
- **Psl** Salado Formation
- **Pc** Castile Formation
- **Pat** Artesia Group; includes Tansill, Yates, Seven Rivers, Queen and Grayburg Formations
- **Pcbc** Capitan, Bell Canyon, and Cherry Canyon Formations
- **Psag** San Andres Formation and Glorieta Sandstone
- **Pcov** Cutoff Shale and Victorio Peak Limestone
- **Pyags** Yeso, Abo, Glorieta and San Andres Formations
- **Pct** Cutler Formation
- **Ph** Hueco Formation
- **Pb** Bursum Formation

PERMIAN–PENNSYLVANIAN

- **PIP** sedimentary rocks, undivided
- **PIPsc** Sangre de Cristo Formation

PENNSYLVANIAN

- **IP** sedimentary rocks, undivided
- **IPm** Madera Group
- **IPs** Sandia Formation
- **IPpsl** Panther Seep and Lead Camp Formations

MISSISSIPPIAN–CAMBRIAN

- **M** sedimentary rocks, undivided (Mississippian)
- **MD** sedimentary rocks (Devonian to Mississippian)
- **SOC** sedimentary rocks (Cambrian to Silurian)

PROTEROZOIC

- **YXp** plutonic rocks, undifferentiated; includes dikes (Early and Middle Proterozoic)
- **YXm** metamorphic rocks, undifferentiated (Early and Middle Proterozoic)
- **Xq** quartzite; includes Ortega Quartzite (Early Proterozoic)
- **Xvm** metavolcanic rocks (Early Proterozoic)

- **Tvs** volcaniclastic sedimentary rocks (late Eocene to Oligocene)
- **Trf** rhyolitic lavas and local tuffs (late Oligocene)
- **Turp** pyroclastic rocks and ash-flow tuffs of the Mogollon Group (late Oligocene)
- **Tlrp** pyroclastic rocks and ash-flow tuffs of the Datil Group (late Oligocene)
- **Tla** lava rocks and pyroclastic breccias (Eocene)
- **Tv** volcanic rocks, undifferentiated (late Eocene to early Miocene)
- **Ti** intrusive rocks, undifferentiated; includes dikes (late Eocene to Pliocene)
- **TKav** lava rocks and pyroclastic breccias (Paleocene and Late Cretaceous)
- **TKi** intrusive rocks (Paleocene and Late Cretaceous)
- **OCp** plutonic rocks of Florida Mountains (Ordovician and Cambrian)

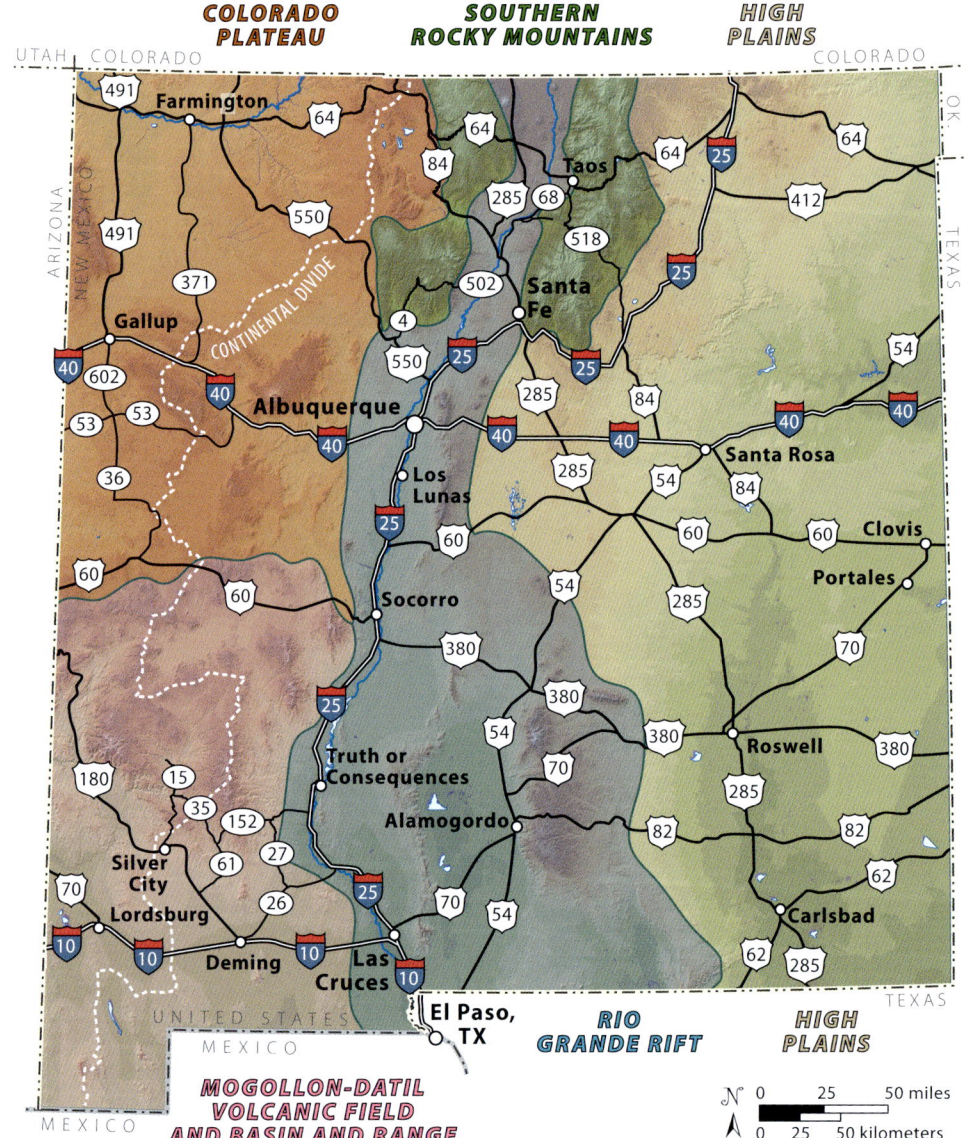

Roads and sections of Roadside Geology of New Mexico.

CONTENTS

Acknowledgments xi
Note to the Readers xi

Geologic History of New Mexico 1
    Types of Rocks 4
    Precambrian Era 7
    Paleozoic Era 11
    Mesozoic Era 15
    Cenozoic Era 18
    The Modern Landscape 27

The Colorado Plateau 31
    I-40: Arizona—Gallop—Grants 35
    US 64: Chama—Farmington—Arizona 39
    US 491: Gallup—Colorado 47
    US 550: Colorado—Cuba—San Ysidro 54
        Aztec Ruins National Monument 56
        Chaco Cultural National Historic Park 59
        Gypsum of White Mesa 63
    NM 602 and NM 53: Gallup—Grants 64
        El Morro National Monument 66
        El Malpais National Monument 69

Rio Grande Rift 73
  Northern Rift and Southern Rocky Mountains 77
    I-25: Las Vegas—Santa Fe 77
        Pecos Canyon State Park 81
    US 64: Raton—Ute Park—Taos 82
    US 64: Taos—Chama 90
    US 285: Santa Fe—Colorado 98
    NM 68: Española—Taos 103
    High Road to Taos Scenic Byway: Pojoaque—Ranchos de Taos 108
    NM 522 and NM 38 (Enchanted Circle Scenic Byway):
        Taos—Questa—Eagle Nest 114
        Rio Grande del Norte National Monument 117

JEMEZ MOUNTAINS   121
   NM 502 and NM 4: Pojoaque—White Rock—San Ysidro   124
      Diablo Canyon Recreation Area   127
      Bandelier National Monument   129
   Bandelier to San Ysidro   131
      Fenton Lake State Park   34
   NM 96: Cuba—Abiquiu   138

Southern Rift   145
   I-25: Santa Fe—Albuquerque   145
      Kasha-Katuwe Tent Rocks National Monument   147
   Albuquerque   150
      Sandia Mountains and the Tramway   153
   NM 14 (Turquoise Trail National Scenic Byway): Tijeras—I-25 near Santa Fe   155
      Cerrillos Hills State Park   158
   I-25: Albuquerque—Socorro   160
   I-25: Socorro—Truth or Consequences   167
   I-25: Truth or Consequences—Las Cruces   171
      Organ Mountains–Desert Peaks National Monument   176
   Las Cruces to Texas on I-10   177
   I-40: Grants—Albuquerque   178
      Petroglyph National Monument and the Albuquerque Volcanoes   182
   I-40: Albuquerque—Clines Corners   183
   US 54: Texas—Carrizozo   189
   US 60: Bernardo—Mountainair—Encino   194
      Salinas Pueblo Missions National Monument   197
      Manzano and Quarai Ruins   199
   US 70: Las Cruces—Alamogordo   200
      Aguirre Spring Recreation Area   204
      White Sands National Park   206
   US 380: Carrizozo—San Antonio (I-25)   209
      Trinity Site National Historic Landmark   213

The Peaceful Plains   215
   I-25: Colorado—Raton—Las Vegas   220
      Sugarite Canyon State Park   223
      Canadian River Canyon in Sabinoso Wilderness   226
   I-40: Clines Corners—Santa Rosa   229
   I-40: Santa Rosa—Tucumcari—Texas   233

US 54 Carrizozo—Santa Rosa   238
US 64: Oklahoma—Clayton—Raton   242
   Clayton Lake State Park and Dinosaur Trackways   244
   Capulin Volcano National Monument   247
US 70: Clovis—Roswell   248
US 82: Alamogordo—Artesia   251
   Sunspot Scenic Byway   256
US 285: Santa Fe—Encino   260
US 285: Vaughn—Roswell—Carlsbad   263
   Carlsbad Caverns National Park   268
US 380: Texas—Roswell   272
   Bottomless Lakes State Park   275
US 380: Roswell—Carrizozo   276

The Volcanic Southwest   284
   I-10: Arizona—Lordsburg—Deming   287
   I-10: Deming—Las Cruces   292
      Rockhound State Park and Spring Canyon Recreation Area   294
   US 60: Socorro—Datil   297
   US 60: Datil—Arizona   302
   US 180: Deming—Silver City—Arizona   306
      City of Rocks State Park   308
      Gila Cliff Dwellings National Monument   310
      The Catwalk   314

Glossary   317

References   328

Index   341

Pool at El Morro National Monument.

Acknowledgments

We wish to thank those who have helped us prepare this book. First is the original author, Halka Chronic, who, with Lucy as her field assistant, drove many miles for research, drafted maps and figures, and typed on what would now be considered an antique computer. We want to thank Felicie Williams, who helped with the book proposal and many of the figures. We would like to thank Magdalena's husband and three children, who accompanied her on many road trips during the updating of this text. We want to thank all geologists whose combined effort has helped us understand New Mexico's geology. We would especially like to thank the people who thoughtfully reviewed the manuscript: Diane Agnew, Lewis Land, Marisa Repasch, Jason Ricketts, and Cal Ruleman. Finally, we would like to thank the staff at Mountain Press, particularly our editor, Jennifer Carey.

Note to the Readers

The first chapter summarizes the geologic history of New Mexico, beginning with the oldest rocks and ending with the modern erosional setting. This chapter gives context before you head out on the roads. The rest of the book is divided into four geographic regions: the Colorado Plateau, the Rio Grande rift and the Rocky Mountains, the volcanic fields of southwestern New Mexico, and the Great Plains of eastern New Mexico. Each section starts with an introduction to the geology of the area, which we recommend reading before traveling and referring back to as you travel. A geologic map accompanies each road guide.

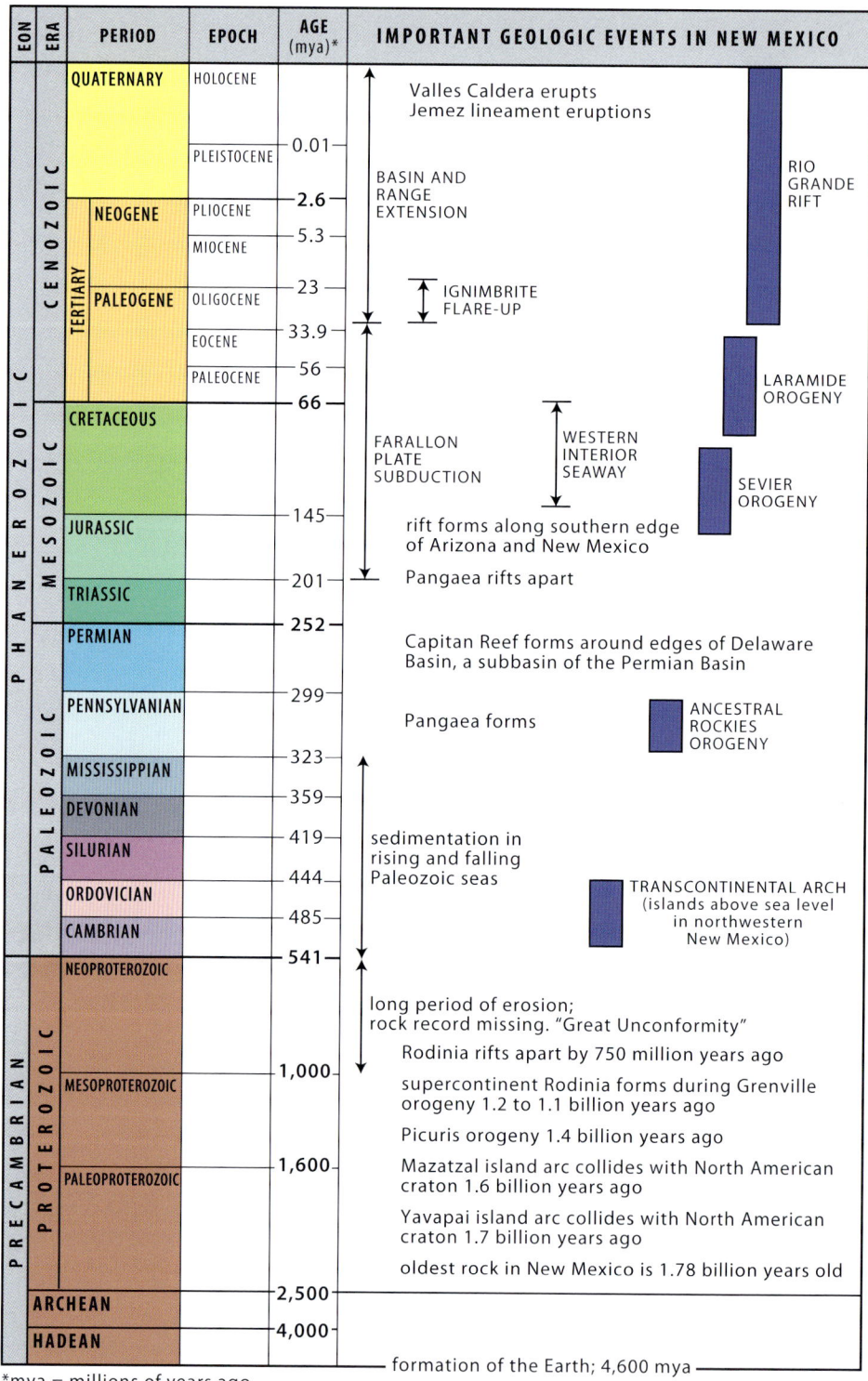

Geologic time scale and important events in the geologic history of New Mexico. —Dates used are from the Geologic Time Scale Version 6.0, published by the Geological Society of America in 2022

Geologic History of New Mexico

New Mexico's geologic history extends nearly 2 billion years and includes continental collisions, rising and falling oceans, and mountains growing and eroding away. Volcanoes erupted, precious minerals were deposited, and caverns formed. These dynamic geologic processes continue to this day. For example, the Rio Grande rift—a 600-mile-long tear in Earth's crust—formed as the North American plate stretched apart, causing the land surface along the centerline to drop in elevation over time. The central sag now houses the fourth-longest river in North America. The volcanic outpouring of two supervolcanoes that erupted 1.6 and 1.2 million years ago in Pleistocene time formed the Jemez Mountains. The complex geologic history of these caldera eruptions is vividly exposed in the massive, vertically walled canyons that radiate from the collapsed caldera center. The second-largest known magma body on Earth lies 12 miles beneath Socorro, forming a pancake-shaped blob of molten rock estimated to be about 130 cubic miles in volume. Occasionally, small earthquakes can be felt, though most of those that arise from the slowly inflating magma body are too small to feel and are detected only by seismometers.

For more than a century, field geologists have made observations and gathered data about New Mexico's rocks and landscapes. Our understanding of the Earth has increased and become more nuanced as we've gained information through remote sensing, computational modeling, geophysics, seismology, geochemistry, and other analytical methods. For example, geologists now isolate grains of zircon (a durable mineral that forms in igneous rock) from much younger sedimentary rock and use radiometric dating to determine the age of the zircon. Zircon contains the radioactive element uranium, which decays to lead, so it can be dated through the measurement of uranium and lead isotopes. If the zircon is 1.43 billion years old, then the scientists know the sediment containing the detrital zircon eroded from an area of 1.43-billion-year-old igneous rock. Geologists use this information to reconstruct ancient drainages and estimate when a particular-age rock was high in the landscape and eroding. Zircons in river sediments tell us from where the river flowed. Many of the sandstones in New Mexico have been studied for their detrital zircon chronology, supporting our understanding of New Mexico's geologic history.

To frame the geologic history of New Mexico, however, we must begin with a brief overview of how the Earth was formed. Our universe was born 13.8 billion years ago, in a supersized expansion of matter and energy called the Big Bang. Energy and matter expanded in all directions in the first microseconds of existence, cooling from initially extreme temperatures and pressures. Tiny variations in the concentration of matter from place to place caused variations in gravity. Slightly denser areas attracted nearby space material, concentrating this early matter into gas clouds and, eventually, stars and galaxies. Such a collection of particles, held together by gravity yet kept apart by their whirling motion, became our galaxy. Our solar system formed about 4.6 billion years ago, emerging from the collapse of a cloud of interstellar gas and dust.

The Earth formed 4.54 billion years ago. At first, much of the Earth remained molten, constantly bombarded by additional cosmic collisions. Eventually, the planet cooled enough to separate into a layered sphere with a solid inner core of nickel and iron. Surrounding the inner core was a fluid outer core, a mostly solid mantle, and a thin, solid crust. The oldest known whole rock formations on Earth, more than 4 billion years old, are found on continental cratons and contain important clues about Earth's early history. These ancient cratons are found in Canada, Australia, and Africa. The oldest dated terrestrial materials are 4.4-billion-year-old zircon grains taken from a metamorphosed sandstone conglomerate in the Jack Hills of western

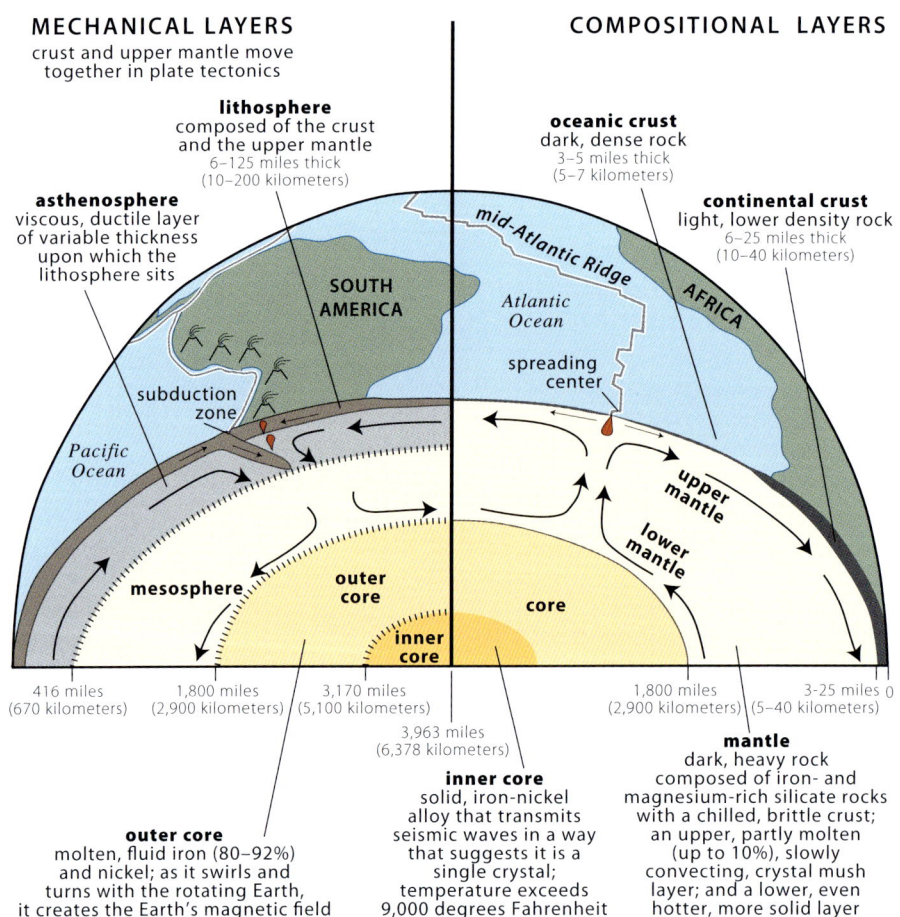

Cutaway of the Earth. The inner core, about 760 miles in diameter, consists of a solid nickel-iron alloy. Temperatures exceed 9,000 degrees F. The fluid outer core is approximately 1,500 miles in diameter, also composed of iron and nickel. The mantle, the thickest layer of our Earth, extends from 4 miles to about 1,800 miles depth and is divided into the upper and lower mantle layers. It is composed of iron- and magnesium-rich silicate rocks and is largely solid. While the mantle is solid, it still convects, or flows, over very long geologic time scales. The outer part of the mantle is called the asthenosphere, a viscous, ductile layer of variable thickness, upon which sits the lithosphere. The outermost layer of the solid Earth, called the lithosphere (litho means "rocky" in ancient Greek), includes the crust and the uppermost part of the mantle.

Australia. The oldest rocks in New Mexico, 1.78-billion-year-old metavolcanic rocks in the Taos Range, are much younger but still very old. As dating techniques and methods grow, so does our ability to refine the ages of the materials that compose the Earth.

During the Earth's early evolution, a process began in which sections (plates) of the thin crust move over Earth's surface. Known as plate tectonics, the process is responsible for building mountains, generating volcanoes, and triggering earthquakes. The Earth is broken into seven major plates and several minor plates, all rigid pieces of the thin lithosphere. These plates float on and through the still solid but more elastic asthenosphere. Convection of the mantle drives motion of the tectonic plates.

Plates are composed of two sorts of crustal materials: continental crust and oceanic crust. Continental crust ranges from 20 to over 40 miles in thickness, with an average of about 25 miles thickness. Continental crust is composed largely of silica-rich intrusive rock types like granite. Oceanic crust, only about 4 miles thick, is predominantly iron- and magnesium-rich rocks formed at oceanic spreading ridges. Oceanic crust composes the floors of the major oceans. Continental crust has a lower density (2.7 grams per cubic centimeter) and can float above the denser (3.5 grams per cubic centimeter) oceanic crust. A lot of important geologic history happens where plates interact with neighboring plates. In subduction zones, denser oceanic crust is pulled below less-dense continental crust.

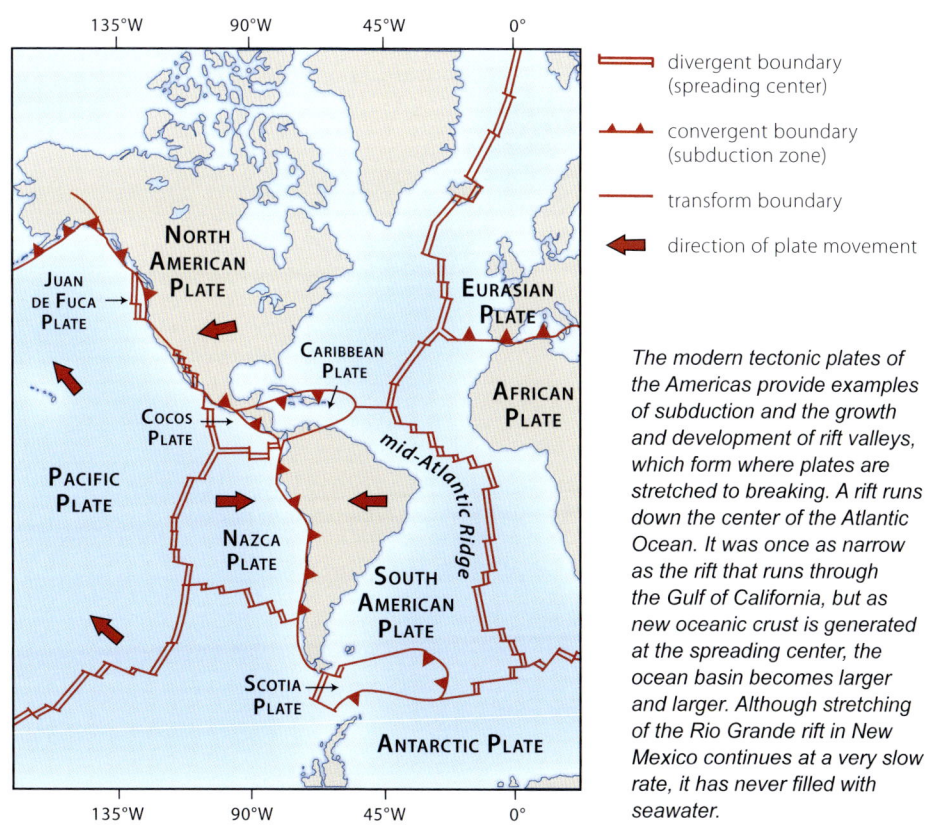

The modern tectonic plates of the Americas provide examples of subduction and the growth and development of rift valleys, which form where plates are stretched to breaking. A rift runs down the center of the Atlantic Ocean. It was once as narrow as the rift that runs through the Gulf of California, but as new oceanic crust is generated at the spreading center, the ocean basin becomes larger and larger. Although stretching of the Rio Grande rift in New Mexico continues at a very slow rate, it has never filled with seawater.

TYPES OF ROCKS

There are three main types of rocks: sedimentary, igneous, and metamorphic. This section showcases only the first two types because metamorphic rocks, which form when high heat and elevated pressure cause profound chemical and physical changes to existing rocks, are not commonly seen along the roads in New Mexico. Most metamorphic rocks in the state are Proterozoic gneisses found in the Rocky Mountains.

COMMON ROCKS IN NEW MEXICO

ROCK TYPE	ROCK NAMES	ROCK DESCRIPTION
igneous intrusive	gabbro	very dark, coarsely crystalline, with visible individual grains of dark minerals
	diorite and monzonite	medium-colored, coarsely crystallized rocks, intermediate between gabbro and granite; visible crystals include feldspar, black mica, and hornblende
	granite	common, light-colored, coarse-grained rock with visible crystals of quartz and feldspar, and small amounts of mica or hornblende
igneous extrusive, volcanic	basalt	very fine-grained, gray or black; forms widespread flows or cinders
	andesite - dacite	dark-colored, intermediate in composition between basalt and rhyolite
	rhyolite	extrusive equivalent of granite, light-colored and often pinkish; occurs as thick flows and tuff or ash beds
sedimentary clastic (from pieces)	shale, siltstone	very fine-grained, made of silt and clay cemented together; usually breaks into flat slabs
	sandstone	sand grains cemented together
	arkose	sandstone with at least 25 percent of the grains composed of feldspar; therefore, deposited close to a rapidly eroding granitic or metamorphic source
	conglomerate	sand, pebbles, and gravel deposited then cemented together; may include cobbles and boulders
	breccia	broken, angular rock fragments cemented together; smiliar to conglomerate
	tillite, diamictite	ice-deposited rock composed of completely unsorted material
sedimentary biologic	limestone	gray or white in color, made primarily of calcite (calcium carbonate), often from shells and skeletons of marine organisms; fossils are often visible
	coal	black, carbon-rich layers derived from plant material
sedimentary chemical	evaporite	formed by the evaporation of mineral-rich water; includes gypsum, halite, potash, and phosphate from seas and lakes, and travertine in limestone caves and hot springs
metamorphic non-foliated	marble	recrystallized limestone, often with visible calcite crystals; often white or gray in color
	quartzite	sandstone recrystallized so it breaks through the grains instead of around them
metamorphic foliated	schist	stripey, medium-grained rock with abundant parallel mica grains, hornblende, or both; tends to split along the mica partings
	gneiss	coarse-grained with alternating bands of aligned flaky and granular minerals; more intensely metamorphosed than schist

Sedimentary rocks originate from sediment, the material that remains after the breakdown or weathering of rocks. Sediment can be transported by water, wind, ice, and gravity. Sedimentary rocks are almost always layered, or stratified, because their components are deposited and then lithified.

Volcanic breccia contains angular fragments of rock broken by the eruption and encased in a finer-grained matrix. Oxidation has colored this breccia from the Valles Caldera red.

The Madera Group includes classic limestone with jointed gray layers weathered by the elements, photographed here at the crest of the Sandia Mountains.

The Zuni Sandstone at El Morro National Monument weathers into pockets and knobs called tafoni.

Igneous rocks take two forms: intrusive rocks that cool within the crust and extrusive rocks that are erupted and cool on the surface. Intrusive rocks, such as granite and gabbro, cool slowly over thousands or millions of years because surrounding crustal rocks insulate their magma chambers, and this slow cooling allows large crystals to form. Extrusive rocks, such as lava and volcanic tuff, cool rapidly after eruption, resulting in a glassy texture with microscopic crystals. Some igneous rocks have a porphyritic texture in which visible crystals are surrounded by a groundmass of crystals invisible to the naked eye. The heat of the erupting material sometimes is so great that the erupted ignimbrite flow—the ash, pumice, and other rock fragments—becomes "welded," or highly dense and massive.

The small pockets, or holes, in this basalt are vesicles, places where gas bubbles were trapped as the lava was cooling. Flow banding, the stretched, stripy texture in the middle of the photo, forms due to friction as the flowing magma cools.

The Amalia Tuff shows porphyritic texture in which larger crystals that crystallized while the magma was underground are surrounded by a light matrix of microscopic crystals that cooled nearly instantly when the material was erupted. —Photo by Kent Budge

The Sandia Granite is rich in feldspars (pink), quartz (clear to white), and biotite (black). The large crystals indicate it cooled slowly deep underground.

PRECAMBRIAN ERA
4,600 to 541 million years ago

The name Precambrian was given to the oldest rocks, once thought to have formed before there was life on Earth. Scientists now know that primitive life forms, similar to modern algae, stromatolites, and flagellates, flourished as early as late Precambrian time, but did not grow hard, readily preserved shells. These stromatolite mats and microfossils preserve the beginning of life around 3.5 billion years ago and set the stage for the explosion of life forms that developed in the Paleozoic Era. The Precambrian Era, divided into the Hadean, Archean, and Proterozoic Eons, encompasses 4 billion years, roughly 80 percent of the Earth's existence. All of New Mexico's Precambrian history occurred in the Proterozoic Eon, which began 2.5 billion years ago.

Ancient crystalline rocks underlie the entirety of New Mexico, deep beneath the layered sedimentary rocks, volcanic fields, and alluvium. These rocks outcrop at the surface in only about 5 percent of the state, mostly in the mountains. Known as basement rock, these crystalline rocks are present in many iconic ranges of New Mexico: the Sangre de Cristos towering above Santa Fe and Taos, the Sandia and Manzano Mountains near Albuquerque, the San Andres Mountains rising to the west of White Sands National Park, and the Gila and Zuni Mountains. Proterozoic rocks in the latter are scattered amidst volcanic rocks.

The oldest rocks in New Mexico formed 1.8 to 1.7 billion years ago as parts of island arcs, chains of volcanic islands that develop at convergent plate boundaries. The collision of island arcs is one of the main ways that continents form and grow. Many of these oldest rocks are plutons, large masses of igneous rock that cooled from magma that rose into and pooled within Earth's crust. Other old rock was originally volcanic: lavas, ash, and volcanic fragments. No matter the origin, these rocks were all extensively metamorphosed over the ensuing nearly 2 billion years. Many have been through multiple metamorphic events, and deciphering their history is challenging.

About 1.7 billion years ago during the Yavapai orogeny, a series of island arcs collided with the Wyoming Province, which was part of Laurentia, the original

Gneiss, a metamorphic rock, displays folds from a complex history of deformation. The white rock at left is a vein of quartzite. Photo taken in the Sandia Mountains.

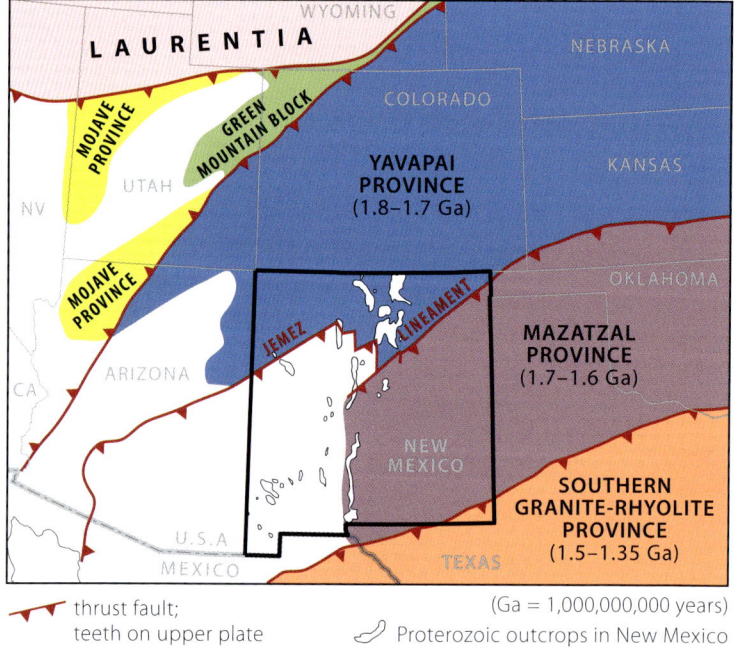

The basement rock of New Mexico is composed of the ancient island arcs of Yavapai and Mazatzal, with a tiny part of the Southern Granite-Rhyolite Province in the southeastern corner. The basement rock is almost entirely buried beneath younger rock. The outcrop areas of basement rock, mostly in uplifted mountain ranges, are shown in white. —From Brandes, 2021

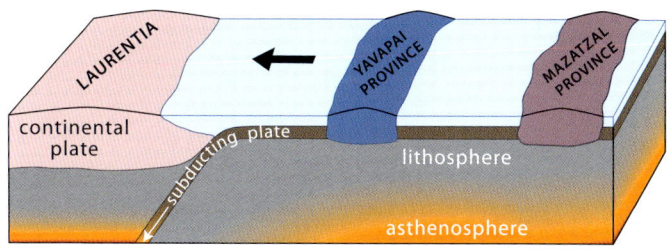
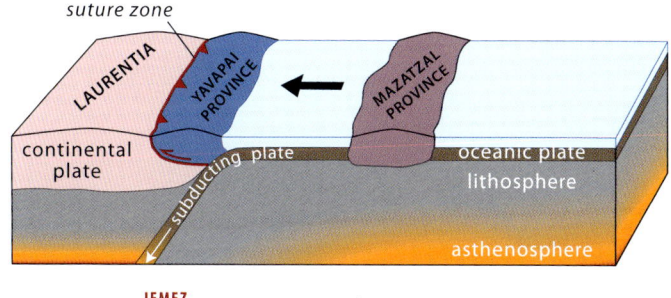
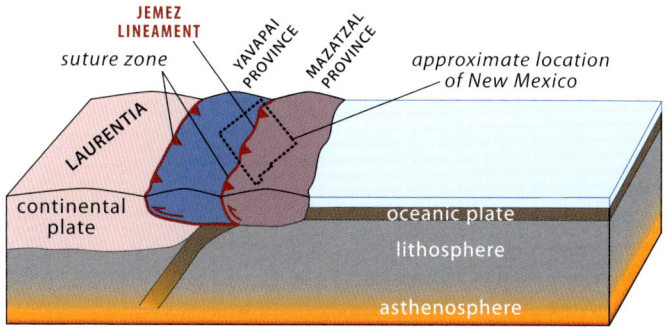

The process of subduction was responsible for the Yavapai and Mazatzal tectonic episodes in Proterozoic time.
—From Brandes, 2021

continental craton of North America. These island arcs were pushed up against the margin of the Wyoming Province, creating a band of new crust and enlarging the North American continent in a process called accretion. The rocks of these ancient terranes, which stretched across Arizona and New Mexico, extended north to Utah and Colorado and east to Kansas and Nebraska. Today, units of this origin in New Mexico include the Vadito Group and Ortega Quartzite, which outcrop in the Tusas, Picuris, and Truchas Mountains.

A second period of continental growth occurred at about 1.6 billion years ago during the Mazatzal orogeny, when another island arc was added to North America. The remains of this island arc include metavolcanics presently exposed in the Los Pinos Mountains northeast of Socorro and the Sandia and Manzano Mountains near Albuquerque.

From 1.6 to 1.5 billion years ago, during a period of quiescence following Mazatzal mountain building, sediments accumulated in intermontane basins. The Vadito Groups contain sediments that accumulated in these basins but are now metamorphosed rock.

Another major metamorphic event, the Picuris orogeny, occurred 1.49 to 1.45 billion years ago. Named for the Picuris Mountains south of Taos, the orogeny is not well understood. It may be related to the development of the Southern Granite-Rhyolite Province that extends from southeasternmost New Mexico into Texas and Oklahoma, or it may have been another suturing event related to converging land at the edge of the continent. It is mainly known from batholiths—enormous groupings of igneous plutons—intruded into the crust. One example of these is the granite of the Sandia Mountains east of Albuquerque.

For most of the next nearly 1 billion years, New Mexico remained above sea level, gradually eroding. To the east, the Grenville orogeny occurred around 1.2 to 1 billion years ago and affected rocks that stretch from modern day southern New Mexico through the Appalachians, Adirondacks, and into Canada. The Grenville orogeny was part of the amalgamation of continents into the supercontinent Rodinia, which existed until the supercontinent broke up into smaller continents about 750 million years ago.

The North American continent that existed at the time was beveled to an extensive flat surface, called a peneplain. Sediments were transported toward distant oceans, and a major section of the rock record—more than 1 billion years in places—is missing from this area. This contact where the section of rock is missing is called the Great Unconformity and is found throughout the southwestern United States. It usually takes the form of an eroded surface of ancient Precambrian basement rock upon which layered sedimentary rocks of Paleozoic age were deposited. In New Mexico, Paleozoic sedimentary rocks were deposited directly on an eroded surface of schists, gneisses, and granites from the Paleoproterozoic. Other unconformities also exist in New Mexico, representing shorter periods of time during which there was no deposition or when erosion removed some of the rock before younger sediments were deposited. Other than a few dikes (intrusions of magma that followed fractures), New Mexico does not contain any rock from the last 500 million years of the Proterozoic Eon.

The Great Unconformity on the east side of the Sandia Mountains, where the Paleozoic-age Sandia Formation and Madera Limestone overlie an eroded surface of Proterozoic-age Sandia Granite. This exposure is directly east from the Doc Long Picnic Ground (milepost 2) on the Sandia Crest National Scenic Byway.

PALEOZOIC ERA
541 to 252 million years ago

The Paleozoic Era was a time of continental change, with much of North America covered by shallow seas. In Cambrian time, the Sauk Sea inundated the entire continent except for the Transcontinental Arch, a series of islands that extended from New Mexico to the Great Lakes region. These islands, a series of basement-controlled structural highs and basins, or sags, formed a continental-scale, platform-like feature that remained elevated until Mississippian time.

The sedimentary record from this time starts with the Sauk sequence, the earliest of six cratonic sedimentary sequences recorded on the North American continent. A cratonic sequence is the sedimentary record of a complete marine transgression (rising sea level) and regression (falling sea level). Sea levels have risen and fallen for various reasons throughout geologic time. For example, the lack of land near the poles meant less water would be tied up in polar ice caps, and a warm climate meant higher sea levels because less water was tied up in ice caps and glaciers. In New Mexico, shallow water deposits during the Sauk transgression are the sandstones and clastic carbonates of the Cambrian Bliss Formation. Retreating sea levels exposed these shallow water deposits when the sea regressed.

During Cambrian to Ordovician time, a widespread magmatic event emplaced many alkaline igneous bodies in what is now New Mexico and Colorado. A failed rift, known as the New Mexico aulacogen, stretched the crust here, possibly related

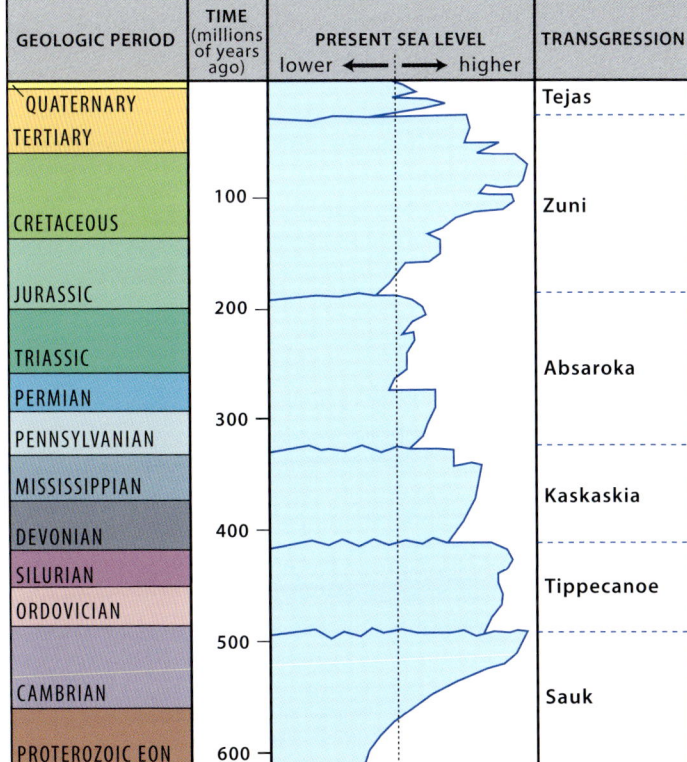

The Sauk was the first of six major transgressions during Paleozoic time. All but the last transgression affected New Mexico.
—Brandes, 2021

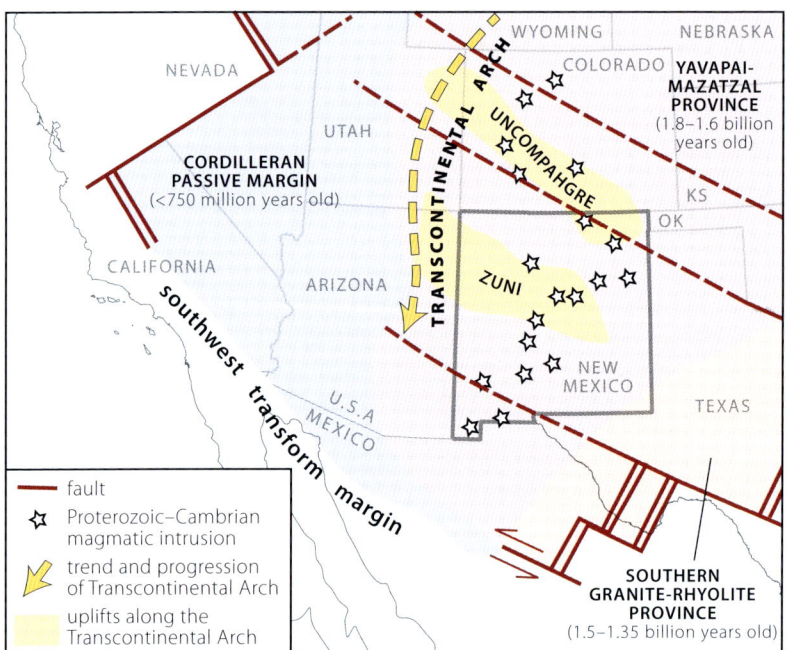

In Cambrian to Ordovician time, the northwestern corner of New Mexico consisted of small islands elevated along the Transcontinental Arch and surrounded by a vast ocean. To the south of the arch, the crust stretched, allowing magmatic intrusions to rise into the crust. —Simplified from McMillan and McLemore, 2004

to this magmatic event. Carbonatites, syenites, monzonites, and alkaline granites all intruded the crust, forming the Iron Hill carbonatite complex in Colorado and similar deposits in the Lemitar and Chupadera Mountains, Lobo Hill, and Monte Largo in New Mexico. These intrusions are associated with rare earth deposits including uranium, thorium, niobium, zirconium, and other elements.

The Paleozoic seas that covered parts of North America provided scientists with ample rocks and fossils for developing the geologic sequence. Sediments that accumulate in oceans are deposited at or near horizontal. As layer after layer is deposited and buried, these sediments eventually lithify, or turn into sedimentary rock. The fossils preserved within these rock layers provide geologists with a way of determining the relative age of a rock. Layers create an order of deposition because younger layers stack on top of older layers. Animals evolved through time, so fossils were used to correlate distant sedimentary units, creating a *relative* geologic timeline. The Mississippian and Pennsylvanian Periods of the Paleozoic Era were named for sequences of rocks in North America. *Absolute* ages, gained by radiometric dating of uranium, potassium, and other elements, have been incorporated in the earlier dateless relative geologic timeline. As scientific methods improve, these dates undergo frequent revision and refinement. In this way, a combination of relative and absolute ages helps us understand the history of Earth.

Sedimentary rocks were deposited in what is now New Mexico during the Ordovician, Silurian, and Mississippian Periods, but many of the sedimentary units were eroded away when the land was uplifted during mountain-building events. During Mississippian time, New Mexico was at the edge of a broad, shallow carbonate shelf that crossed what is now North America from northeast to southwest. The shallow seas were full of life, and rocks from this time have abundant fossils, including conodonts, crinoids, ammonoids, and other invertebrates. By the end of Mississippian time, the rocks were exposed on land and developed soil, and limestone, a carbonate rock, dissolved to form sinkholes and caves, features known as karst.

The Ancestral Rockies, a series of Pennsylvanian- and Permian-age uplifts in Colorado, New Mexico, Texas, and Oklahoma, rose in response to a plate collision far to the east as the continents of Africa and Europe collided with North America. The uplifts, including the Zuni, Sierra Grande, and Pedernal (from north to south) in New Mexico, are not the modern mountains seen today. Modern ranges rose much later in the same place and along many of the same faults and fault zones that these ancient ranges formed along. The topographic expression of the Ancestral Rockies had eroded by the time the Western Interior Seaway covered the region in Late Cretaceous time.

This Pennsylvanian collision of continents sutured the landmasses into the supercontinent Pangaea. The fossiliferous limestones of the Madera Group were deposited in areas that remained submerged under seawater from Pennsylvanian to early Permian time. The layered beds of the Madera Formation limestone now form the striking cap of the Sandia Mountains above Albuquerque. At the same time, the rising Ancestral Rockies shed debris from their flanks into adjacent basins. These clastic sedimentary deposits are preserved in the conglomerates, sandstones, siltstones, and shales of the Sangre de Cristo and Flechado Formations of northern New Mexico. These two units also preserve diverse fossils, including crinoids and brachiopods, testament to the interactions between land and water.

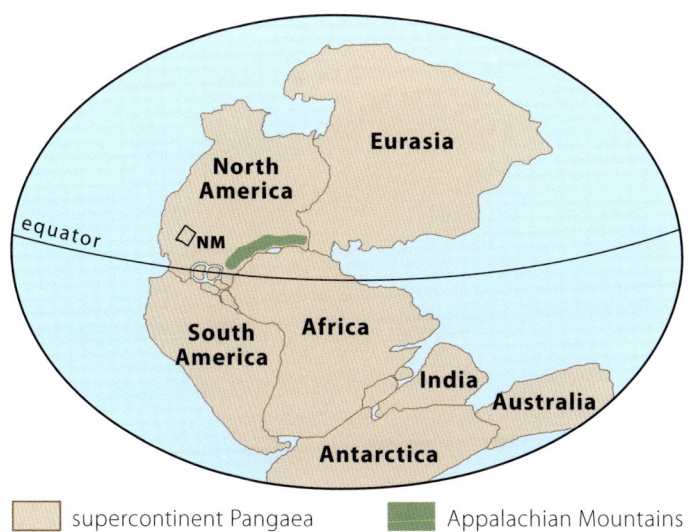

Despite New Mexico's position well within the interior of the North American continent, it was affected by the plate tectonic collision far to the east that created the supercontinent Pangaea and formed the Appalachian Mountains.

The uplift of the Ancestral Rocky Mountains brought New Mexico above sea level, so sedimentary sequences show a transition from marine to continental deposition. In Early Permian time, rivers flowing south toward the sea deposited the bright-red beds of the Abo Formation. The Abo also preserves evidence of the reptiles, amphibians, and synapsids of the tetrapod species that evolved in the Permian Period.

While northern New Mexico was rising, the sea still reigned in southern parts of the state. The Capitan Reef ringed the Delaware Basin of southeastern New Mexico and western Texas. This massive fossil reef, which extends across state lines and is present in multiple national and state parks and monuments, consists of 1,000 to 2,000 feet of limestone and dolomite. The reef, when actively being constructed

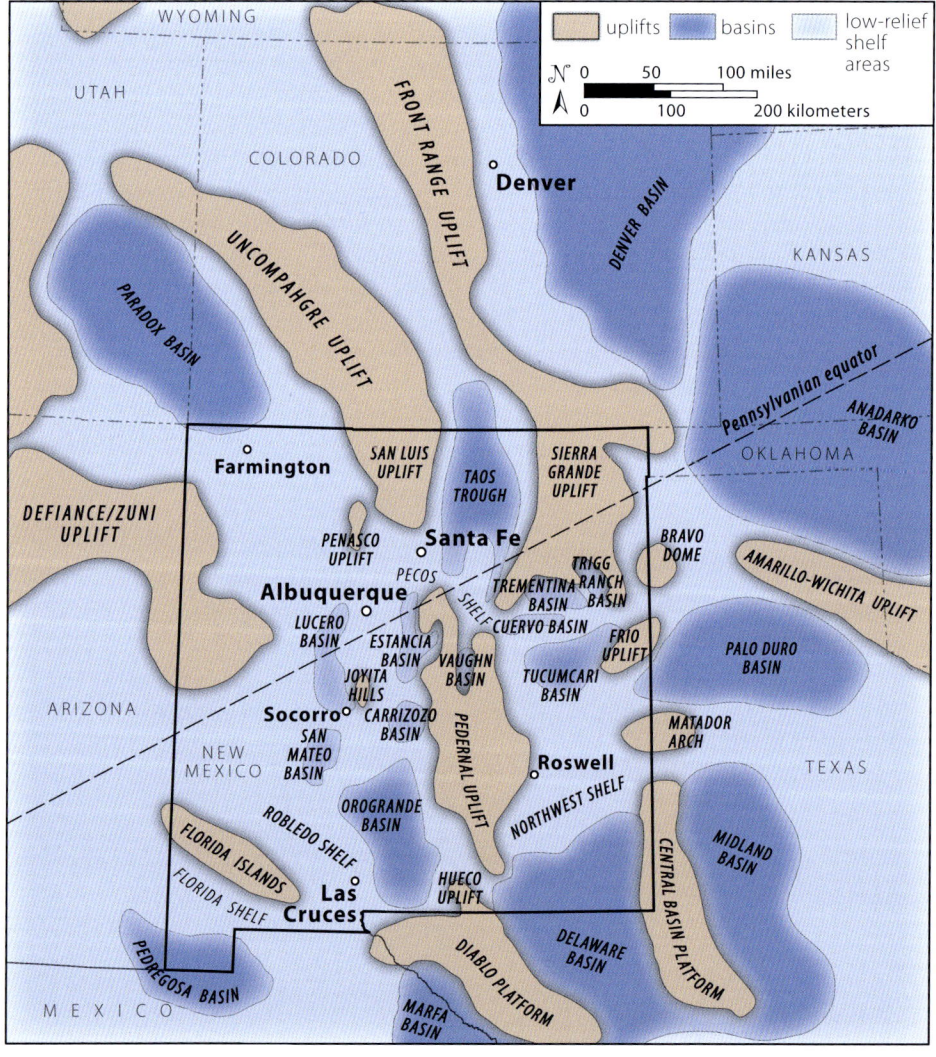

Uplifts and basins of the Ancestral Rockies developed in Pennsylvanian time in response to distant continental collisions. —Modified from Kluth and Coney, 1981; Schroder and others, 2021

by living organisms, may have stood several hundred feet above the shallow ocean floor. Fusulinids and brachiopods, diagnostic of the Guadalupian Epoch (named for the Guadalupe Mountains on the Texas border) of the middle Permian Period, are exposed in the spectacular Capitan Formation southwest of Carlsbad. The Delaware Basin is part of the larger Permian Basin of Texas, a major structural basin that contains enormous quantities of oil and gas. The petroleum contained in the Permian Basin accumulates in the Permian limestones, which are riddled with caverns and passages of all sizes.

Limestone of the San Andres Formation was also deposited during the Permian Period. The San Andres has the most surface exposure of any Paleozoic unit in New Mexico. Sinkholes, dissolution features characteristic of the karst topography common in limestone bedrock, are common where the San Andres is at or close to the surface. See the Peaceful Plains section for more information about the San Andres Formation.

MESOZOIC ERA
252 to 66 million years ago

When early geologists established the geologic time scale, they located the boundaries of major divisions where life forms and abundance changed dramatically. The Mesozoic Era begins with the Permian-Triassic extinction event (P-T extinction). The Permian is the last period of the Paleozoic Era and the Triassic Period is the first of the Mesozoic Era. More than 80 percent of marine species, 70 percent of terrestrial vertebrate species, and nearly 60 percent of biological families became extinct at the end of the Permian Period. One major catalyst for this global die-off was the eruption of the Siberian Traps, a large region of flood basalts that emitted huge amounts of carbon dioxide when they erupted. A mantle plume (similar to what presently exists beneath Hawai'i) breached the Siberian Craton and erupted basaltic lavas for more than 2 million years. More silicic eruptions occurred at the same time in South China and likely contributed to the extinction event. The eruptions elevated global temperatures, triggered ocean acidification, and reduced the oxygen in global ocean water, conditions unsuitable for supporting life. It took several million years for life to recover and reach the same level of diversity.

During the beginning of Triassic time, New Mexico lay in the interior of Pangaea and the climate was arid. Rocks were mostly eroding rather than being deposited, so the geologic record in the early Triassic is largely missing. The oldest Triassic units in New Mexico, from the Middle Triassic (247 to 237 million years ago), record a shift to a more temperate climate. The low-energy river and coastal plain sandstone, mudstone, and shales of the Moenkopi Formation outcrop extensively throughout the southwestern United States, including western New Mexico. Sandstones and conglomerates of the Anton Chico Formation are seen along the canyon walls of the Pecos River and near Santa Rosa in eastern New Mexico.

The Chinle Group unconformably overlies the Moenkopi and is another regionally extensive unit. These Late Triassic sandstones are the deposits of a continental-scale river system, which included lake and windblown deposits. They thicken to more than 1,700 feet in places and records a lush, wet environment of equatorial Pangaea. The Chinle is famous for its scenic outcroppings at Ghost Ranch, which is also the location of the quarry for the *Coelophysis*, one of the earliest dinosaurs. This small, bipedal, ground-dwelling, carnivorous dinosaur was a speedy runner. The fossils were first found at the Ghost Ranch location in 1887. The Chinle Group contains some volcanic ashes, now bentonite clay layers, erupted from faraway volcanoes associated with subduction zones. The Uncompahgre uplift of the Ancestral Rocky Mountains was the sediment source for some of the Chinle deposits. Petrified wood and fragmented fossilized teeth and scales of amphibians, ostracods, bivalves, and phytosaurs are reported in the Ojo Huelos Member (lower San Pedro Arroyo Formation of the mid Chinle Group).

At the end of Triassic time, Pangaea began to split apart. A rift formed that eventually opened into the Atlantic Ocean basin. One or more subduction zones to the west of North America pulled the continent west, away from the new spreading ridge in the Atlantic. At the same time, the Gulf of Mexico began to open to the south.

The Jurassic Period brought a return to an arid environment. The Entrada Sandstone, which forms the tan-buff-orange cliffs at Ghost Ranch and near Gallup, was originally deposited in an enormous sand sea, similar to today's Sahara Desert in Africa. Later, in the Middle Jurassic, sea level briefly rose. Drying of the sea resulted in the deposition of the gypsum and limestones of the Todilto Formation, visible in the striking White Mesa outcrops near San Ysidro. The Morrison Formation, famous for its extensive large dinosaur fossils and brilliant red coloring, was deposited from 156 to 146 million years ago in the Late Jurassic. Mountains were building to the west in Utah during the Sevier orogeny, and rivers deposited the Morrison into the basin forming to the east of the mountains.

To the south, the Mexican Border rift system developed because of complex relationships between the subduction to the west and the opening of the Gulf of Mexico to the east. Lavas dated at 155 to 145 million years ago erupted from the rift in southwestern Arizona.

Steamboat Butte, along the Dry Cimarron River in far northeastern New Mexico, displays an angular unconformity. The angled rocks in the lower half of the butte are Triassic red beds. These beds were uplifted, tilted, and then leveled off by erosion before the sediments in the upper half, the Entrada Sandstone, were deposited in Jurassic time. The mesa in the distance is capped with young basalt.
—Photo by Alexandra Priewisch

Depositional environments changed early in the Cretaceous Period. After more than 100 million years of the continent being above sea level, a shallow sea moved in. The Western Interior Seaway submerged the interior of North America from the Arctic Ocean to the Gulf of Mexico, including much of New Mexico during the Late Cretaceous. The Dakota Formation, a unit of sandstones, shales, and clays, was deposited in beaches, bays, and lagoons along the coast as the Western Interior Seaway spread across the continent. Numerous formations from Late Cretaceous time contain tracks, skin impressions, and skeletal remains of dinosaurs that lived along the coastline and along rivers that flowed into the sea. The Kirkland and Fruitland Formations of the San Juan Basin have been particularly fruitful for dinosaur hunters.

The Western Interior Seaway covered about two-thirds of New Mexico during the Cretaceous Period, although sea level fluctuated over time.

The Mancos Shale, a body of thick, sticky-when-wet, gray-black clays, was deposited at the bottom of the seaway. The Mancos Shale is prominent in New Mexico throughout the San Juan Basin, near Heron and El Vado Lakes in north-central New Mexico, and along US 550 west of the Nacimiento fault zone that runs along the base of Jemez Mountains. Many of the units deposited during this time are of extreme economic importance because of the oil and natural gas they contain.

Mesozoic time came to a close with a changing landscape that is preserved in the multiple formations of the Mesaverde Group, another regionally extensive sequence notable for its coal-bearing units. That group is present across much of the Southwest, with the Point Lookout, Menefee, and Cliff House units in the San Juan Basin of northwestern New Mexico and the Point Lookout Sandstone, Crevasse Canyon Formation, and Gallup Sandstone in southern New Mexico, particularly on the flanks of Sierra Blanca.

The Western Interior Seaway, which dominated much of the Mesozoic, receded as the Laramide orogeny began to elevate topography, beginning about 75 million years ago. In the 1970s, geologists hypothesized that a plate subducting beneath the western edge of North America became coupled to the overlying plate. This process, called flat-slab subduction, transferred collisional stress inland, across the continent, resulting in a major folding, faulting, and piling up of the crust far from the subduction zone. Geologists are now debating whether this was indeed the case or whether a microcontinent colliding with the west coast and then moving north caused the compression. The roots of the Sierra Nevada Mountains extended deep into the Earth at that time, decreasing the likelihood of the coupling of a low-angle slab. Flat-slab subduction, however, might have occurred from a southwest-trending subduction zone near the US–Mexico border and may have extended northeastward and underplated the Colorado Plateau. Regardless of the mechanism, the southern Rocky Mountains, which extend from New Mexico through Colorado and into Wyoming, were born in the Late Cretaceous and continued building into the Cenozoic Era.

CENOZOIC ERA
66 million years ago to the present

The Cenozoic Era begins with a definitive event that has a globally recognized geologic signature: the K–Pg boundary, the time boundary between the Cretaceous and Paleogene Periods. A band of rocks with iridium present at hundreds of times the normal amount marks this boundary (formerly known as the K-T boundary). The impact of a large meteorite 66 million years ago provided the iridium and formed the Chicxulub Crater in what is now the Yucatán Peninsula in Mexico. This impact triggered tsunamis, caused the formation of sulfuric acid in the atmosphere that resulted in global acid rain, and changed the global temperature. The result was the extinction of 75 percent of plant and animal life on Earth, including all but avian dinosaurs. This high-iridium layer is exposed in the Raton Formation in northeast New Mexico and between the Ojo Alamo and Kirtland Formations of the San Juan Basin. To learn more about the layer, including where to see it up close, go to the road guide for I-25: Colorado—Las Vegas in the Peaceful Plains section.

The Laramide orogeny, which began 75 million years ago in Late Cretaceous time, continued into the Cenozoic Era. Crystalline basement rocks that formed long ago were uplifted in large blocks along faults to (or near) the surface. These uplifts are unusual in that they occurred 400 to 1,000 miles east of the western convergent margin, well within the interior of the continent. These uplifts include what we know today as the Wind River Range in Wyoming, the Front Range in Colorado, and the Sangre de Cristo Range in New Mexico.

Though the mountains of the Laramide orogeny are visibly striking and form spectacular topography, the structural basins that formed in tandem with the mountains as the crust was buckled and warped are just as interesting to geologists. The basins' sedimentary units preserve the history of the changing landscape. Rivers carried sediment from the uplifts and deposited it in adjacent basins. The early sediments were eroded from overlying sedimentary rocks; eventually, the crystalline bedrock from deep below was exposed and eroded. The ages and thicknesses of these units provide geologists with the rate of sediment accumulation and thus the rate of uplift and geometry of surrounding areas. Some of these units that are particularly visible include the San Jose Formation of early Eocene age, beautifully eroded in the San Juan Basin (US 550) and the Galisteo Basin south of Santa Fe (NM-14/Turquoise

GEOLOGIC HISTORY OF NEW MEXICO   19

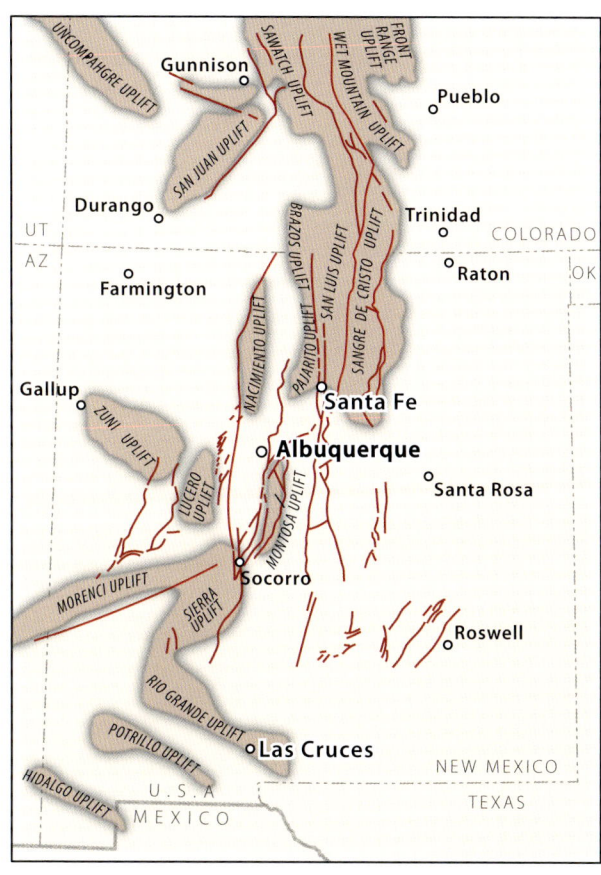

Many of the same areas that were uplifted during the Ancestral Rockies orogeny were uplifted again during the Laramide orogeny, using many of the same faults. The faults shown here formed before the faults of the Rio Grande rift broke the crust beginning in Oligocene time. —Modified from Cather, 2004

In the Late Cretaceous, Laramide deformation and uplift resulted in increased erosion, with sediments accumulating in adjacent basins. By Paleocene and Eocene time, erosion had exposed the older rocks in the core of the uplifts, and the uplifts had disrupted drainages, resulting in closed basins. —Modified from Cather, 2004

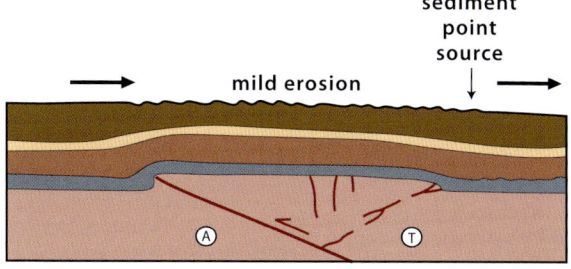

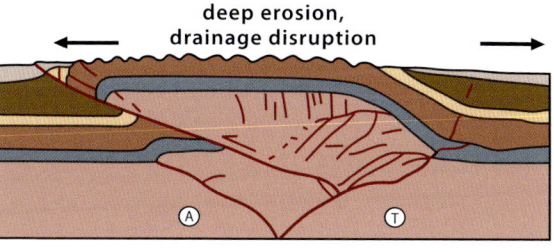

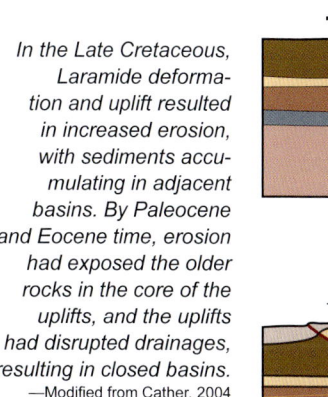

fault movement
A=away
T=toward

Trail), and the Eocene Baca Formation in the Baca Basin west of Socorro (US 60). The basins' accumulated sands, conglomerates, and silts preserve both the evolution of topography and the remains of mammals and other animals that lived there.

Laramide deformation is especially visible at the edges of the basins, where tilted sedimentary rocks form hogbacks and flatirons that parallel the uplifted mountain ranges. The San Juan Basin, a large structural basin formed from the broad downwarping of rock layers, filled in with thick sequences of sedimentary rocks and is now among the highest-producing regions of natural gas and oil in the US. The basin includes both the source material for the oil and gas and the reservoirs they exist in now. Petroleum and natural gas are fossil fuels, meaning they are substances produced by slow decomposition of plant and animal material. Intense heat and pressure over millions of years causes the decomposition and formation of hydrocarbon fuels. The oil and gas rise through the rocks until an impermeable layer or fault traps them. Then they collect in what is called a reservoir, a porous and permeable rock. Reservoir rocks of the San Juan Basin are Cretaceous and Paleogene sandstones.

The basin units also preserve ash from the earliest volcanic activity of the next major player in shaping the landscape: the Oligocene ignimbrite flare-up. From 40 to 25 million years ago in the late Eocene to Oligocene Epochs, some of the most voluminous volcanism in geologic history dominated the region from Nevada to western Colorado, triggered by hot mantle material rising close to the surface. Volcanism of this time formed the San Juan volcanic field of southwestern Colorado and the Indian Peak and Central Nevada volcanic fields in Nevada. In New Mexico, the Mogollon-Datil volcanic field was active, as were the Latir volcanic field and intrusions of the Ortiz porphyry belt. The latter two are part of the San Juan volcanic field but were separated from the main field by extension of the Rio Grande rift. Volcanism was triggered as the oceanic slab, subducted beneath the west coast, peeled off, rolled back, and began sinking into the mantle. This rollback allowed hot mantle material to rise close to the surface. Ignimbrites, volcanic tuffs solidified from pyroclastic flows (hot, dense, gassy flows of material that rush along the ground after being ejected from volcanoes) erupted from calderas. Ignimbrite is derived from Latin and means "fiery

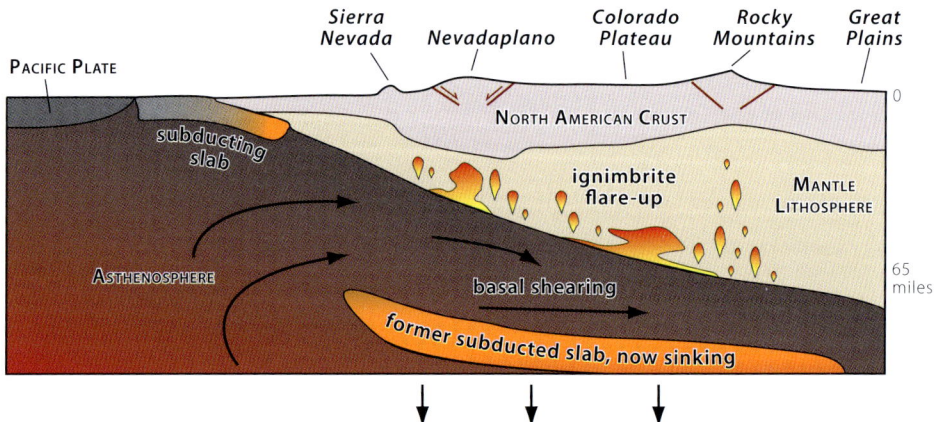

Hot mantle welled up from below, initiating the melting that triggered the Oligocene ignimbrite flare-up. —Modified from Bahadori and others, 2022

The vertical walls of a dike extend out from Ship Rock, an Oligocene intrusion in northwestern New Mexico. The enclosing Chinle Group sediments eroded away from the more resistant igneous dike.

rock dust cloud." Note that the Valles Caldera in the Jemez Mountains, one of the most well-known calderas in the world, is not from the Oligocene ignimbrite flare-up. It erupted much more recently.

While the extrusive products of this volcanism are highly visible, the intrusive volcanism that occurred during this time resulted in major mineral and precious metal deposition as hot fluids associated with intruding magmas interacted with existing bedrock and groundwater. Gold and silver are found near the Ortiz and Magdalena Mountains and near Silver City, copper near Santa Rosa and Silver City, and molybdenum near Red River and Questa. For more on the Mogollon-Datil volcanic field, see the introduction to the Volcanic Southwest section.

Following hard on the heels of massive volcanism was the opening of the Rio Grande rift, a spreading seam in Earth's crust that extends from central Colorado south into Texas and northern Mexico. The rift divides New Mexico, with the Colorado Plateau to the west and the Great Plains to the east, and it now houses the Rio Grande, the fourth-longest river in North America. The river's length is due in part to the connecting of basins as the rift opened. Rifting began synchronously along its entire length approximately 35 to 30 million years ago. The extension is still ongoing today, although very slowly. The stretching of the crust is related to the same forces that thinned the crust in the Basin and Range Province to the west. The Basin and Range Province stretches all the way through Arizona and Nevada; this landscape is characterized by a series of narrow ranges separated by larger basins, all lined up for hundreds of miles in places. In fact, the Rio Grande rift forms the eastern boundary of the Basin and Range Province in the far southwestern corner of New Mexico.

The extension stems from changing dynamics of tectonic plates far to the west in southern California. The San Andreas fault zone developed as the Pacific Plate contacted the North American Plate and replaced the compressive subduction zone with a strike-slip fault zone. The new right-lateral, strike-slip faults in California and Nevada added a clockwise rotation to the Great Basin. The complexity of heat flow

VOLCANISM

New Mexico is one of the most volcano-rich states in the United States. Volcanoes are dramatic landscape features. Mt. Taylor, for example, is a dormant stratovolcano near Grants. Young, recently active edifices are on display at Capulin Volcano National Monument, El Malpais National Monument, Valley of Fires Recreation Area, and Valles Caldera. The eroded remains of older volcanoes are everywhere, testament to New Mexico's fiery past.

Volcanoes form where molten magma reaches the Earth's surface. They vary in size from small spatter cones to immense mountains or fissures. The shape and eruptive behavior of volcanoes depend on the chemistry of the rocks that built them. The amount of silica (silicon dioxide) determines the fluidity and explosivity of the magma. The more silica there is, the more viscous and sticky it becomes. Silica-rich magma doesn't flow easily when it erupts, so it piles up to form domes of rhyolite or dacite. When gases get trapped in the thick, silica-rich magma, it explodes violently, forming calderas. When silica-poor magma erupts, it is more fluid and flows along the surface as basalt lava.

Types of volcanoes.

in the Earth's mantle that led to the continent splitting apart here in New Mexico is still being studied.

Below the surface, the rift is a series of asymmetrical halfgrabens, or down-dropped blocks, with a major normal fault on one side, while the other side acts as a hinge, lowering the sediment-filled rift more gradually. The relief on these rift-bounding faults can be massive: in places, Proterozoic basement is offset more than 11,000 feet along a fault zone. The east-tilted, flat-topped, ramp-like Sandia and Manzano Mountains near Albuquerque were uplifted along rift faults.

As the crust pulled apart, major basins formed in New Mexico: the San Luis Basin (from Alamosa, Colorado, south to Taos), the Española Basin, and the Albuquerque Basin, as well as numerous additional basins in southern New Mexico. Between

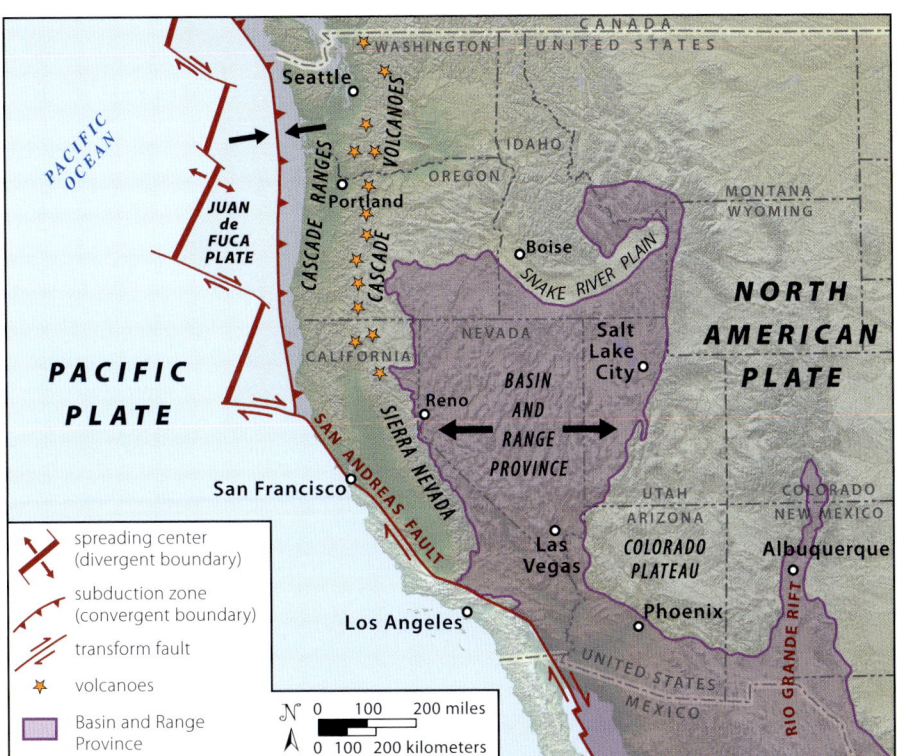

As the San Andreas strike-slip fault developed in southern California, the brittle upper crust of the continent to its east was stretched and broken by normal faults, developing the Basin and Range, as well as the Rio Grande rift.

major basins are numerous subbasins as well as transfer zones, which are areas where strain is transferred from one feature to another. Transfer zones are complex zones of faulting and folding that accommodate variations in movement between adjacent basins. Sediments were continuously washed into the basins from bordering mountains and by local and regional tributaries, always adding to the sediment fill of the rift. At times, these rift basins did not have a through-going river, so surface runoff collected in low-elevation places, forming lakes. Together, the highly variable sedimentary units, which include conglomerates, sandstones, siltstone, shale, evaporites, and lava flows and ashes of diverse volcanic origin, form the Santa Fe Group, which remains an important aquifer today. The Santa Fe Group erodes into stunning tan, buff, white, and orange badlands. The volcanics mixed in with the sediments are mainly from rift-related volcanism. The thinning crust allowed magma to rise from the mantle, spilling out as young basaltic lava, such as that erupted from the Albuquerque Volcanoes. For more information about the rift, see the introduction to the Rio Grande rift section, beginning on page 73.

A series of Miocene volcanic centers, most younger than 10 million years, extends from Arizona into northeastern New Mexico, forming along an alignment called the Jemez lineament. These volcanic centers include the spectacular Zuni-Bandera and Mt. Taylor volcanic fields in west-central New Mexico; the Jemez volcanic field,

FAULTS

Faults, fractures in Earth's crust along which blocks of rock on either side have shifted, are classified by the angle of the fault surface and by the relative movement of the two sides. Three main types of faults are found in New Mexico: normal, reverse, and strike-slip. Normal faults occur in extensional settings where the crust is stretched, and the two sides of the fault move apart from each other. Normal faults generally occur at 60-degree angles, and the rock overlying the fault surface moves down relative to the underlying rock. The Rio Grande rift is defined by normal faults. The Basin and Range is also characterized by normal faults that drop basins and lift ranges. Thrust faults, low-angle reverse faults, occur in areas of compression, where plates are pushed together. The mountain fault blocks of the Laramide orogeny were uplifted along reverse faults. In strike-slip faults, the third major fault type, the sides slide past each other, and motion is primarily horizontal rather than vertical. The enigmatic Picuris-Pecos fault zone is a major strike-slip fault in northern New Mexico that may have originated during Proterozoic time but may also have moved during Laramide deformation or more recently. The Embudo fault zone contains multiple oblique-slip fault strands that connect the northern end of the west-tilted Española Basin to the south and the east-dipping San Luis Basin to the north.

faults formed by extension or stretching
the hanging wall moves *down* along the fault

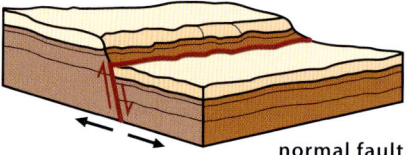

normal fault

Normal, reverse, thrust, and strike-slip, faults are found in New Mexico.

faults formed by compression or pushing
the hanging wall moves *up* along the fault

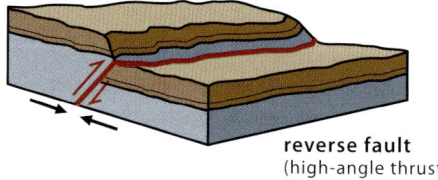 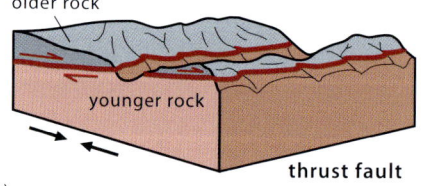

reverse fault
(high-angle thrust)

thrust fault

faults formed by shearing force
the fault blocks move horizontally alongside one another

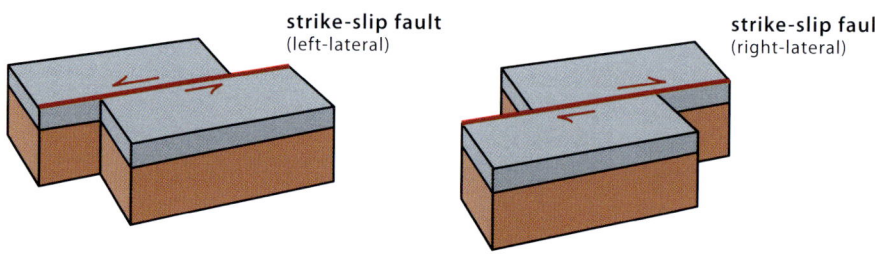

strike-slip fault
(left-lateral)

strike-slip fault
(right-lateral)

including the Valles Caldera; the Taos volcanic field; the Cerros del Rio volcanic field; and the Ocate and Raton-Clayton volcanic fields. The origin of this chain of volcanic centers is thought to be a zone of weakness associated with an ancient subduction or continental suture zone where the Yavapai and Mazatzal Provinces smashed together 1.7 billion years ago. In this region of thinning crust, magma exploits the crustal weakness and rises to the surface. The intersection of the north-south-trending Rio Grande

Neogene volcanic centers are aligned along the Jemez Lineament, a suture zone between the Proterozoic-age Yavapai and Mazatzal provinces. Rising magma exploited the weakness in the crust, especially where the lineament and the Rio Grande rift intersect. The suture zone is mostly hidden beneath the young volcanic rock. —Modified from CD-ROM Working Group, 2002; Nereson, 2013

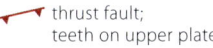

Cabezon Peak, a volcanic neck south of Cuba, intruded into Cretaceous-age sediments about 2.6 million years ago as part of the Mt. Taylor volcanic field. When magma begins to cool, it contracts and solidifies, fracturing the columns, as displayed here by the vertical fractures. Cabezon Peak is just one of the Rio Puerco necks, smaller edifices that also formed as part of the Mt. Taylor volcanic field.

rift with the northeast-slanting ancient scar of the Jemez lineament is the location of one of New Mexico's most spectacular volcanic sites: the Jemez Mountains. While early volcanism forms the base of the Jemez Mountains, caldera volcanism steals the show now. Two major calderas existed here nearly on top of each other: the older, 1.6-million-year-old Toledo Caldera and the younger, 1.2-million-year-old Valles Caldera. The huge volumes of ignimbrite (hot, rapidly moving slurry of ash, gas, and particulate) that poured from these calderas formed the massive pile of volcanic tuff, which is more than 1,000 feet thick in places. These repeated ash flows completely changed the topography.

With the onset of the Pleistocene Epoch, a cooler, wetter climate initiated glaciation in the mountains of New Mexico. Glaciers and small ice caps were present on the Sangre de Cristo Range of northern New Mexico as well as on Sierra Blanca, which today towers above White Sands National Park. These alpine environments feature tumbled glacial moraines, carved cirques, and outwash terraces. The glaciation of New Mexico mountains was accompanied by incision and the development of the river systems seen today.

During the cooler climate, several lakes occupied closed drainage basins. Ridges, beaches, spits, and bars mark the lakes' former shorelines, and lake deposits exist in the subsurface beneath the modern playa lakebeds. The youngest of these lakes existed between 6,000 and 3,000 years ago, in the Holocene Epoch. The Valles Caldera was filled with a long-lived lake in Pleistocene time. Sediment cores taken from the Valles Caldera contain mudstones that preserve ancient pollens and oxygen isotope data that help scientists better understand the glacial-interglacial cycles and ancient climates.

High in the Sangre de Cristo Mountains, Lake Katherine sits in a cirque, a three-sided bowl of steep bedrock eroded by an alpine glacier during the Pleistocene ice ages. Late Holocene moraines form the hummocky approach to the lake.

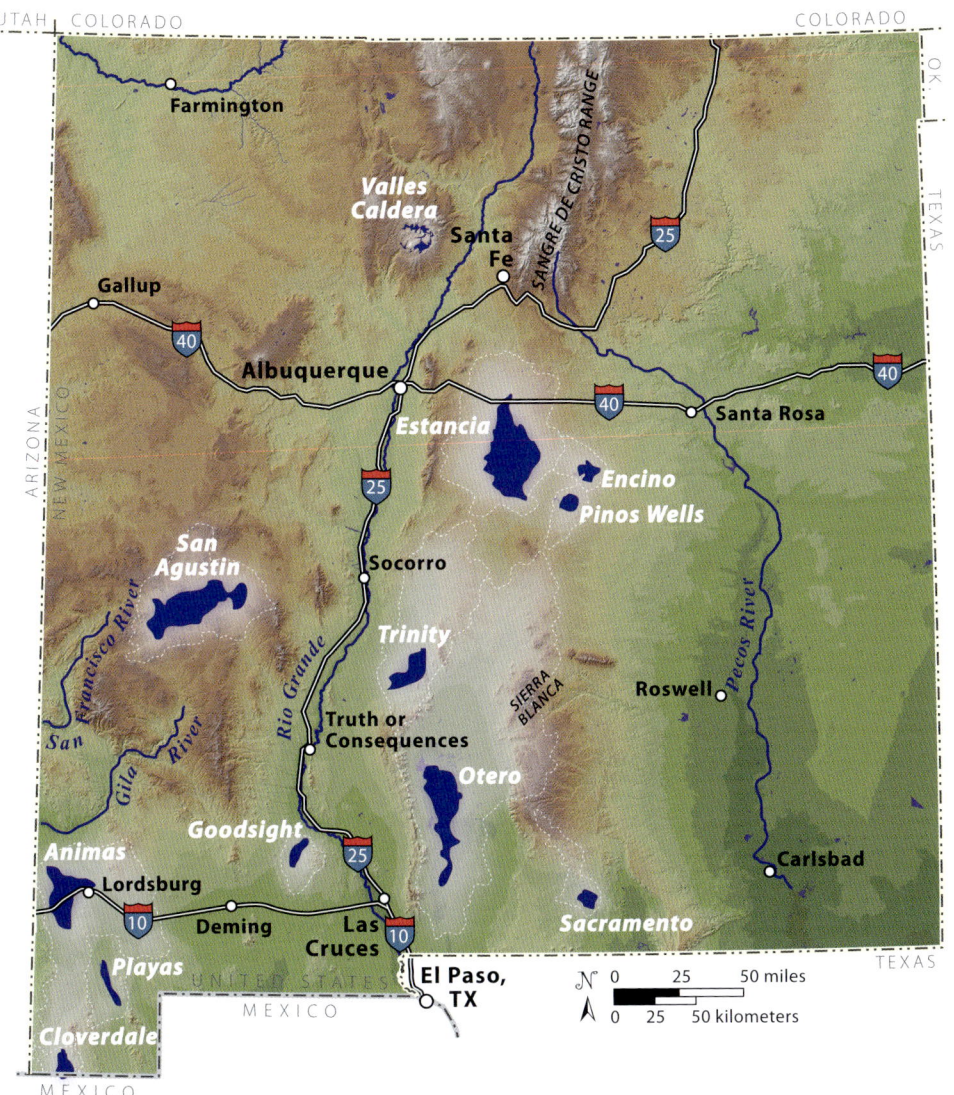

The extent of Pleistocene lakes in New Mexico. The dashed lines show the lakes' drainage basins. South of the Jemez Mountains and Santa Fe, the Rio Grande and Pecos River were through-flowing streams during the ice age, so lakes did not form in their drainages.
—Modified from Allen, 2005

THE MODERN LANDSCAPE

New Mexico's diverse landscape occurs in part because it sits at the juncture of five physiographic provinces: the Great Plains, the Rio Grande rift, the Basin and Range, the Rocky Mountains, and the Colorado Plateau. The provinces overlap to varying degrees along their edges. For example, the southern half of the Rio Grande rift in New Mexico is also part of the Basin and Range. Geologists like to draw neat boundaries on maps and charts, but the natural world does not conform to these boundaries.

More than 80 percent of the state is above 4,000 feet in elevation, though elevations range from a low of 2,842 feet above sea level near Red Bluff Reservoir on the Pecos River near the Texas state line to a high point of 13,161 feet at Wheeler Peak in the Sangre de Cristo Mountains. With the state's high elevation, it is no surprise that the Continental Divide crosses it from north to south. The Rio Grande, with its tributaries of the Pecos River, Canadian River, Rio Chama, and Rio Puerco, drains the collective basins of the Rio Grande rift and forms the national border with Mexico before reaching the Gulf of Mexico in the Atlantic Ocean. The San Juan and

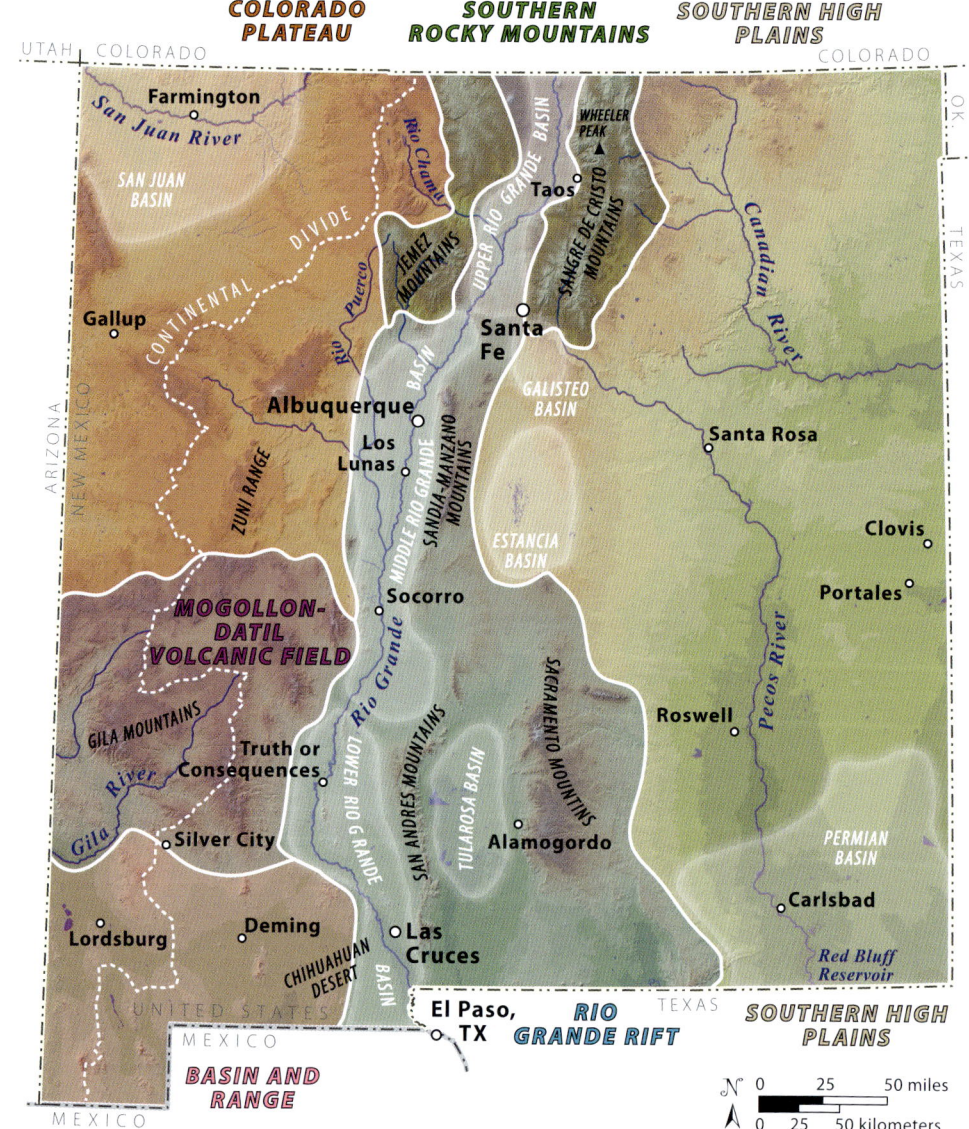

The Continental Divide snakes through three of New Mexico's five physiographic provinces.

Gila Rivers join the Colorado River, draining to the Gulf of California in the Pacific Ocean. These major rivers flow year-round and sustain wildlife, recharge groundwater aquifers, provide for human residential and industrial use, and are critical for agriculture. Most other streams in New Mexico, however, are ephemeral, filling with water only seasonally or during storms. With the exception of forested and alpine habitats in the higher elevations, New Mexico is a dry region of mesas, badlands, arroyos, and playas. The Chihuahuan Desert, the largest desert in North America, extends north from Mexico into southern New Mexico.

When New Mexico's limited rainfall arrives in heavy storms, it can quickly erode the land. Different rock types respond to erosion in different ways. Soft sedimentary rock, covered with little vegetation to produce and hold soil, easily erodes, forming canyons and badlands. When sedimentary units are removed completely, the harder, underlying crystalline bedrock is exposed. This bedrock forms the resistant peaks of many mountain ranges. Volcanic outflows can protect underlying sedimentary rocks, forming a caprock. In this case, a hard, rigid basalt flow might protect the underlying sandstone, and this combination can form a cliff. Wind also contributes to erosion, gathering sand and depositing it in dunes where the air currents slow.

Alluvial fans, cone- or fan-shaped deposits composed of coarse sand and rock fragments, are common along the edges of recently uplifted mountain ranges. The point, or apex, of the fan points toward the source area of the sediments. Alluvial fans form when a confined water channel moves into a more open area, such as where a narrow canyon reaches the mountain front. The abrupt opening of the channel triggers a drop in stream power, or energy, resulting in the loss of the stream's ability to transport sediment. Large sediment falls out of suspension first, with finer sediments being carried farther down the length of the fan. This produces a characteristic "fining out" of the fan deposits; coarser grains settle near the mountain front, and finer grains are carried toward the end, or toe, of the fan. Alluvial fans are very common in arid and semiarid regions. The flanks of the Rio Grande rift are dotted with alluvial fans that vary in size from very small to 10 or more miles across.

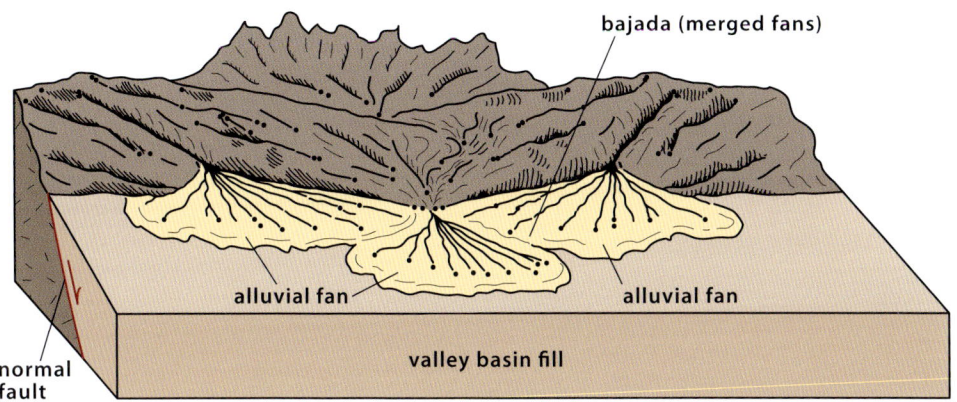

Alluvial fans form at the mouths of canyons. Bajadas, or alluvial aprons, form as alluvial fans grow and interweave with their neighbors and as mountain faces are eroded back from the faults along which they were uplifted.

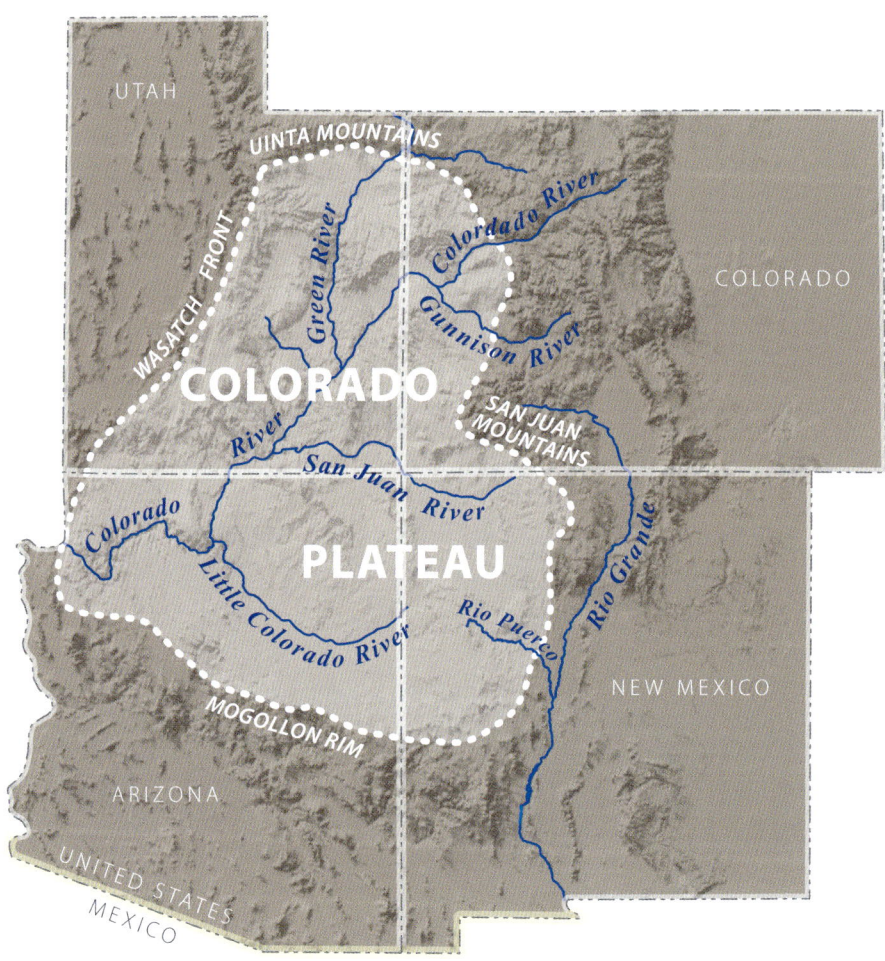

The Colorado Plateau, an uplifted region of mostly flat-lying sedimentary rock, spans four states, including northwestern New Mexico.

The Colorado Plateau

The Colorado Plateau physiographic province, the region west of the Rio Grande rift and Rocky Mountains in northwest New Mexico, is often identified by the iconic horizontal red sedimentary rock layers made famous by the Arizona, Utah, and Colorado national parks to the west and north, such as the Grand Canyon and Arches. The plateau region in New Mexico, the southeastern corner of the great Colorado Plateau, is equally spectacular but less seen. The physiographic province of the Colorado Plateau is more than 130,000 square miles in size and derives its name from the Colorado River, which crosses the region and cuts magnificent canyons through the flat-lying sedimentary rocks.

In New Mexico, the Colorado Plateau is an erosional landscape shaped by the action of water and wind. Tributaries of the Colorado River and Rio Grande dissected the region into fluted mesas separated by flats and badlands. Many of the stark mesas here are capped with black basalt flows. An abundance of modern springs and seeps exist, and numerous large-volume travertine deposits are located at the southeastern edges of the Colorado Plateau, at Mesa del Oro (south of Acoma) and near Springerville over the border in Arizona. These deposits have accumulated throughout the Quaternary, attesting to the longevity of these springs.

Uplifts of crystalline basement rocks punctuate the broad, weakly deformed sedimentary rocks, forming the cores of the Zuni Mountains, the Defiance uplift (including the Chuska Mountains), and the Nacimiento uplift. Compression during the Laramide orogeny from Cretaceous to Paleogene time produced these basement-cored uplifts, but the basement rock is visible only where deep erosion has removed the overlying sedimentary material. These Laramide uplifts are fault-bounded, and the sedimentary rock bordering them is often deformed or draped to form monoclines—enormous warps or kinks in the otherwise flat layers. Along the Arizona–New Mexico line, the rocks of eastern Arizona fold down sharply off the Defiance uplift in a prominent monocline. These steeply tilted beds are prominent as hogbacks along I-40 near Gallup. They drape at least 7,000 feet from their highest point at the crest of the Defiance uplift to their lowest position to the east in what is known as the Gallup sag. The sedimentary layers drop much more gently down to the west off the Defiance uplift west of Window Rock, Arizona. This relief is not generated by one single fault, but rather a series of faults that the sedimentary cover has been carried over like a rug over stairsteps.

In the Colorado Plateau of New Mexico, Pennsylvanian- and Permian-age rocks sit directly on the Proterozoic basement rocks. Cambrian through Mississippian rocks were either not deposited or were eroded away before the deposition of the Pennsylvanian rocks, forming a gap in the geologic rock record so notable it is called the Great Unconformity. Near Gallup, for instance, the Permian-age Supai and Coconino Formations sit directly on the Proterozoic granite, gneiss, and quartzite basement rocks. The Great Unconformity is famously exposed in the Grand Canyon and can also be found throughout parts of New Mexico, including in the Sandia Mountains near Albuquerque and in the Jemez Mountains northwest of Santa Fe.

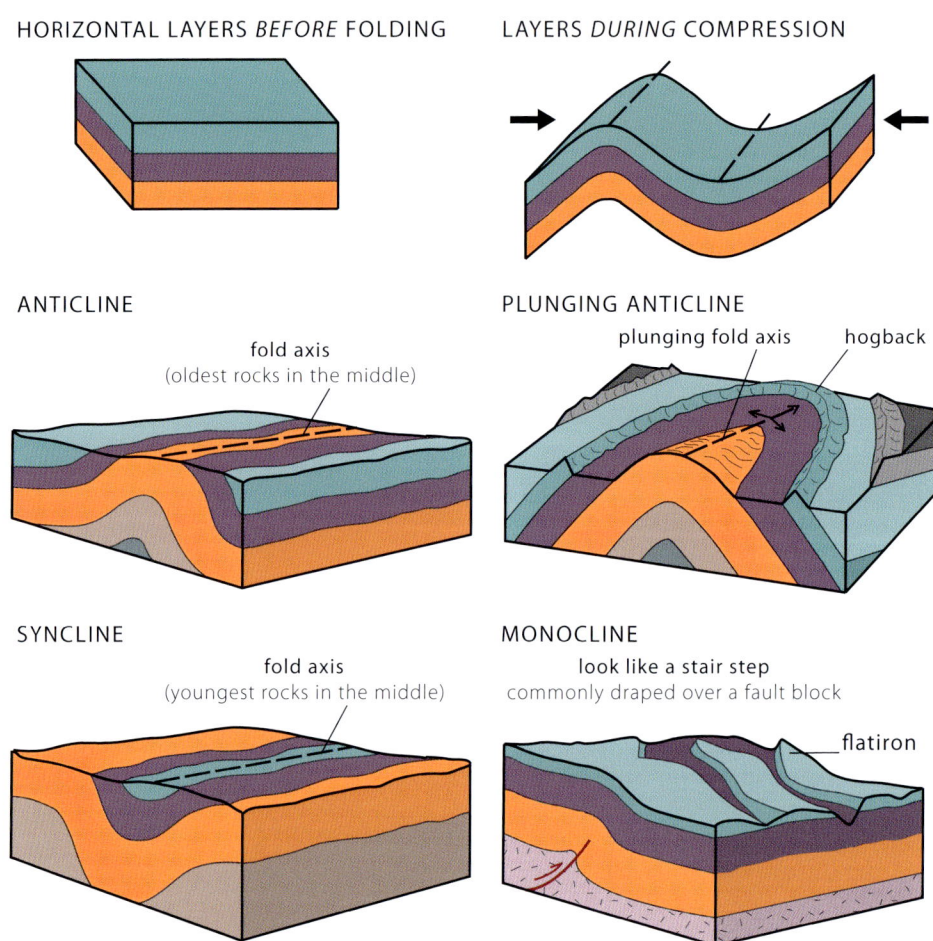

Sedimentary beds can be folded into anticlines, synclines, or monoclines.

Northeast of Gallup, the stack of plateau sedimentary rock layers dips below and into the San Juan Basin, a downward fold known as a structural basin. More than 7,500 square miles in surface area, the San Juan Basin extends into neighboring Arizona, Utah, and Colorado. This asymmetrical structural basin holds Paleozoic to Paleogene rock layers that dip gently from the southwest side of the basin. On the northeast side, where the basin abuts the San Juan Mountains, the rock layers dip more steeply and rise sharply to the surface. The more resistant sandstone layers stand out as hogback ridges bordering the Brazos and Nacimiento uplifts. These blocks of Proterozoic and Paleozoic basement rock were thrust upward along faults during Laramide continental compression. Sedimentary rocks bounding these cores have been severely deformed and faulted.

The San Juan Basin is home to the largest coal-bed methane field in the world, and the region has been extensively studied. Oil and natural gas from coal-bed methane have been extracted here since the early twentieth century. More than 300 oil fields hosting over 40,000 wells have produced more than 42 trillion cubic feet of natural

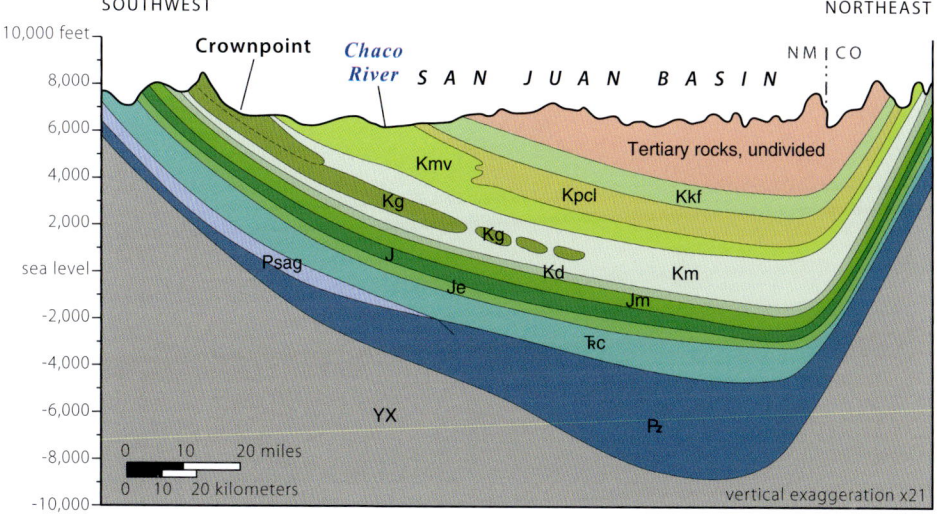

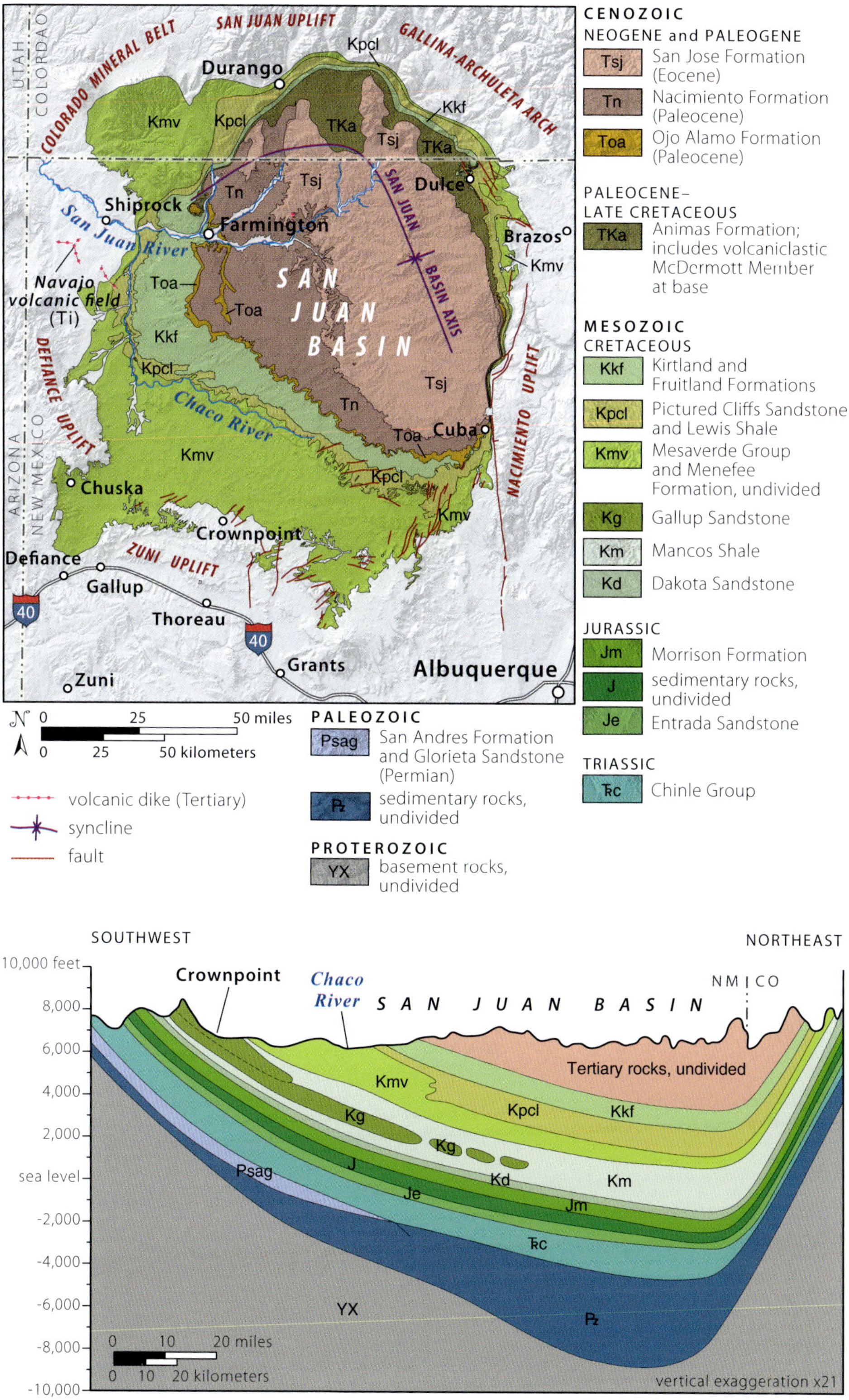

Map and cross section of the San Juan Basin. —Modified from Stone and others, 1983

COLORADO PLATEAU STRATIGRAPHIC COLUMN

ERA	PERIOD (mya)*	Rock Unit	Rock Color/Composition	Depositional Environment	Occurrence in NM Parks
CENOZOIC	QUAT. — 2.6 —	recent sediment, lavas / Santa Fe Group		soil-covered sediments deposited in basins associated with the Rio Grande rift zone	
	NEOGENE	Mt Taylor volcanics		eruption of Mt. Taylor	
	— 23 —	Bidahochi Formation		lake sediments deposited in one or more large lakes	
	TERTIARY PALEOGENE	Chuska Sandstone		broad windblown dune field	CHACO CULTURE NAT. HISTORICAL PARK / EL MORRO N.M. / EL MALPAIS NATIONAL MONUMENT
		San Jose Formation		soil-covered plains with slow moving rivers and shallow lakes	
		Nacimiento Formation		large soil-covered plains with occasional shallow lakes, rivers	
	— 66 —	Ojo Alamo Fm / Animas Fm		alluvial stream deposits	
MESOZOIC	CRETACEOUS	Mesaverde Group: Cliff House Sandstone, Menefee Formation, Point Lookout Sandstone		shoreline beaches, lagoons, and rivers	NAVAJO LAKE STATE PARK / BISTI/DE-NA-ZIN NATIONAL MONUMENT / AZTEC RUINS NATIONAL MONUMENT / EL MORRO NATIONAL MONUMENT
		Mancos Shale: Crevasse Canyon Formation, Gallup Sandstone, Tres Hermanos		ocean with volcanic ash blowing in from the west and a few shoreline sands	
	— 145 —	Dakota over Cedar Mountain–Burro Canyon		rivers, beaches, lagoons	
		Morrison Formation		floodplain with rivers, streams, and lakes	
	JURASSIC	San Rafael Group: Salt Wash Member		tidal flats, near-shore rivers	BLUEWATER LAKE STATE PARK
		Bluff Sandstone / Zuni Sandstone		sand dunes in east; sand washed by rivers or ocean in west	
		Summerville Formation		broad tidal flats	
		Todilto Formation		sand dunes	
		Entrada Sandstone		streams washing dune sand / sand dunes	
	— 201 —				
	TRIASSIC	Chinle Group / Shinarump conglomerate		continental streams, floodplains and volcanic ash	GHOST RANCH AND VICINITY / RED ROCK STATE PARK
PALEOZOIC	— 252 —	San Andres Formation		marine	
	PERMIAN	Glorieta Sandstone		shoreline sand dunes, nearshore marine sand and mud	
		Yeso Formation (south and east)		coastal sand, mostly dunes, also worked by rivers and waves	
		Abo Formation (south and east) / Cutler Formation (arkose)		alluvial fans sloping down to the west from eroding Uncompahgre Mountains	
	— 299 — PENN.				

*mya = millions of years ago

~~~ unconformity
· · · river/beach sandstone
· · · dune sandstone
= = = mudstone or shale
≡ ≡ siltstone
▭ limestone
■ coal
△▲△ volcanic ash
∘∘∘∘ conglomerate

Common rock units in the Colorado Plateau of New Mexico.

gas and 380 million barrels of oil. More than 90 percent of wells drilled in the San Juan Basin are in New Mexico, and energy extraction from these largely Cretaceous-age rocks is a major economic resource for New Mexico.

The San Juan Basin formed during the Laramide orogeny. Beginning approximately 95 million years ago, pressure from a plate collision along the western edge of North America resulted in the folding of continental crust throughout the Rocky Mountain region. Initially, a broad area of subsidence extended from the north to the south across the continent, flooding to form the Western Interior Seaway.

The depth and depositional setting of the Western Interior Seaway fluctuated over the millions of years it existed. At times, sedimentary units record a deep, quiet-water depositional setting; in others, shallow waters and influences from the nearby shores are seen. These varied settings led to the varied types of rocks we see across the San Juan Basin. The first major transgression of the Western Interior Seaway inundated the area around 95 to 80 million years ago. The first marine deposit was the Mancos Shale, preserved here and in other basins across central North America. This was followed by a drop in sea level and the deposition of the Mesaverde Group. A return to watery environments marks another sea transgression and the deposition of the Lewis Shale and the coal-rich Pictured Cliffs and Fruitland Formations. Deposition of marine units ceased as compression and tectonic deformation of the Laramide orogeny began to exert additional stress and much more intense mountain range uplift and basin subsidence to the region, bringing the final regression of the Western Interior Seaway.

The uplift of any mountain range subjects its rocks to erosion. The sediments that fill the San Juan Basin came from many directions. Volcanic debris sourced from early San Juan volcanic field eruptions in southwestern Colorado entered the basin from the north in the form of the conglomerates and sandstones of the Ojo Alamo Sandstone and Animas and Nacimiento Formations. The youngest basin fills are the Eocene-age sandstones and shales of the San Jose Formation.

The western and southeastern parts of the San Juan Basin are dotted with the stark volcanic necks of the Navajo volcanic field. The most famous of these is Ship Rock, west of Farmington, rising about 1,580 feet above its surroundings, with radiating dikes forming walls and ridges across the desert. These volcanic necks and the dikes that connected and distributed magma were once part of an ancient volcano's underground plumbing. Formed 27 million years ago and nearly 3,000 feet below the ground surface, the volcanic rock cooled from magma that was moving through the sedimentary rocks. In the time since eruption and cooling, the softer sedimentary rocks through which the volcano originally erupted have eroded, leaving the more resistant volcanic rocks standing in high relief.

I-40
Arizona—Gallup—Grants
79 miles

Following the valley of the Puerco River, I-40 enters New Mexico between cliffs of colorful Jurassic sandstone that weathers to smooth surfaces. Many cavities and arching alcoves have formed where unsupported rock fell away. Above the main cliffs are younger Mesozoic rocks, notably the Dakota Group, a stack of sandstone,

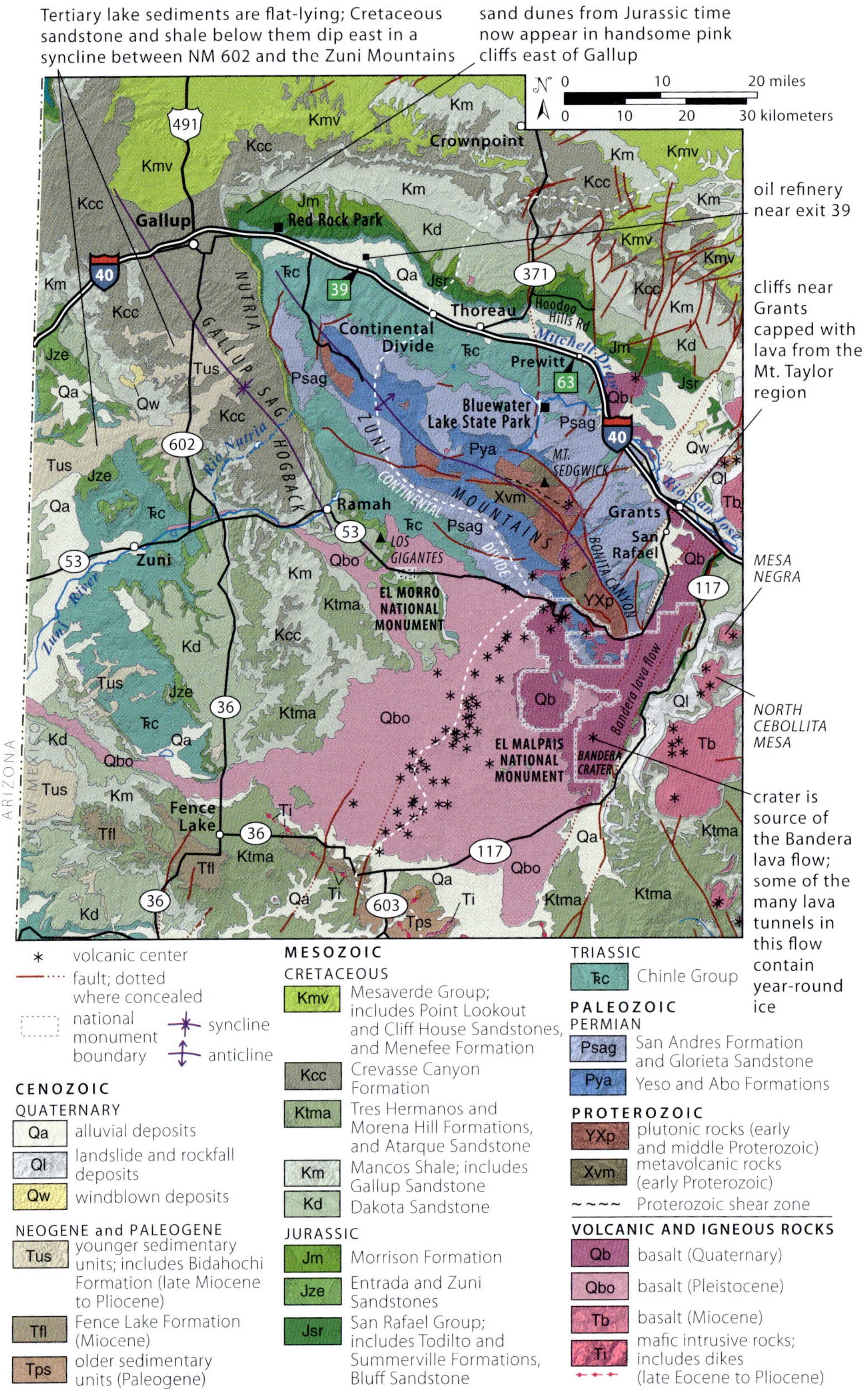
Tertiary lake sediments are flat-lying; Cretaceous sandstone and shale below them dip east in a syncline between NM 602 and the Zuni Mountains
sand dunes from Jurassic time now appear in handsome pink cliffs east of Gallup
0 10 20 miles
0 10 20 30 kilometers
oil refinery near exit 39
cliffs near Grants capped with lava from the Mt. Taylor region
MESA NEGRA
NORTH CEBOLLITA MESA
crater is source of the Bandera lava flow; some of the many lava tunnels in this flow contain year-round ice
Crownpoint
Gallup
Red Rock Park
Thoreau
Continental Divide
Prewitt
Bluewater Lake State Park
MT. SEDGWICK
Grants
San Rafael
Ramah
Zuni
LOS GIGANTES
EL MORRO NATIONAL MONUMENT
EL MALPAIS NATIONAL MONUMENT
BANDERA CRATER
Bandera lava flow
Fence Lake
NUTRIA
GALLUP SAG
HOGBACK
ZUNI MOUNTAINS
CONTINENTAL DIVIDE
BONITA CANYON
Hoodoo Hills Rd
Mitchell Draw
Rio Nutria
Rio San Jose
Zuni River
ARIZONA
NEW MEXICO
491
40
39
371
63
602
53
117
36
603
volcanic center
fault; dotted where concealed
national monument boundary
syncline
anticline
CENOZOIC
QUATERNARY
Qa alluvial deposits
Ql landslide and rockfall deposits
Qw windblown deposits
NEOGENE and PALEOGENE
Tus younger sedimentary units; includes Bidahochi Formation (late Miocene to Pliocene)
Tfl Fence Lake Formation (Miocene)
Tps older sedimentary units (Paleogene)
MESOZOIC
CRETACEOUS
Kmv Mesaverde Group; includes Point Lookout and Cliff House Sandstones, and Menefee Formation
Kcc Crevasse Canyon Formation
Ktma Tres Hermanos and Morena Hill Formations, and Atarque Sandstone
Km Mancos Shale; includes Gallup Sandstone
Kd Dakota Sandstone
JURASSIC
Jm Morrison Formation
Jze Entrada and Zuni Sandstones
Jsr San Rafael Group; includes Todilto and Summerville Formations, Bluff Sandstone
TRIASSIC
TRc Chinle Group
PALEOZOIC
PERMIAN
Psag San Andres Formation and Glorieta Sandstone
Pya Yeso and Abo Formations
PROTEROZOIC
YXp plutonic rocks (early and middle Proterozoic)
Xvm metavolcanic rocks (early Proterozoic)
Proterozoic shear zone
VOLCANIC AND IGNEOUS ROCKS
Qb basalt (Quaternary)
Qbo basalt (Pleistocene)
Tb basalt (Miocene)
Ti mafic intrusive rocks; includes dikes (late Eocene to Pliocene)

Geology along I-40 between the Arizona border and Grants.

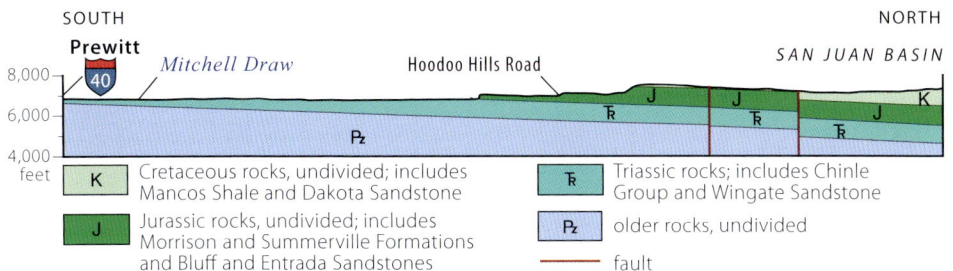

Cross section extends 10 miles to the northeast from I-40 near Thoreau. Note the rock layers dipping north into the San Juan Basin. —Modified from Rawling and others, 2015

mudstone, and shale deposited in the Late Cretaceous in warm, wet river and coastal environments at the edge of the Western Interior Seaway. Except for several small anticlines, these rock layers all dip eastward into a broad syncline called the Gallup Sag, with rocks becoming younger to the east. Within a few miles of the Arizona border, the massive sandstones of the Dakota Group disappear, and reddish Late Cretaceous Gallup Sandstone tops the cliffs.

Roadcuts near milepost 22 in Gallup expose the cross-bedded Gallup Sandstone. The pink, well-bedded sandstone has small layers of shale and siltstone and thin beds of coal. In places, sand-filled channels cut into the coal beds. The coal formed from plant material deposited in marshy or swampy lagoons along Cretaceous shores, which overlying sediments then buried and compacted. Layers of both sandstone and coal thin, thicken, and disappear—the irregularities expected in the coastal environment in which these rocks were deposited. Coal deposits near Gallup were among the factors considered when the Santa Fe Railroad was routed through here.

Cretaceous and older rocks bend up suddenly into a hogback near milepost 25, just east of Gallup, so that from this point eastward, successively older rocks—Cretaceous,

Cliffs of Jurassic sandstones rise above the Puerco River along the Arizona–New Mexico border. A tributary of the Colorado River, this ephemeral river flows in a deep arroyo with high, steep walls of fine, tan, poorly consolidated silt.

then Jurassic, then Triassic—border the highway. The tilted rocks are on the northwestern edge of the Zuni uplift, a northwest-trending anticline. Cretaceous rocks and underlying older Jurassic, Triassic, and Permian layers are faulted here to form the Nutria monocline, locally called the Hogback. Two ridges parallel each other and define this western side of the larger Zuni Mountain anticline. The rock layers tilt steeply, with dips commonly more than 60 degrees, and in places become nearly vertical. The faults that formed this monocline have moved nearly 4,000 feet here to set Permian rocks (to the east) against the Cretaceous Dakota Sandstone to the west. Ten to fifteen miles to the south of the highway, the Zuni Mountains rise above the sedimentary rocks. The Proterozoic granites and metamorphic rocks of the Zuni Mountains were originally amalgamated during continental accretion and then uplifted during the Ancestral Rocky and Laramide orogenies.

A sharp hogback just east of Gallup near milepost 25 marks the western edge of the Zuni uplift.

Tall buttresses of red Entrada Sandstone continue to rim the area to the north of the highway east of Gallup. The foreground valley, eroded in weak Triassic and Jurassic shales and mudstones, is floored with fine silt and etched with deep gullies.

At exit 39, the highway passes an oil refinery. When the road crosses the Continental Divide near milepost 48, it leaves the drainage basin of the Colorado River, which drains westward to the Pacific Ocean, and enters the drainage basin of the Rio Grande, which eventually drains into the Gulf of Mexico and the Atlantic Ocean. The highway runs to Grants in a valley eroded by the Rio San Jose, which cuts into the soft Triassic mudstones and siltstones of the Chinle Group. Some of the dark-red and gray Chinle appears in highway roadcuts. Ahead to the northeast is the graceful volcanic cone of Mt. Taylor. Bluewater Lake State Park is located to the south from exit 63. The popular fishing reservoir impounds water on the limestone of the San Andres Formation, surrounded by sculpted cliffs of Triassic Chinle Group.

The red cliffs of Entrada Sandstone to the north of the highway end abruptly southeast of Prewitt at milepost 65, where a large, rift-related fault has dropped the sedimentary units down to the east. To the east, cliffs of tan sandstone and gray shale of the Mesaverde Group are capped with Cenozoic basalts. These sedimentary

THE COLORADO PLATEAU   39

Huge cliffs of Entrada Sandstone form the backdrop for Red Rock Park, visible to the north from milepost 29. Large-scale cross-beds of fine, well-sorted sand grains indicate this formation originated as desert sand dunes in Jurassic time.

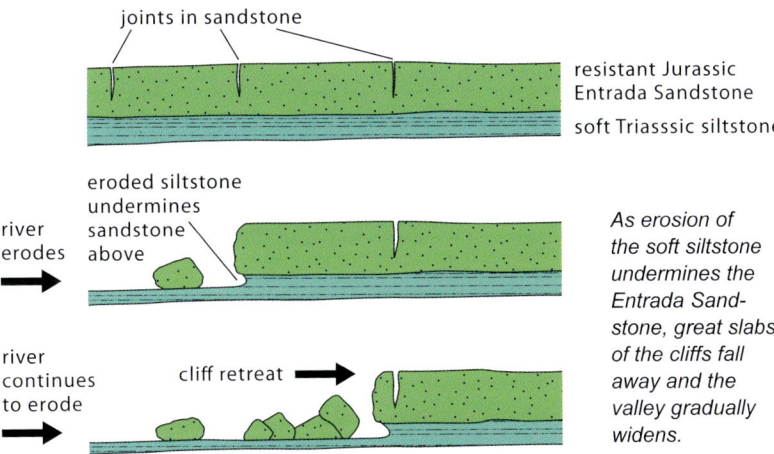

units are warped into shallow anticlines and synclines. Just southeast of Prewitt, the highway crosses a Quaternary lava flow, the youngest rock eastbound travelers will have seen along this route between Arizona and Grants.

US 64
Chama—Farmington—Arizona
161 miles

A short distance south of Chama, US 64 turns westward, skirting tall bluffs of Mancos Shale capped with sandstone of the Mesaverde Group. The Mancos Shale is at the surface of the Chama Basin, which extends south for about 30 miles. This basin warped downward during the Laramide orogeny about 75 to 40 million years ago, when compression bowed up the mountain ranges of the Rocky Mountains and folded down basins between them.

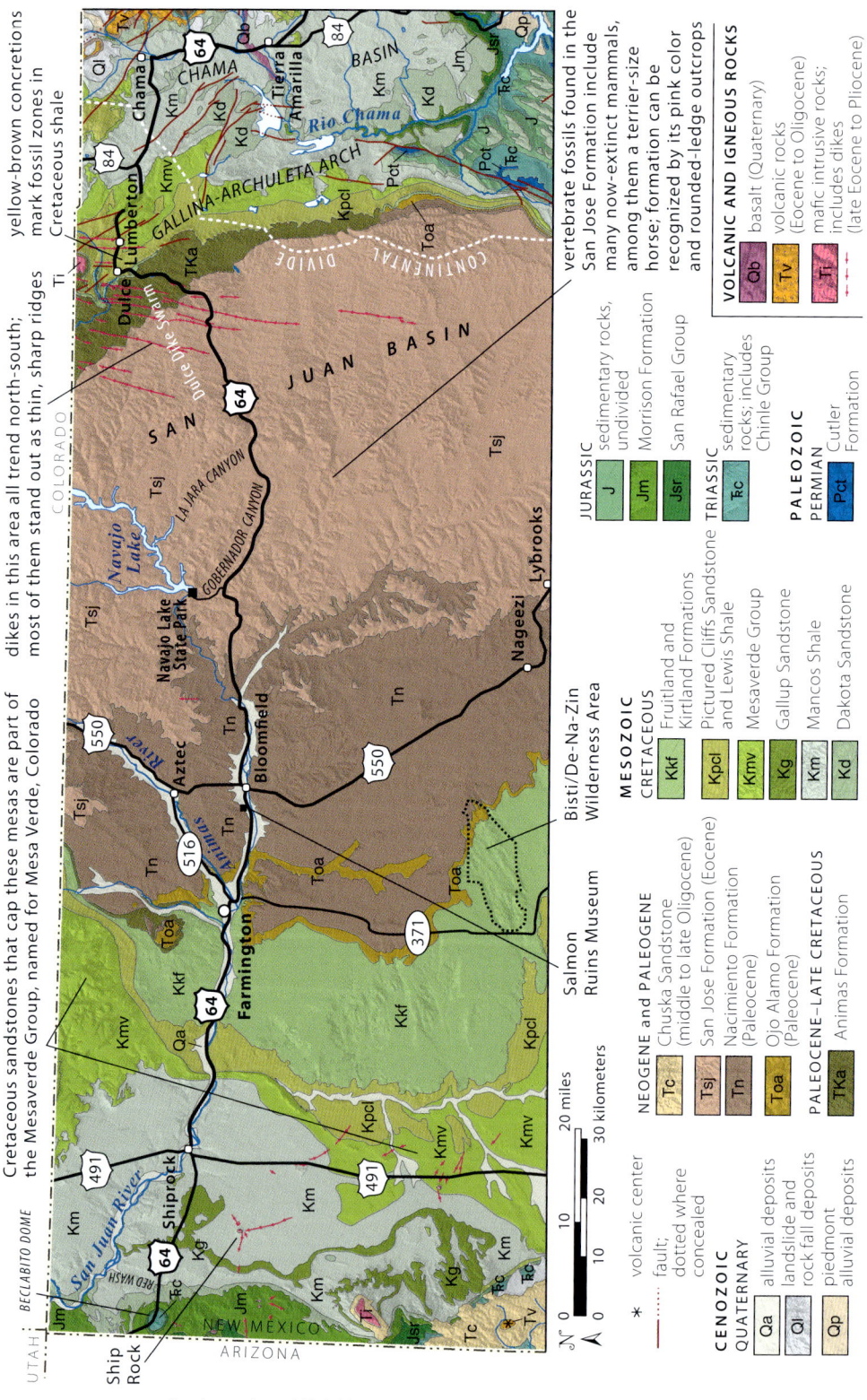

Geology along US 64 between Chama and the Arizona border.

Near milepost 153, US 64 crosses the Continental Divide, leaving the Rio Grande drainage basin—the water of which ultimately ends up in the Gulf of Mexico—and entering the drainage of the San Juan River, which flows into the Colorado and onto the Pacific. Elevation of the divide here is 7,275 feet.

West of the junction with US 84, at about milepost 147, a mesa to the north boasts a prominent cliff of mostly flat-lying Point Lookout Sandstone, the lower sandstone of the Mesaverde Group. Farther west, however, the sandstone starts to dip toward the road, indicating the highway has crossed the crest of the Gallina-Archuleta arch, a north-south-trending anticline that separates the Chama syncline from the much deeper San Juan Basin to the west. Rock layers now become younger as the route proceeds westward into the San Juan Basin.

The San Juan Basin, a geologic subprovince, extends into Colorado to the north and is nearly 100 miles north to south and 90 miles east to west. The San Juan Basin's stratigraphy and tectonic history started in the Late Cretaceous when a collision along the western edge of the North American plate resulted in a downwarping of continental crust. The San Juan Basin filled as it subsided and now contains a vast array of sedimentary sand, gravel, clay, and volcanic ash deposits sourced from the San Juan Mountains to the north and the southern tip of the Rockies to the east.

Coal was mined at Monero (milepost 145) from beds in the Menefee Formation, the shaly middle unit of the Mesaverde Group. The Menefee Formation was deposited in river floodplains and coastal swamps. Coal forms from organic matter, largely plant matter that collects in swamps, marshes, and peat bogs, and is then compacted and heated as overlying sediments accumulate. The coal here is soft bituminous coal rather than the hard anthracite found in the eastern United States.

In this roadcut just west of milepost 146 along US 64, the layers of the lower Mesaverde Group sandstone are streaked with coal. The beds dip toward the San Juan Basin because this outcrop is on the west side of the Gallina-Archuleta arch.

At milepost 139 near Lumberton, the road crosses west onto the Lewis Shale. This olive-gray marine shale is similar in appearance to the Mancos Shale but is younger, deposited about 75 million years ago on the western slopes of the Western Interior Seaway.

Sharp ridges on either side of milepost 138 are two of the many basalt dikes that cut north-south through this area as part of the Dulce dike swarm. The highway crosses several of these dikes in the next 15 miles. Many of the dikes are only 3 to 6 feet thick but can be traced for nearly 12 miles. The majority were emplaced 26 to 25 million years ago as part of the post-caldera magmatic activity associated with the Platoro caldera in the San Juan volcanic field of southern Colorado. The intruding magma baked and hardened the shale on either side of the dikes. The basalt of the dikes and the baked shale, both harder than surrounding rocks, tend to stand up as sharp ridges.

From Dulce (milepost 136), US 64 heads south into a short canyon walled first by the Cretaceous Lewis Shale, then a cliff of Pictured Cliffs Sandstone, and finally the Animas Formation. The Pictured Cliffs Sandstone, composed of the sands of beaches and nearshore waters, marks the shoreline of the Late Cretaceous sea. The younger Animas Formation, consisting of conglomeratic sandstone alternating with shales and mudstones, contains volcanic clasts and ash brought into the basin in Paleocene time from the rising San Juan volcanic field in southwestern Colorado.

The route curves westward again near milepost 125 at the junction with NM 537, where the road crosses from the Animas Formation west onto the San Jose Formation. These tan-colored beds were deposited in high-energy river channels and floodplains in Eocene time during initial subsidence of the San Juan Basin. They were later folded by movement of the Nacimiento fault, the major fault along the western edge of the Rocky Mountains in New Mexico. The San Jose Formation contains

A basaltic dike of the Dulce dike swarm cuts north-south through horizontal layers of Eocene San Jose Formation rock near milepost 121.

many diverse fossil vertebrates and is largely sandstone with cross-stratification of conglomerates and minor channel deposits that alternate with muddy silts, clays, and fine-grained sandstones that have experienced extensive bioturbation. In places the San Jose Formation erodes into spectacular hoodoos and mushroom rocks, visible in the Bisti/De-Na-Zin and the Ah-Shi-Sle-Pah Wilderness Areas (managed by the BLM). These areas are located south of Bloomfield at milepost 65 of this route. West of milepost 122, the road wends its way through breaks in two prominent, finned-top ridges that are dikes of the linear Dulce dike swarm.

The Dulce dike swarm near Lumberton and Dulce forms ridges and walls stretching for many miles. This one, also near milepost 121, forms a nicely finned ridge.

Near milepost 113 at the Forest Service boundary, wind and rain have sculptured soft sandstone of the San Jose Formation.

The highway crosses several canyons draining northwest into the San Juan River. These canyons house ephemeral streams and see seasonal flash floods, powerful erosive forces in the weak Paleogene sedimentary units. La Jara Creek, a tributary of the San Juan, flows in La Jara Canyon (milepost 107). The San Jose Formation forms irregular ledges and slopes of tan sandstone and shale that weather reddish brown. Oil well sites become more prevalent as the road enters the San Juan Basin natural gas field. Canyon bottoms are favored drilling sites because wells do not have to be as deep. West of La Jara Canyon, the highway drops once more through layered sedimentary rocks into the wider Gobernador Canyon, home to another San Juan tributary, and follows it between mileposts 96 and 88.

Wells and storage tanks are abundant near La Jara Canyon (milepost 107). These wells are the surface markers of rich oil, gas, and coal-bed methane found in the San Juan Basin.

NM 539, at milepost 88, leads northwest to Navajo Lake State Park, home to the second-largest body of water in New Mexico (after Elephant Butte Reservoir). The earth-and-rock dam is 402 feet high and spans 3,648 feet across the San Juan River. Finished in 1962, the reservoir collects water from La Jara Creek and the San Juan, Piedra, and Los Pinos Rivers.

West of milepost 79, surface rocks are the Paleocene-age Nacimiento Formation. From here, rocks become older westward because the road has crossed the deep axis of the San Juan Basin. Beds now dip shallowly down to the east, but the shift in dip direction of the sedimentary layers is barely noticeable.

Between mileposts 75 and 74, US 64 crosses the San Juan River, a major tributary of the Colorado River with headwaters in the eastern San Juan Mountains of Colorado. Bluffs here expose the Animas Formation. US 64 follows the San Juan downstream to the town of Shiprock, and the river water provides irrigation for fertile bottomland soils.

At milepost 70, east of Bloomfield, the Ojo Alamo Formation is exposed. West of downtown Bloomfield, between mileposts 62 and 61, is the Salmon Ruins Museum, which provides access to an ancient Chacoan peoples archeological site. At milepost 60, the soft hills are composed of the brown, coal-bearing shales of the Late Cretaceous-age Kirtland and Fruitland Formations.

Farmington, at milepost 54, is built upon the Quaternary alluvial deposits at the confluence of the San Juan and Animas Rivers. These old river terraces record the incision of the river into the valley floor. Terraces originate as floodplains whose river remains in a stable environment, migrating back and forth and depositing layer after layer of sediment. When the river system becomes more erosive, it can incise into this floodplain, eroding its own sediments, and leaving flat terraces to either side of the modern course. The shifts between depositional and incisional river systems are tied to changes in the base level of the rivers' destination. Integration with other major rivers or the rising or lowering of a sea can change the local base level. Climate also plays an important role in geology. Long-term drought or periods of glaciation can result in drastic changes to water volume and sediment availability and transportation.

On the south side of Farmington, NM 371 climbs out of town past Farmington's Dune Vehicle Recreation Area, where dunes and rough badlands are edged by friable cliffs of the Ojo Alamo Formation. This Cretaceous to Paleocene–age unit contains numerous fossils, including dinosaurs, crocodiles, and turtles.

Small, abandoned coal mines can be seen in some of the rocks north of the highway near Farmington. The Four Corners Generating Station west of Farmington burns Cretaceous coal from larger mines. Smoke plumes from this plant have been blamed for polluting the air of the Four Corners region and endangering the health of those who live downwind from the facilities. The Arizona Public Service has begun decommissioning of the plant, which should be complete by 2031.

Badlands in the Ojo Alamo Formation on the south side of the San Juan River valley at Farmington.

At milepost 31, Cretaceous rocks rise sharply, forming cuestas and hogback ridges. The Menefee Formation shales form the center part of the photo; the ridge to the left is the Point Lookout Sandstone; the ridge to the right is the Cliff House Sandstone.

The highway quickly crosses more than 9,000 feet of Cretaceous sedimentary rock west of Farmington: the Kirtland and Fruitland Formations form the surface between mileposts 44 and 38 as the road passes through their namesake towns of Fruitland and Kirtland. Then comes the Pictured Cliffs Sandstone (milepost 35), and Lewis Shale (milepost 33). The Cliff House Sandstone, Menefee Formation, and Point Lookout Sandstone are crammed together and passed within 1 mile, exposed in a hogback at milepost 31. This hogback along the western edge of the San Juan Basin is known simply as the Hogback and continues for many miles north and south of the highway. West of this striking feature is a wide expanse of Mancos Shale.

The highway diverges from the San Juan River west of the town of Shiprock, but both continue through the Mancos Shale and its light-colored silty coating. In the

Along the San Juan River near Shiprock, bluffs of Mancos Shale show that the Cretaceous rocks have leveled out on the western side of the San Juan Basin.

distance to the south, you can see Ship Rock, a volcanic neck and major landmark. Ship Rock is discussed in the road guide for US 491: Gallup—Colorado.

Between mileposts 10 and 7, US 64 crosses the eroded valley of Red Wash, filled with sediments derived from the many nearby red rocks. The red coloring comes from the iron oxide that cements the grains of sandstone together. The highway then climbs onto the Beclabito Dome, cresting it near milepost 4. It is thought to be domed by an underlying igneous intrusion, although no igneous rocks are exposed. The uplifted, reddish Morrison Formation and San Rafael Group sandstones are visible from the road. In the distance to the west rise the Carrizzo Mountains of Arizona.

US 491
Gallup—Colorado
110 miles

US 491 at Gallup lies on the Crevasse Canyon Formation, a Cretaceous-age, coal-bearing unit. The Western Interior Seaway, from which the Mancos Shale and Mesaverde Group were also derived, encroached into New Mexico in the Late Cretaceous, uniting shallow seas that existed to the north and south. The seaway rose and fell numerous times, and these environmental changes are recorded in the varied rock types that were deposited at different seawater depths. The withdrawal of the seas is part of what defines the end of the Cretaceous time period. All younger sedimentary rocks in this area, including the thick Paleogene deposits that fill the center of the San Juan Basin to the northeast of Gallup, are continental, having been deposited on land.

North of Gallup, at milepost 4, the highway crosses into surface exposures of the shale and sandstone of the Menefee Formation, part of the Mesaverde Group. This

The alternating hard and soft layers of the Menefee Formation near milepost 10 result in the ledge-slope-ledge landscape so common in the Colorado Plateau.

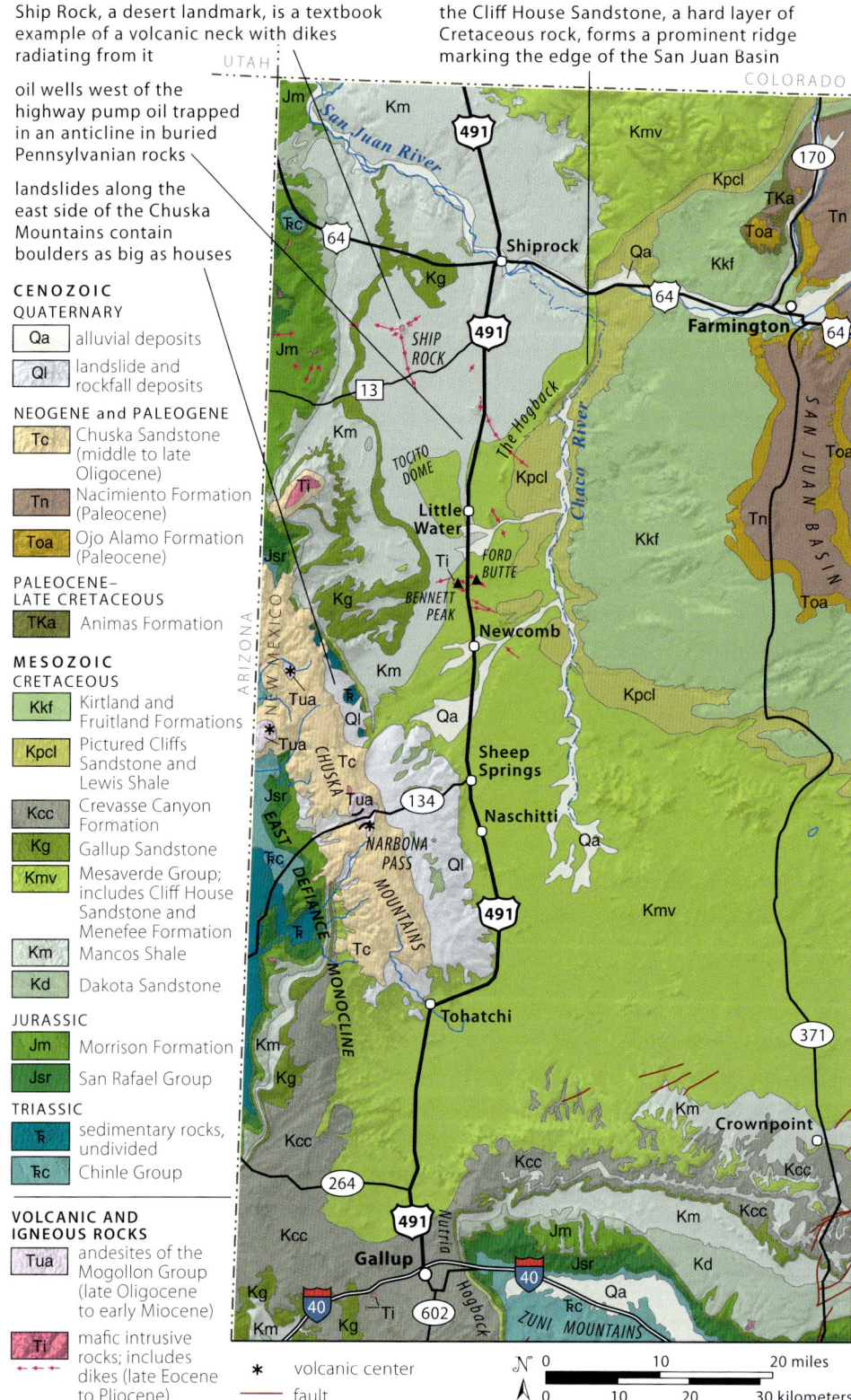

Geology along US 491 between Gallup and the Colorado border.

The eastern flank of the Chuska Mountains is covered with landslide deposits—boulder-studded debris on a hummocky slope that contrasts with the precisely defined slopes and mesas east of the highway.

unit consists of light-colored, often cross-bedded, fairly resistant sandstone layers that alternate with dark, slope-forming shale and mudstones, with seams of coal near the bottom and top of the formation. The unit forms many picturesque erosional features: buttes, finger-like pinnacles, and mushroom rocks.

West of the highway, older rocks rise to the surface in a long, deeply carved escarpment. These rocks dip eastward as the eastern slope of the great fold of the Defiance uplift, centered just west of the Arizona state line. The asymmetrical monocline is approximately 35 miles wide and 95 miles long from north to south. The Chuska Mountains, the toe of which we curve around at Tohatchi near milepost 25, are part of that long ridge-like anticline and straddle the Arizona–New Mexico state line. The mountains are topped with the 33- to 27-million-year-old Chuska Sandstone, which sits unconformably above the underlying, tilted Cretaceous, Jurassic, and Triassic rock layers. The Chuska Sandstone is an erosional remnant of a once much larger expanse of windblown sand deposited in an erg, a regional sand sea similar to that seen in the Sahara Desert. The unit is very pale in color—almost white when freshly broken, then weathering to yellowish gray. The Chuska Mountains are fringed with landslides where younger rocks slid down over soft, slippery shales that wall the range.

In the Chuska Mountains west of US 491 at Sheep Springs (milepost 47), the Narbona Pass maar volcano is well-exposed on the Defiance uplift. This volcano is one of about eighty volcanic features of the Oligocene to Miocene Navajo volcanic field. A maar is a broad crater that forms when rising magma contacts groundwater and heats the water enough to form steam. This confined steam and magma then violently erupt, forming a shallow maar crater at the surface of the Earth, which is surrounded by a blanket of ejected volcanic material. The Narbona Pass maar feature consists of a sequence of tuffs capped with lava flows, the oldest of which erupted 25 million years ago. Continued pyroclastic surges and fallout deposits were capped with the lava flows and plugs. There are also pillow lavas, bulbous features that form when lava enters water. The maar can be reached by driving west on BIA-32 (Tribal Road 134) about 10 miles.

Bedded pyroclastic deposits erupted from the Narbona Pass maar volcano form a forbidding wall on the eastern side of the Chuska Mountains near milepost 9 on BIA-32.

Columnar volcanic rock forms a noticeable cliff above BIA-32 just to the west of Narbona Pass. The column-shaped jointing occurred as the lava cooled and contracted.

THE COLORADO PLATEAU

A. Laramide time (approximately 90–45 million years ago)

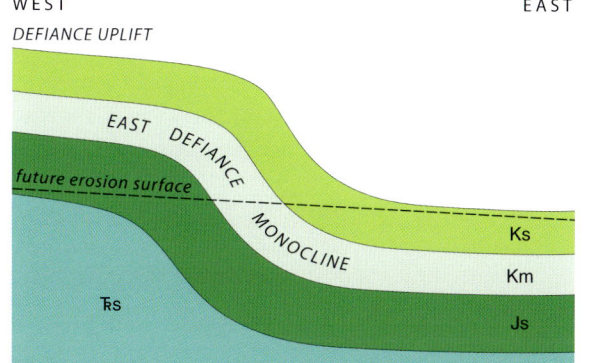

B. period of erosion (approximately 35 million years ago)

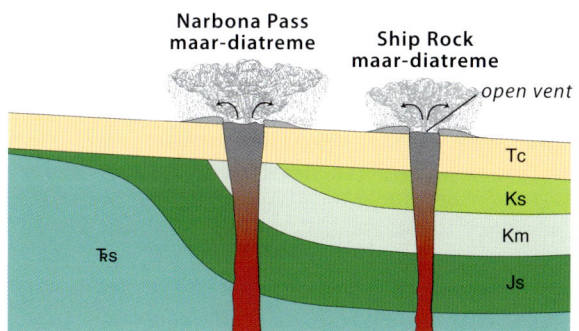

During the Laramide orogeny, Mesozoic rock layers were bent into a monocline, forming the Defiance uplift. Erosion leveled the landscape, and the Chuska Sandstone was deposited during the Oligocene Epoch. During Oligocene magmatic activity, volcanic diatremes intruded and exploded into the landscape. Millions of years of erosion removed most of the volcano, but much of one remains at Narbona Pass. —Modified from Brand and others, 2009

C. mid-Oligocene (approximately 25 million years ago)

D. present

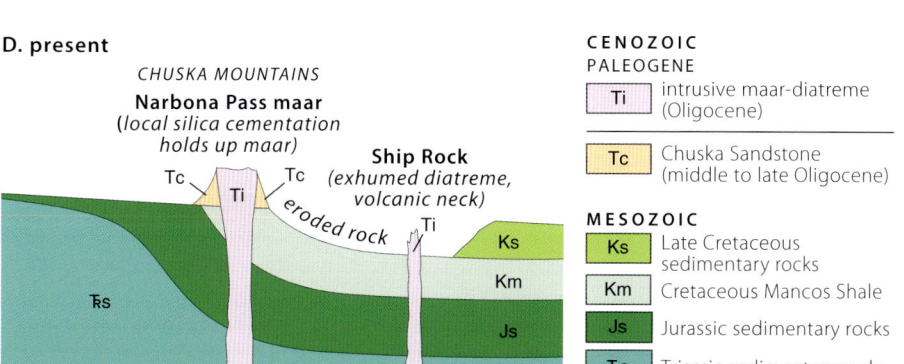

CENOZOIC
PALEOGENE
- Ti — intrusive maar-diatreme (Oligocene)
- Tc — Chuska Sandstone (middle to late Oligocene)

MESOZOIC
- Ks — Late Cretaceous sedimentary rocks
- Km — Cretaceous Mancos Shale
- Js — Jurassic sedimentary rocks
- ₸Rs — Triassic sedimentary rocks

North of Newcomb (milepost 57), the route enters a land of volcanic necks, dikes, and ubiquitous prairie dog burrows. The western and southeastern parts of the San Juan Basin are dotted with gaunt volcanic necks of the Navajo volcanic field. These volcanic necks and connected dikes distributed magma as part of an ancient volcano's underground plumbing. In the time since eruption and cooling, the softer sedimentary rocks through which the volcanoes originally erupted have eroded, leaving the more resistant volcanic rocks standing in high relief. Bennett Peak, west of the highway at milepost 63, and Ford Butte to the east are small mafic intrusions. Dikes radiate outward from them, forming cross-country ridges that can be traced in some cases for many miles. Volcanic dikes, diatremes, and plugs are characteristic features of the Navajo volcanic field, which was active from 28 to 19 million years ago.

To the east beginning at milepost 68, the long ridge of the Hogback marks one flank of a small monocline, a fold formed of steeply tilted Cliff House Sandstone, part of the Mesaverde Group.

To the west of milepost 70, oil wells dot the crest of an anticline called Tocito Dome. An anticline is one type of geologic structure that forms a trap for oil. Because oil floats on water, it tends to work its way upward through porous, water-saturated rock—in this case sandstone—until it reaches an impermeable layer. Unless the sandstone is perfectly horizontal, the oil will then migrate up-dip as far as it can. Here, impermeable rock layers arching over the crest of the anticline trap upward-migrating petroleum. There it remains, in a sort of upside-down pool. Anticlines on the surface commonly reflect anticlines at depth, so wells such as these perch on top of them. Oil may also be trapped along faults where a permeable layer happens to abut impermeable rock, or in areas where sandstone or other porous rock layers thin and pinch out. Structural petroleum traps of all these types exist in the San Juan Basin.

Near milepost 71, the highway leaves the Menefee Formation and crosses north onto the Mancos Shale, a dark-gray marine shale also of Cretaceous age, but somewhat older than the Menefee. At milepost 79, the highway passes close to the chubby finger of another volcanic neck with dikes radiating from it. At milepost 80, look northwest for a fine view of Ship Rock rising high above the Mancos Shale that now forms the surface of the plain and is often coated with ephemeral silt and sand. Ship Rock, part of the Navajo volcanic field, is an erosional remnant of the throat of a volcano that erupted about 27 million years ago. It is composed of volcanic breccia and minette, a black igneous rock rich in potassium. The throat seen today originated more than half a mile below the land surface. More resistant to millennia of erosion than the soft sandstone and shale country rock through which the volcano erupted, the hard volcanic remnants stand starkly in the landscape. Six dikes, which originated as magma that flowed away from the main conduit through cracks in the host bedrock, radiate from the center of the volcano. Closer views of Ship Rock are accessible via BIA 13, which goes west from just south of the Shiprock airstrip at milepost 85.

The Four Corners Generating Station east of Ship Rock, just visible to the east from milepost 89, burns coal mined from the nearby Mesaverde Group. Here, coal comes from accumulated plant material laid down in coastal swamps and marshes of Cretaceous time, which was later buried and metamorphosed. The Arizona Public Service has begun decommissioning of the plant, which should be complete by 2031.

US 491 crosses the San Juan River, one of the main tributaries of the Colorado River, just south of the town of Shiprock near milepost 92. The town sits on silty hills of Mancos Shale in the well-watered valley. The road climbs north out of the valley through yellow bluffs of Mancos Shale onto a broad tableland surfaced with the formation's gray silt. The ground is thinly coated with pebbly desert pavement, small stones left when wind blows away finer sediment.

In the distance to the north rise the steep flanks of Mesa Verde, capped with thick sandstone layers of the Mesaverde Group, among them the Cliff House Sandstone seen in the Hogback. These sandstone layers represent beaches and bars along the edge of

Ship Rock, one of New Mexico's most famous landforms, towers 1,581 feet above the surrounding desert. The long ridge to the left is a dike, a wall-like ridge of solidified magma that flowed through a crack radiating from the volcano.

Note the brick-like jointing of the igneous rock in the dike that runs south of Ship Rock, here only a few feet thick. When the rock cooled, shrinking caused joints in both dikes and necks.

the shallow Cretaceous sea. High mountains visible farther north over Mesa Verde are the San Juan Mountains of southwestern Colorado, headwaters of both the San Juan River and the Rio Grande. The main mass of Mesa Verde, as well as Mesa Verde National Park, is in Colorado. West of Mesa Verde is the Sleeping Ute Mountain in Colorado. Its highest peak forms the Ute's folded arms. The mountain is composed of laccoliths, whose intruding magma domed up overlying sedimentary rocks in Cenozoic time.

View to the northeast near the Colorado border shows sandstone cliffs of the Mesaverde Group above eroding slopes of Mancos Shale. Mesa Verde is visible in the distance. Note the beautifully precise sculpturing on the steep slopes, the treelike branching of gullies, and the sharpness of ridges—all common arid-climate erosional features.

US 550

COLORADO—CUBA—SAN YSIDRO
Including Aztec Ruins National Monument and Chaco Cultural National Historic Park
112 miles

US 550 enters New Mexico in the broad Animas River valley, whose floodplains and water support agriculture and industry, as well as the many small towns of northwestern New Mexico. During the late Pleistocene, the Animas Glacier was one of the largest masses of ice in the Rockies and flowed through the San Juan Mountains as far down the Animas River valley as Durango, Colorado, about 20 miles north of the New Mexico border. Meltwater flowed downstream, depositing glacial outwash in the valley in New Mexico.

The canyon is walled with the 64.5- to 61-million-year-old Nacimiento Formation, whose interbedded sandstone, mudstone, conglomerate, and shale indicate fluctuating environments of deposition at the beginning of the Paleogene Period. The sediment deposited in floodplains, rivers, and lakes in the gradually bowing-down San Juan Basin came from the San Juan Mountains to the north and the Sangre de

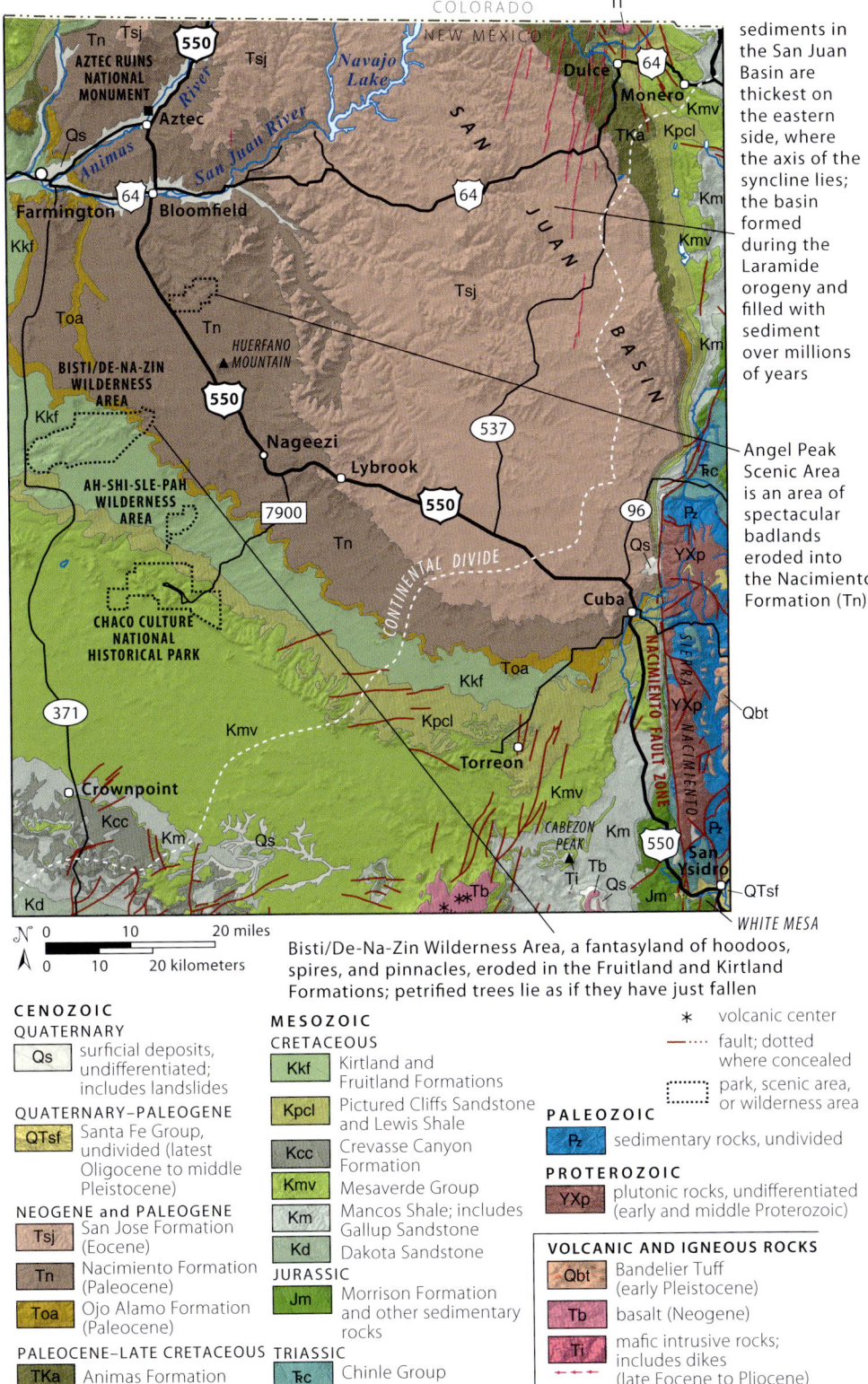

Geology along US 550 between the Colorado border and San Ysidro.

AZTEC RUINS NATIONAL MONUMENT

Aztec Ruins National Monument, a settlement that was occupied between AD 1050 and AD 1150, was excavated and partially restored during the 1920s. The site includes a large restored kiva, several smaller kivas, and numerous smaller rooms, many of which are connected into one large building. The settlement was located between Mesa Verde and Chaco Canyon, two of the largest known settlements in the southwestern United States. The Aztec Ruins structures are built on the outwash plain on the west side of the Animas River. Below the glacial and younger stream deposits, bedrock consists of Nacimiento Formation sandstones and shales.

Twelve large limestone disks held up four large pillars supporting the ceiling of the large kiva. The disks, weighing around 355 pounds each, were quarried at an unknown location many miles from here; rocks similar to this do not outcrop nearby.

The partially restored structures at Aztec Ruins National Monument. The rock used for the original construction of the structures was sourced from nearby Nacimiento Formation outcrops. The Nacimiento consists of layers of many-colored shale, siltstone, and sandstone, which were used to make stripes and patterns in the walls.

Cristo uplift to the east. The cobble layers in the Nacimiento stand out—the cobbles are well rounded and increase in size to the north and proximity to the San Juan uplift. Near Durango, Colorado, cobbles are up to 1 foot in diameter. The Nacimiento Formation contains abundant fossils that record life after the extinction event at the K–Pg boundary, 66 million years ago. These fossils provide an important glimpse of which lineages of vertebrates survived the extinction and how they evolved afterward. Mammals, for instance, became more abundant and diverse after the extinction event.

The Animas River swings west away from US 550 at Aztec (milepost 160), where ancestral Puebloan people—not Aztecs—lived. The ruins of their villages are highlighted in Aztec Ruins National Monument. South of Aztec, the road rises onto the Nacimiento Formation, then drops into the floodplain of the San Juan River.

South of Bloomfield (milepost 151), the road rises again onto the Nacimiento Formation, then heads southeast across the San Juan Basin, the largest oil, gas, and coal source in New Mexico. Thousands of feet of sediment (in places nearly 15,000 feet) accumulated in the growing basin, deposited over millions of years in river valleys and along shorelines. These moist settings produced abundant vegetation, which then became buried and heated, eventually converting to the fossil fuels the region is famous for. The Eocene-age San Jose Formation is the youngest unit in the basin and marks the end of the regional depositional setting. The sediments deposited in the subsiding basin kept the basin floor relatively level at the surface, but the sediments are thousands of feet thick along the axis of the basin to the east of US 550.

County Road 7175, a good dirt road just south of milepost 137, heads east to Angel Peak Badlands Overlook. Angel Peak and the other high pinnacles are capped by the more resistant San Jose Formation, here a river-deposited sandstone that overlies the Nacimiento.

The true beauty of the Nacimiento Formation is revealed at Angel Peak Scenic Area, where its colored layers are incised into spectacular badlands. —Photo by Sandy Partridge

Huerfano Mountain, to the east at milepost 128, is also capped with the San Jose Formation. County Road 7500, a dirt road just south of Huerfano Mountain at the Huerfano Trading Post, accesses the eastern side of Bisti/De-Na-Zin Wilderness, another large area of badlands. The area can be more easily accessed from NM 371 to the west. In Bisti, many trails offer ever-changing badland scenery to the adventurous hiker.

US 550 continues south through occasional outcrops of Nacimiento Formation. Near Lybrook (milepost 105), the highway passes east onto the sandstones and shales of the San Jose Formation, upon which it travels nearly to Cuba. At the eastern edge of the San Juan Basin, just west of Cuba, the road goes through a quick succession of Paleocene to Cretaceous rocks.

The swelling and shrinking of the clays at Bisti/De-Na-Zin Wilderness Area create a popcorn texture on the surface. —Photo by Jon Knudson

The badlands in the Bisti/De-Na-Zin Wilderness echo the ever-changing coastal environments in which its sediments were deposited. Here, coal-rich sediment, deposited in a lagoonal environment, gives way to sandy sediment deposited along a sand bar or beach. The softer dark sediments undermine the more resistant sandstone.

CHACO CULTURE NATIONAL HISTORIC PARK

County Road 7900, a paved route to Chaco Culture National Historic Park, heads south at milepost 113. Chaco Canyon is famous for prehistoric sites constructed between AD 850 and AD 1150. Set in a broad canyon walled with cliffs of Mesaverde Group sedimentary rock layers, the numerous dwellings represent the peak of pueblo culture. Trails led out from Chaco to surrounding pueblos, and trading avenues reached south into Central America. The majority of the main buildings, including the Pueblo Bonito great house, were built near the northern canyon wall, which is composed of Cliff House Sandstone. The sandstone was deposited as a sandbar at the edge of the Western Interior Sea in Cretaceous time and contains evidence of the life of the seaway in its walls. The canyon itself is fairly young, having begun incising about 2 million years ago as a small river (now an ephemeral stream) cut into the sedimentary bedrock.

This trace fossil in the Cliff House Sandstone at Chaco is a burrow called Ophiomorpha. It belonged to what is interpreted to be a crustacean and is evidence of the animal's activity, not the actual animal itself. Similar Ophiomorpha occur as far back as the Permian.

Late Cretaceous shell coquinas, once part of a sandy beach, can be found in the Cliff House Sandstone along the Pueblo Alto Trail above Pueblo Bonito.

An overhanging cliff, known as Threatening Rock, fell onto Pueblo Bonito at Chaco on January 22, 1941, causing considerable damage to the structure.

The mountain range in view to the east, the Sierra Nacimiento, forms the eastern edge of the Colorado Plateau. The Nacimiento fault zone on the western face of the mountains includes reverse, normal, and strike-slip faults, all serving to uplift the block of basement rock during the Laramide orogeny. These faults are thought to be continuous with the Pajarito fault to the south that helps define the western flanks of the Jemez Mountains. The Sierra Nacimiento is cored with 1.7-billion-year-old granodiorite and gabbro, which were once part of a juvenile volcanic arc, and 1.4-billion-year-old granites and gneisses. Sedimentary units, severely deformed and tilted by this extreme tectonic activity, form spectacular ridges that abut the basement core of the range. Displacement along the Nacimiento fault zone ranges from 0.5 to 18 miles and slip is greater in the northern parts of the mountains. For more on the Sierra Nacimiento, see the Jemez Mountains in the Rio Grande rift section.

Mesas west of the highway between Cuba (milepost 64) and about milepost 48 are capped with the Cliff House Sandstone Member of the Mesaverde Group. These rock units were deposited along the shores of the Western Interior Seaway, which covered this area in Cretaceous time. They dip gently toward the northwest, into the San Juan Basin. South of Cuba, both Cretaceous and Jurassic rocks are turned upside down and appear to dip east.

US 550 follows the valley of the Rio Puerco south from Cuba to about milepost 50. In several places across the valley, deep arroyos cut into Pleistocene- to Holocene-age brown, silty terrace and channel deposits of the river. The dramatic arroyos that cross the floodplain may result from the stripping of vegetation by overgrazing by horses, cattle, and sheep, as well as a drying climate and aridification of the general environment. Though it flows only seasonally, the Rio Puerco is one of the major sources of sediment to the Rio Grande. Some of the arroyos show looping meanders developed before downcutting began, when the stream swung lazily across a nearly flat plain.

At milepost 42, the eastern mountainous skyline is composed of Proterozoic basement rocks of the Sierra Nacimiento. These granites and gneisses have been faulted

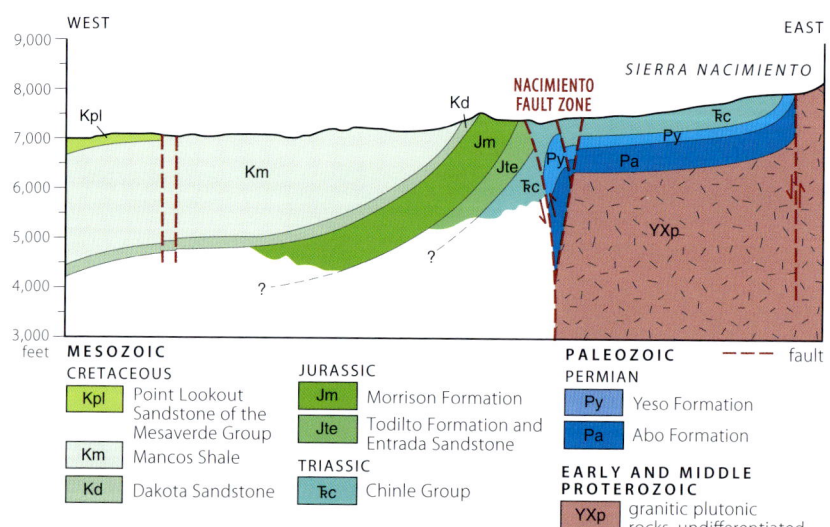

Cross section shows sedimentary layers tilted up near the Nacimiento fault, which lies at the base of the Sierra Nacimiento. —Modified from Woodward and others, 1972

up along the Nacimiento fault, a major fault that runs nearly exactly north-south for the length of the entire mountain range.

Geologic patterns along this edge of the San Juan Basin are complicated by solution and distortion of salt and gypsum layers. Most of the salt has long since dissolved and washed away, contributing to collapse of sandstone and siltstone layers above it. The gypsum remains as thick white beds sandwiched between layers of sedimentary rock. Gypsum flows like glacial ice under pressure. When overlying rocks press down or canyons cut into the rock layers, the gypsum will flow out and around, distorting overlying rock layers.

A massive ledge of resistant, gray-white gypsum of the Middle Jurassic Todilto Formation stands high in the landscape to the northwest near milepost 33. The less-resistant, fine-grained sandstone below will gradually undermine the gypsum, leading to collapse of the cliff face. Chunks of tumbled gypsum litter the cliff base.

Cenozoic volcanic ash (from the San Juan and Jemez regions) has quite an influence on the scenery here, too. Volcanic ash decomposes into clays that expand when wet and shrink as they dry, discouraging plant growth. These swelling clays form gray mats that blanket eroding hillsides in places.

Cabezon Peak, visible to the southwest from many parts of this highway, including at milepost 39, is a prominent, 2.6-million-year-old volcanic plug. It is the most visible of several volcanic necks in the area. A volcanic neck, or plug, is a formerly lava-filled conduit. In the case of Cabezon Peak—and nearby Cerro Chafo, Cerro Cuate, Cerro Santa Clara, and Cerro Guadalupe—these are the inner workings of cinder cones. These necks and nearby basalt-capped Mesa Prieta and Mesa Chivato are part of the Mt. Taylor volcanic field, active from 4 to 2 million years ago. These necks are so well exposed because of the huge amount of erosion that has occurred within the Rio Puerco valley. The ephemeral Rio Puerco transports this eroded sediment to the Rio Grande just south of Belen.

Cabezon Peak, visible to the west from milepost 38, is a large volcanic neck, part of the Mt. Taylor volcanic field. The neck is the cooled plumbing of an ancient cinder cone and preserves striking hexagonal cooling columns (inset). Cretaceous sediments form the shoulders of Cabezon Peak. The Rio Puerco valley below it to the north is actively eroding.

View to the west of several other volcanic necks from the top of Cabezon Peak.

GYPSUM OF WHITE MESA

White Mesa, southwest of San Ysidro, is a popular mountain bike area and home to a well-exposed anticline. To get to the White Ridge Bike Trails, head south 2.4 miles on US 550 from the intersection with NM 4 in San Ysidro, then turn right (west) on Cabezon Road and go another 4.5 miles. South-bound travelers on US 550 will see White Mesa to the southwest from milepost 23. It is topped with Jurassic Todilto Formation, a whitish, gypsum-rich unit. Below it lies a vertical cliff band of Entrada Sandstone, which in turn overlies the reddish Triassic Chinle Group sediments.

The bike trails are on an anticline to the west of White Mesa. The Entrada peters out there, so the Todilto overlies the Chinle Group. The bright-red beds visible in the eroded core or middle of the anticline are the Chinle Group sedimentary units, including the aptly named Painted Desert Member. Exposed on the outer sloping faces of the dipping fold limbs are the Jurassic Todilto, Summerville, and Morrison Formations, and the Cretaceous Dakota Sandstone. The variable robustness of these units results in a ridged anticline, with more resistant units standing high, while softer units have been deeply eroded.

The Todilto Formation is an evaporite sequence with basal shale and limestone (the Luciano Mesa Member), and an upper gypsum deposit known as the Tonque Arroyo member. The upper gypsum member is as much as 60 feet in thickness and forms a hummocky, barren topographic surface. It is thought that the unit precipitated in brine pools that would have dotted the surface of an evaporating body of saline water. There are marine fossils in the lower limestone sections of the Luciano Mesa Member, but these fossils are rare and lack diversity, suggesting the environment in which this evaporite sequence occurred was isolated, hypersaline, lacked oxygen, and perhaps experienced periodic marine flooding. The Tonque Arroyo Member is being actively mined for gypsum, as it provides base material used in drywall.

The reddish Chinle Group lies in the eroded core of the anticline, with the draping white Todilto beds in the foreground.

Inset: Gypsum grows as flat, sheetlike crystals, forming "books" in the Todilto Formation.
—Photos by Samantha Wyman

The pink-and-green, varicolored siltstone and sandstones that come into view at mile 22 on US 550 are the 150-million-year-old Brushy Basin Member of the Jurassic Morrison Formation. The bones of the large *Camarasaurus* dinosaur and gastroliths (stones in dinosaur stomachs that are used for digestion) were excavated from near here. The *Camarasaurus* was a large, four-legged herbivorous sauropod commonly found in the Jurassic-age units. A *Camarasaurus* skeleton excavated from rocks near San Ysidro, along with a *Seismosaurus*, are on display at the New Mexico Museum of Natural History in Albuquerque.

NM 602 and NM 53
Gallup—Grants
Including El Morro and El Malpais National Monuments
99 miles

See map on page 36.

South of Gallup, NM 602 rises onto a mesa capped with Crevasse Canyon Formation, a coal-bearing, cross-bedded sandstone and mudstone unit deposited along the shore of a Cretaceous sea. Coal beds are typically only 1 to 2 feet thick, though the cliff-forming Crevasse Canyon Formation can be up to 700 feet thick. As the shoreline fluctuated with changes in the level of the sea, sand collected on beaches and offshore bars, while gray mud, which would later become shale, accumulated in quiet lagoons. Dead plant material, later transformed into coal by the pressure and heat of burial, collected in nearshore swamps. The offshore bars, beaches, lagoons, and swamps of Cretaceous time may have resembled the present Georgia and Carolina coasts.

Modern valleys along the highway are filled with fine, silty stream deposits dissected by deep arroyos. Overgrazing, aridification due to warming climate, and stream integrations drive rapid gullying across the region.

South of milepost 21, the highest parts of the pinon- and juniper-covered tableland are surfaced with remnants of a younger rock, the Bidahochi Formation. This cliff-forming Miocene to Pliocene unit was deposited in a large lake or series of ephemeral lakes known as Lake Bidahochi. This large lake once extended from here well into eastern Arizona and is estimated to have been approximately the same size as the Great Salt Lake in Utah. The Bidahochi Formation contains alluvium, lake sediments, windblown sand, and spring deposits, as well as windblown ash and even a few basalt flows from nearby volcanism.

The Bidahochi beds are flat lying; in contrast, Cretaceous units dip eastward into a syncline, called the Gallup Sag, that lies between NM 602 and the Zuni Mountains, just out of sight to the east. The Laramide tectonic forces that drove the structural warping of the older sedimentary units and also helped uplift the Zuni Mountains had stopped prior to the deposition of the Bidahochi Formation.

The Zuni Mountains are a slipper-shaped, basement-cored anticline. Erosion has stripped away most of the sedimentary rocks that used to cover the Zuni Mountains, exposing their core of 1.7- to 1.6-billion-year-old granites and gneisses. In the Zunis, Pennsylvanian sedimentary rocks were deposited directly on these Paleoproterozoic rocks. Older sedimentary rocks—Cambrian to Mississippian—once existed here but were eroded away during Pennsylvanian time, when this region rose as part of the Ancestral Rockies. Mt. Sedgwick, the highest peak in the Zuni Mountains, is 9,256 feet in elevation.

At milepost 4, MN 602 moves south off the mesa capped with Bidahochi lake deposits and onto the Cretaceous Mancos Shale, a gray, valley-forming unit. The Jurassic-age Zuni and Entrada Sandstones form cliffy tops of the buttes to the west. One of the most easily recognized Cretaceous units is the Gallup Sandstone, a ledge- and cliff-former distinguished by its reddish color. Underlying the Mancos is the Dakota Sandstone, the thin but widespread beach-deposited sandstone that marks the bottom of the Cretaceous sequence.

At about milepost 2, the highway crosses the Rio Nutria, an ephemeral stream that feeds into the Zuni River. The highway sits on soil-covered Dakota Sandstone at the junction with NM 53. Near the junction, look to the west for good glimpses of the distinctive reddish or salmon-pink cliff-former, the Zuni Sandstone, which underlies the Dakota Sandstone. Erosion of another layer of shale below undermines the pink- and cream-colored Zuni Sandstone cliffs, causing them to break down into tumbled blocks of rock. The formation varies in color and resistance to erosion and stands as steep cliffs.

NM 53 East to Grants

This road guide follows NM 53 east. Ramah, between mileposts 33 and 34, lies at the southern end of the great Nutria Hogback of bent up Cretaceous sedimentary rock along the west edge of the Zuni Mountain uplift. The hogback extends almost 30 miles to the north, to just northeast of Gallup. The Zuni Mountains record two periods of uplift. The first, Laramide deformation, occurred here around 80 million years ago. The second, between 20 and 10 million years ago, records regional uplift of much of the Southwest.

NM 53 continues southeastward along a valley between ridges of bent-up sandstone. The valley is on a layer of softer Cretaceous shale. Visible to the east from mileposts 39 to 40 are Los Gigantes, pillars in the Zuni Sandstone, accessed via County Road 135. Between Los Gigantes and El Morro (milepost 44), the highway crosses a low divide into another valley, this one floored with Triassic shale. At milepost 42, NM 602 crosses a grass-covered basaltic to andesitic lava flow.

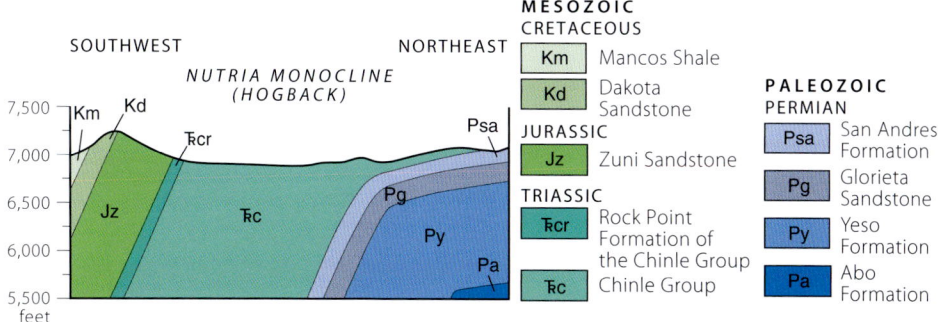

Cross section of tilted rocks in the Nutria Hogback. —Modified from Drakos, Riesterer, and Bemis, 2013

EL MORRO NATIONAL MONUMENT

At El Morro National Monument, between mileposts 44 and 45, a mesa of Zuni Sandstone juts to 7,544 feet in bold, weather-streaked cliffs on which passing travelers—dating back to when the Spanish were present—have inscribed their names. The cross-bedded, 165- to 155-million-year-old Zuni Sandstone is unconformably capped with 95-million-year-old Dakota Group sandstone and shale. The contact between the two is visible as a whitish band, where the upper few feet of the Zuni Sandstone was bleached when it was an ancient weathering surface prior to deposition of the Dakota Group. Inscription Rock, for which the monument is famous, is composed of the Zuni Sandstone. The cross-beds that etch angles in the unit show the large scale of the windblown sand dunes that were once regionally extensive. The Zuni is equivalent to the Entrada Sandstone, famous in the National Parks of Arizona and Utah. The lower parts of the Dakota, seen in the highest part of the mesas, are also cross-bedded sandstones. Gray shales and mudstone, deposited by low-energy streams on the western side of the Western Interior Seaway in Late Cretaceous time, form the dark, crumbly layers in the Dakota and were used as construction material in pueblo ruins.

Inscription Rock is composed of Zuni Sandstone. The whitish band at the top of the monolith is the bleached weathering surface at the top of the unit.

A short walk brings visitors atop the tall bluffs at El Morro National Monument, where they can inspect the white weathering surface that separates the top of the Zuni Sandstone from the overlying Dakota Sandstone.

Los Gigantes (the Giants) are spectacularly shaped pinnacles and balanced rocks carved into the Zuni Sandstone.

The Zuni-Bandera volcanic field is the source for the basalts seen along this route, both here at El Morro and at El Malpais National Monument. Located along the Jemez lineament west of the Rio Grande rift, the Zuni-Bandera is the second-largest volcanic field of the Basin and Range Province. It covers approximately 950 square miles in a blanket of multiple basaltic lavas that in places reach close to 500 feet in total thickness.

A few miles east of El Morro, east of another headland of Zuni Sandstone, the highway converges with the southeast end of the Zuni Mountains and enters a volcanic realm, signaled by hummocky, sage-covered lava flows. A few islands of pale-gray Permian San Andres Formation limestone jut through the lava. The limestone is quarried near milepost 51 and is exposed in the roadcuts at milepost 59.

Between mileposts 60 and 61, NM 53 crosses the Continental Divide, which is subtle here despite its 7,882-foot elevation. It marks the line between west-flowing Colorado River drainage and east-flowing Rio Grande drainage. Just east of the divide, the small mountains south of the highway are cinder cones of the Zuni-Bandera volcanic field in El Malpais National Monument.

East of Bandera Crater, at about milepost 63, the highway runs close to the contact between reddish Proterozoic granite in the core of the Zuni Mountains to the north and the rough lava of the approximately 25-mile-long and 10-mile-wide Bandera flow. Bonita Canyon, a striking feature to the northwest at milepost 70, is a fault-controlled canyon. To the west of the canyon lie the Proterozoic granites and granitic gneisses of the crystalline basement rocks. To the east, canyon walls consist of Permian sedimentary rocks and limestones of the San Andres, Abo, and Yeso Formations, and Glorieta Sandstone.

At about milepost 63, the dark Bandera basalt flowed over an uneven surface of the underlying red granite.

At milepost 67 is a large roadcut through Proterozoic granite.

Lava-capped North Cebollita Mesa and Mesa Negra are in the distance to the southeast. On the mesas' flanks, Cretaceous units, including the thick Gallup Sandstone and Crevasse Canyon Formation, are largely blanketed in landslide deposits. Large alluvial fans extend from canyons that drain into the central area of the basalt-filled canyon of the El Malpais.

At San Rafael (milepost 83), Permian sedimentary layers dip northward off the northeast side of the Zuni Mountains. Because lava flows dam the valley, drainage is poor near San Rafael, and alkaline salts collect in whitish crusts on soil surfaces. Mt. Taylor, a dormant stratovolcano, rises northeast of Grants, part of the Mt. Taylor volcanic field. See I-40: Grants—Albuquerque in the Rio Grande Rift section for more information.

At milepost 85, just south of the I-40 interchange, is a low roadcut in reddish, warped layers of Zuni Sandstone.

EL MALPAIS NATIONAL MONUMENT

In El Malpais National Monument, the Zuni-Bandera volcanic field contains the mafic outpourings from at least one hundred vents over the last 0.7 million years. Hawaiʻian-style basaltic lava flows and cinder cones of the Zuni-Bandera volcanic field show both pahoehoe and aa lavas, as well as lava tubes. Visitors can view and hike through the McCartys flow, the youngest flow, which erupted 3,800 years ago from a shield volcano about 17 miles southeast of NM 53.

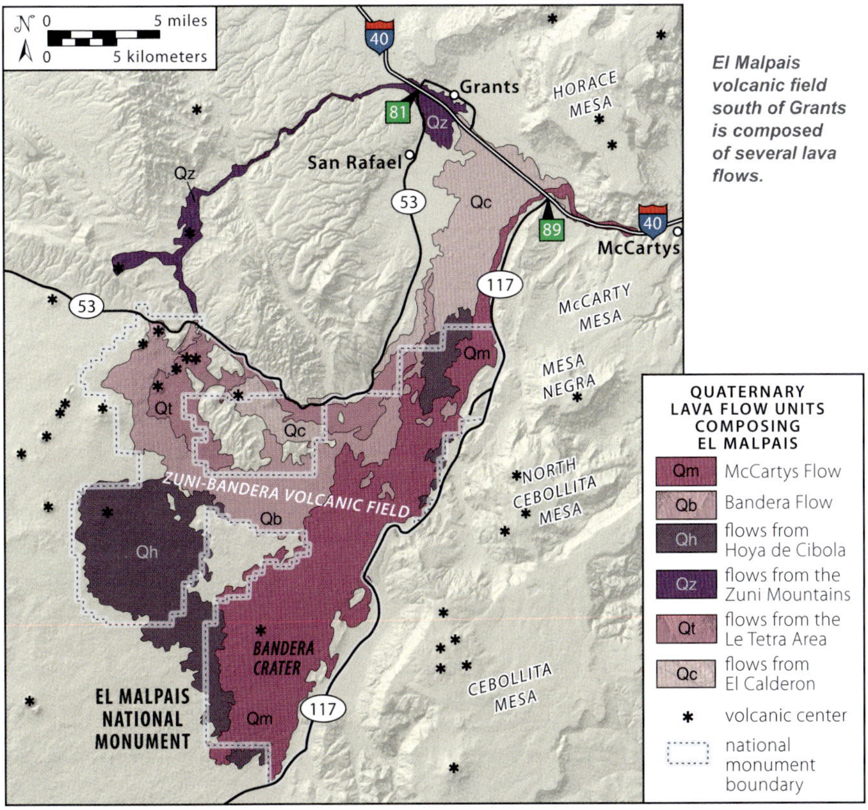

El Malpais volcanic field south of Grants is composed of several lava flows.

QUATERNARY LAVA FLOW UNITS COMPOSING EL MALPAIS

- Qm — McCartys Flow
- Qb — Bandera Flow
- Qh — flows from Hoya de Cibola
- Qz — flows from the Zuni Mountains
- Qt — flows from the Le Tetra Area
- Qc — flows from El Calderon
- ✱ volcanic center
- national monument boundary

Much of the multilayered ice in Ice Cave may be thousands of years old. In this lava tunnel, the ice doesn't see the sun and is green with algea.

Bandera Crater, a large cinder cone, is the source of the Bandera lava flow.

A lava tunnel forms as molten lava flows out from under its own cooling crust. This one was several miles long before portions of its roof collapsed.

View to the south from about milepost 70 of a chasm in a 60,000-year-old basalt flow from the El Caldron vent. Chasms form as the basalt cools and contracts.

The Bandera Crater is a double cinder cone about 500 feet high and 800 feet deep. It is the source of the second-youngest flow in the volcanic field, approximately 10,000 years old. A lava tube extends from the breached crater wall, and the commercially operated tourist destination of the Ice Cave is in a portion of this lava tube. Within the lava tube, in which more than 20 feet of ice has accumulated as precipitation percolates through the porous lavas, temperatures do not exceed 31 degrees Fahrenheit.

Some basaltic magmas contain a high proportion of volcanic gases, and at the start of an eruption, when pressures are released, these gases rise into upper parts of the underground volcanic conduits and spurt outward, similar to how froth rises when a shaken soda is opened. As the frothy lava is released, pressures are further reduced, and more froth forms, bursting upward and then falling to the ground to form a cinder cone. Eventually, as gases are released and the froth is expended, less gaseous magma may flow out on the surface, escaping usually from the base of the loosely consolidated cinder cone. Numerous bubble holes, called vesicles, in the Bandera lava flow show that this lava still contained some gas when cooled.

The Bandera flow is composed of jagged, broken lava, called aa, produced when the surface of the flow cooled and hardened while the underlying lava was still in motion. In places, molten lava splashed out of minor vents to form little spatter cones like the one near the Bandera Crater trail. On the slopes of the main crater are many volcanic bombs tossed out in semi-molten form by the eruption. Many are rounded, or football-shaped, from their journey spinning through the air. Lava showing the less explosive nature of eruption is also visible in elongated, rope-like pahoehoe basaltic lava flows.

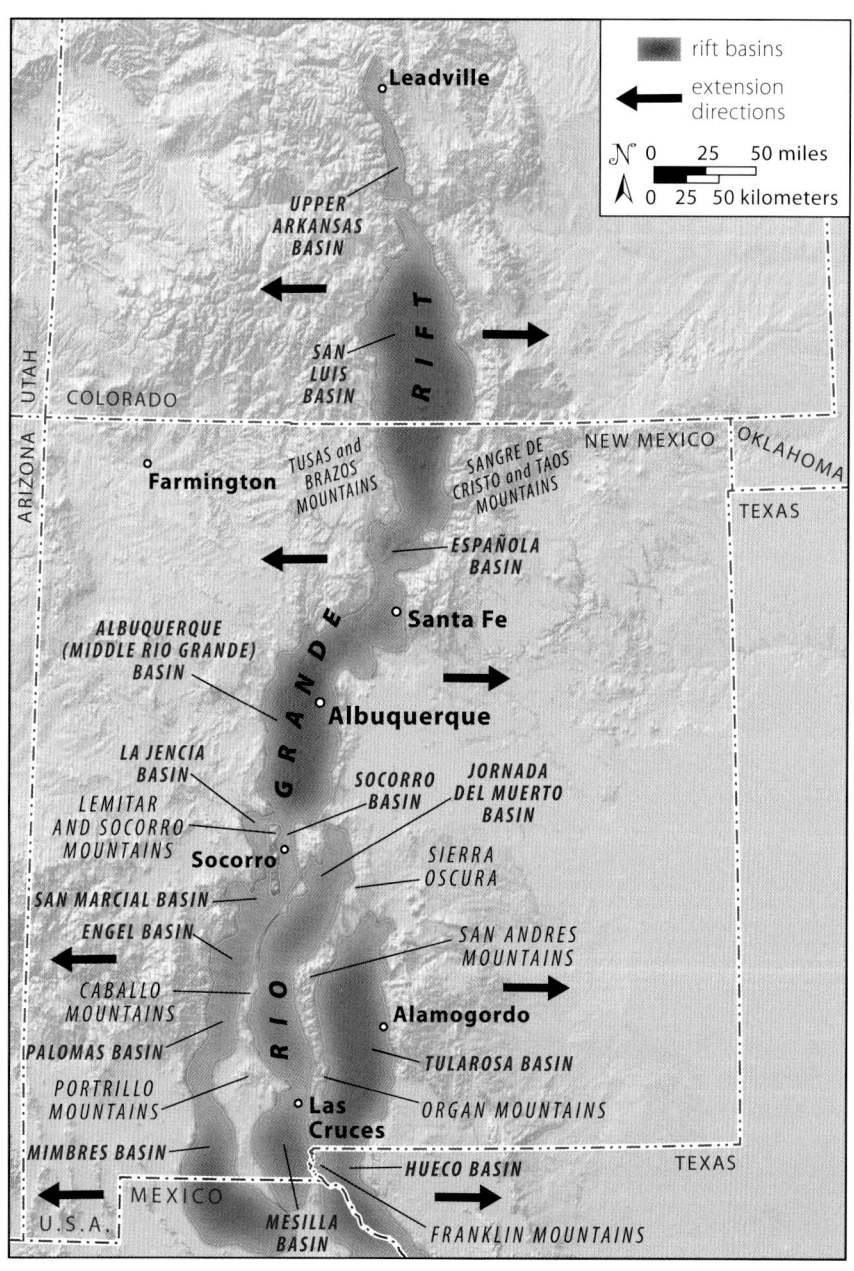

The Rio Grande rift, a series of basins extending from north-central Colorado south to Mexico, formed as the continent began to split apart during Oligocene and Miocene time.

Rio Grande Rift

The Rio Grande rift, a major break in the Earth's crust, starts in central Colorado, cuts south through New Mexico, and extends into the Mexican state of Chihuahua, covering a distance of more than 500 miles. This feature began to form during the waning stages of the Oligocene ignimbrite flare-up as the crust was pulled apart, or extended, from east to west and thinned. Extension accelerated 25 million years ago, resulting in graben development and increased sediment accumulation. Rifting continues today, though at a slower pace. The slow-opening Rio Grande rift is not likely to stretch to the point of reaching sea level, as some rifts such as the Gulf of California or the Red Sea in Africa have. When rifts reach sea level, they either fill with seawater or become scorching, below-sea-level valleys like those of the Dead Sea in Israel or the Imperial Valley, home to the Salton Sea of southern California.

Rift extension was largely synchronous throughout the length of the rift, meaning that the east-west crustal stretching affected the crust for more than 500 miles from Colorado into southern New Mexico at nearly the same time. The mechanisms for this continental-scale stretching are complex but likely included the multistage delamination, rollback, and breakoff of a subducted oceanic slab beneath the North American continent. These massive movements, occurring in the Earth's mantle deep below the surface, drove tectonic activity and stretching of the continent and reshaped western North America.

As the crust extended, the stretching produced a series of staggered, offset basins. In cross section, many basins are asymmetrical half grabens, fault blocks that dropped down along a major fault zone on one side and the other side behaving more like a hinge, with minimal faulting. In Albuquerque, this asymmetry is dramatically visible: the Sandia Mountains form the major rift fault side, while the basin edge is more gradual to the west. The basins vary in total size, with the deepest having dropped some 28,500 vertical feet, or nearly 6 miles, throughout their history, but most did not drop this much.

The Rio Grande rift starts as a narrow sliver in the upper Arkansas River basin near Leadville in central Colorado. Rift basins broaden southward and are about 30 miles wide near Albuquerque. The three largest rift basins are the San Luis Valley in southern Colorado and northernmost New Mexico, and the Española and Albuquerque Basins in New Mexico. Farther south, the basins get wider still and include the Rio Grande Valley, Jornada del Muerto, and the Tularosa Valley to the east. Faults of the same age as the rest of the rift border the latter two. Near Las Cruces, the rift is nearly 100 miles wide and blends with the extensional terrain of the Basin and Range. The basins are connected by transfer fault zones that accommodate tectonic movement.

The Rio Grande rift split the southern end of the Rocky Mountains into two prongs, the Sangre de Cristo Mountains to the east, which include the Taos Mountains, and the Tusas and Brazos Mountains to the west. The same basement rocks found in high peaks in these mountains are thousands of feet belowground within the rift, buried beneath sediments.

Normal faults, with the right side of the fault moving down, in the Peralta Tuff at Kasha-Katuwe Tent Rocks National Monument west of Santa Fe along the Rio Grande.

Numerous volcanic features are found along the length of the rift. Beginning about 38 million years ago, Oligocene ignimbrite volcanism, which formed the Mogollon-Datil and Sierra Blanca volcanic fields, erupted in huge calderas and major ash flows. Younger volcanism continued to occur, with major outflows of basalt near Taos (about 4 to 2 million years ago), the eruptions and collapses of the Jemez calderas near Los Alamos (1.6 and 1.2 million years ago), young volcanic flows near Santa Fe (Diablo Canyon is 3 to 1 million years old), and the Albuquerque Volcanoes (about 200,000 years ago).

The rift first appeared as several closed, internally drained basins. Gradually, these basins filled with sediments eroded from adjacent ranges, along with lava flows and volcanic ash from nearby volcanoes. These sediments and lava are broadly known as the Santa Fe Group and record several million years of basin filling from late Oligocene to Pleistocene time, a span of about 20 million years. The Santa Fe Group contains a myriad of sediments, including those brought in from upstream and those transported from neighboring mountain ranges. This unit varies from place to place because each mountain range supplies unique sediment and each volcanic eruption deposits its own specific lava flows and ash deposits. In general, the sediments of the Santa Fe Group are soft, poorly consolidated, and easily eroded. In many places, continuing movement along rift faults has tilted these sediment layers.

The Santa Fe Group has been studied extensively because units within it contain groundwater aquifers that supply drinking water to Albuquerque and other cities.

The intertonguing alluvial and stream deposits of this group are notoriously variable, especially with the added element of volcanic and windblown deposits. The Santa Fe Group has been divided into several formations differing in rock type and position.

The Rio Grande Valley is not a valley in the usual sense of the word, it was not cut by a river and it doesn't branch upstream the way most river valleys do. The waters of the Rio Grande merely found and followed the preestablished, partly sediment-filled structural grabens of the rift. The developing Rio Grande progressively formed its present, southerly draining beginning around 8 million years ago as it collected waters from the San Juan volcanic field in Colorado. The Rio Chama, which linked the northern rift basins in New Mexico, integrated with the Rio Grande around 5 million years ago, and the through-flowing river made its way to southern New Mexico by 3 million years ago. It incorporated the drainages of the Hueco Basin by 2 million years ago, and the Rio Grande extended to the Gulf of Mexico after 800,000 years ago.

The river, while not forming the valley, markedly influenced the modern shape of this valley floor. The river system alternates between constructive periods of sediment deposition and periods of erosion, during which the river incises into the sediments it previously laid down. We see this in the numerous steplike terraces along the river margins and the unique, complicated intrabasin stratigraphy. At one time, the river flowed across the surface of the Santa Fe Group, blending deposits carried from upstream sources with those in alluvial fans along the mountain front. During dry episodes, roughly corresponding to interglacial stages of the Pleistocene ice ages, erosion rates increased in the highlands and deposition and sedimentation rates increased in basins. The Santa Fe Group was laid down during this increased deposition. Later, driven by climatic change and subsequent integration to the Gulf of

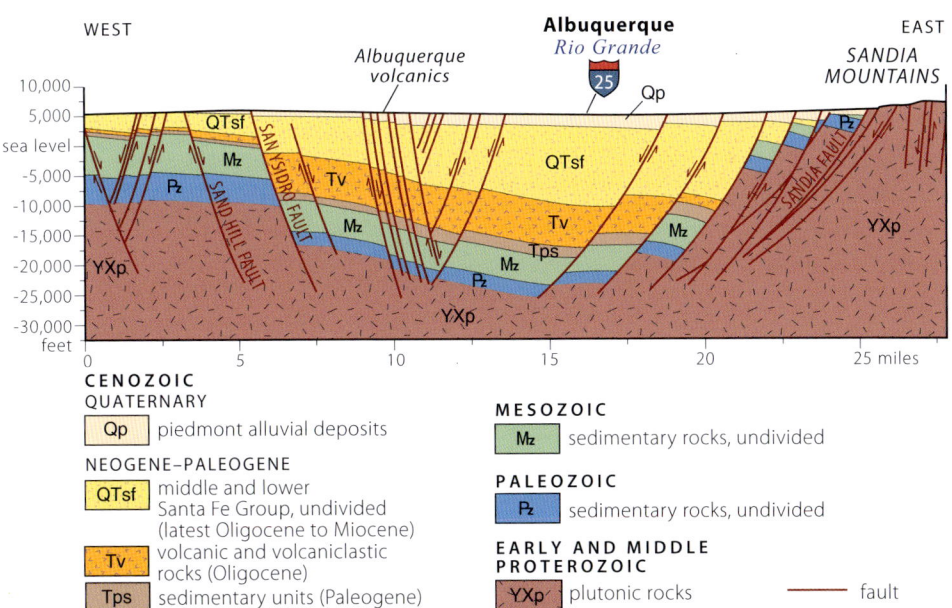

The Rio Grande rift was continuously filled with sediments of the Santa Fe Group as it dropped down along numerous normal faults. —Modified from Grauch and Connell, 2013

Mexico, the river began to incise, or cut through these layers of rift-fill sediment. As a result, the Rio Grande is now entrenched in its valley, which is laterally bound by step-like terraces that represent its previous, higher-elevation positions in the landscape.

While the sediment in alluvial fans is transported short distances from the local mountains, the sediment in river terraces is delivered from a much larger area. The terraces that line the Rio Grande contain sand and gravel brought into the area from all points north, including the main trunk of the Rio Grande, the Chama River, and other tributaries. Distinction between alluvial fans and terraces sounds simple, but there are complications. In some places, the surface gravels are so thick that geologists can't easily determine what underlies them. Continued faulting raised some of the older fans and terrace deposits, and erosion pared them off.

Terraces are conspicuous throughout the rift, the most visible being the Los Duranes Formation (wonderfully exposed to the north near Rio Rancho and Bernalillo), about 100 to 130 feet above the valley floor. This terrace was deposited approximately 150,000 to 100,000 years ago and is over 160 feet thick in places. It represents a major period of basin filling and preserves early bison fossils.

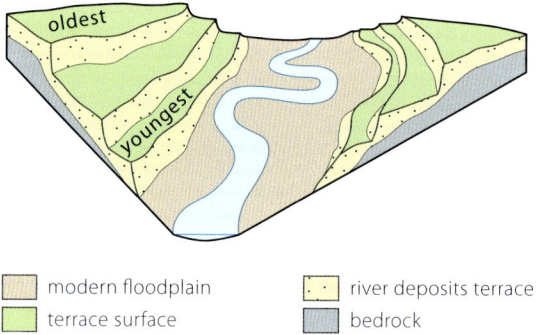

The highest terrace along a river is the oldest.

The same pattern of alternating sedimentation and incision can be recognized today in seasonal microcycles. During times of low discharge, the Rio Grande deposits its sand and sediment load as silty sand and gravel bars. During periods of increased river discharge, it cuts into its established floodplain. Numerous dams control this cycling, and water flow is heavily managed to meet the demands of water users, as well as to conform with requirements of the Rio Grande Compact between Colorado, New Mexico, and Texas, a 1938 agreement that apportions the river water.

In southern New Mexico, the rift and its bordering ranges blend with the Basin and Range region. While the Rio Grande rift is considered a classic example of a narrow rift zone, where the crust has been stretched along a narrow seam, the Basin and Range province has seen much greater extension. Both the Rio Grande rift and the Basin and Range region came about in response to the same great, pulling-apart tensions in the Earth's crust. In both, extension produced a pattern of ranges, which were tilted by bounding normal faults, and intervening valleys built and deepened from late Eocene time forward. In addition, the Basin and Range has an overall clockwise rotation.

Northern Rift and Southern Rocky Mountains

I-25
Las Vegas—Santa Fe
50 miles

Following the route of the Santa Fe Trail, I-25 leaves Las Vegas at milepost 343 to pass through a gap in the lines of hogbacks, ridges of tilted rocks, that edge the southern tip of the Sangre de Cristo Mountains. Uplift of the Rocky Mountains tilted the sedimentary rock during the Laramide deformation. South and east of Las Vegas, the gray, thin-bedded shale and fragmented limestone is the Late Cretaceous Graneros Shale and overlying Greenhorn Formation, known for its oyster and ammonite fossils. The junction with US 84 lies in a valley carved into the Cretaceous Chinle Group. Look for a few red siltstone and shale outcrops of the Chinle to the south.

Southwest of the junction with US 84, the highway cuts through steeply sloping Triassic, Permian, and Pennsylvanian sedimentary rocks, tilted up by a fault in the underlying Proterozoic basement rock. The Triassic and Permian units include, from younger to older in the order in which westbound travelers will encounter them: light-gray and tan sandstones of the Santa Rosa Formation; purplish-red Artesia Group; Glorieta Sandstone; gray San Andres Formation, especially well exposed in the roadcut between mileposts 332 and 331, where faults offset the layered limestones; Glorieta Sandstone again, resurfacing due to faulting; and a thick sequence of Yeso and Sangre de Cristo Formation red beds, which are brick-red siltstone and sandstones.

This magnificent roadcut at milepost 341 exposes tilted, thin-bedded Morrison Formation shale, overlain by light-colored Dakota Group sandstone in one of the hogbacks that border the Sangre de Cristo Mountains.

Proterozoic gneiss is exposed in deep roadcuts at the southern tip of the Sangre de Cristo Mountains (also the southern tip of the Rockies)

ledges and slopes of Glorieta Mesa were formed by erosion of hard sandstone and softer mudstone and siltstone layers

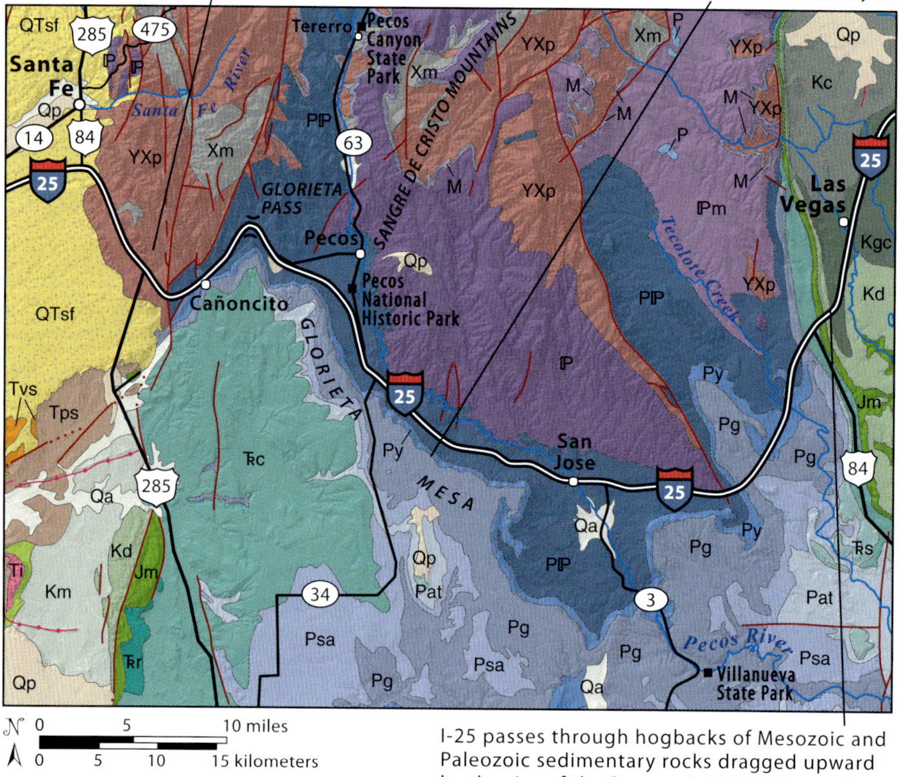

I-25 passes through hogbacks of Mesozoic and Paleozoic sedimentary rocks dragged upward by the rise of the Sangre de Cristo Mountains

CENOZOIC
QUATERNARY
- Qa — alluvial deposits
- Qp — piedmont alluvial deposits (Holocene to early Pleistocene)

QUATERNARY–PALEOGENE
- QTsf — Santa Fe Group, undivided (latest Oligocene to middle Pleistocene)

NEOGENE and PALEOGENE
- Tps — older sedimentary units (Paleogene)

MESOZOIC
CRETACEOUS
- Km — Mancos Shale
- Kc — Carlile Shale
- Kgc — Graneros Shale and Greenhorn Formation
- Kd — Dakota Sandstone

JURASSIC
- Jm — sedimentary rocks undivided; includes Morrison Formation and San Rafael Group

TRIASSIC
- ℞c — Chinle Group
- ℞r — Redonda Formation
- ℞s — Santa Rosa Formation

fault; dotted where concealed

PALEOZOIC
PERMIAN
- P — sedimentary rocks, undivided
- Pat — Artesia Group
- Psa — San Andres Formation
- Pg — Glorieta Sandstone
- Py — Yeso Formation

PERMIAN–PENNSYLVANIAN
- PIP — sedimentary rocks, undivided; includes Sangre de Cristo Formation

PENNSYLVANIAN
- IP — sedimentary rocks, undivided
- IPm — Madera Group and Sandia Formation

MISSISSIPPIAN
- M — sedimentary rocks, undivided

PROTEROZOIC
- YXp — plutonic rocks, undifferentiated (early and middle Proterozoic)
- Xm — metamorphic rocks (early Proterozoic)

VOLCANIC AND IGNEOUS ROCKS
- Tvs — volcaniclastic sedimentary rocks (late Eocene to Oligoene)
- Ti — intrusive rocks; includes dikes (late Eocene to Pliocene)

Geology along I-25 between Las Vegas and Santa Fe.

As sediments were deposited in shallow seas during the Pennsylvanian Period, the Ancestral Rocky Mountains were uplifted, creating ancient sedimentary basins. Sediments of the Sangre de Cristo Formation were deposited in river channels and alluvial fans as these highlands were eroded. The landscape dried in the Permian Period, and the Yeso Formation was deposited on an arid coastal plain. Small streams deposited carbonate rocks in the upper portion of this unit. As the system further developed, large beaches, which would become the Glorieta Sandstone, formed. The encroachment of a shallow sea meant the deposition of the San Andres Formation, a widespread limestone unit that preserves many Permian marine fossils. The continued rising and falling of the shallow sea and the clastic shelf depositional environment of the fine siltstone, salt, and gypsum of the Artesia Group preserve the last of that marine time. An unconformity lies above the Artesia Group. The rock record returns with the river sediments of the Santa Rosa Formation, a pebble conglomerate deposited in Triassic time.

To reach Villanueva State Park, where the Pecos River cut a lovely canyon, take NM 3 (exit 323) south for 13 miles. At the state park, the Yeso Formation forms the lower orange slopes of siltstone and sandstone, deposited in a coastal region of mudflats that periodically dried out. Water left behind in shallow pools became a concentrated brine from which carbonate minerals like selenite, a form of gypsum, were precipitated. Sands in the Yeso were also deposited in windblown environments 268 to 245 million years ago. Look for narrow veins of selenite in the Yeso Formation. The Glorieta Sandstone caps the Yeso here with gray, cross-bedded, cliff-forming exposures that were once large sand dunes. The Pecos River drains the south end of the Sangre de Cristo Mountains and flows southeast toward Carlsbad, providing water to the dry eastern part of New Mexico.

As I-25 curves westward around the southern tip of the Sangre de Cristo Mountains, all these sedimentary layers level out and the upper part of the Permian sequence appears as horizontal layers on the escarpment of Glorieta Mesa, straight ahead for

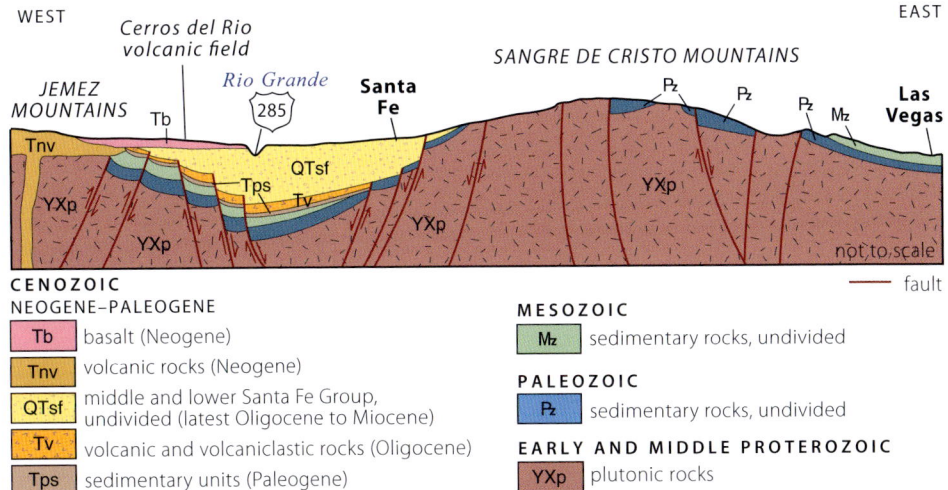

Section across the Sangre de Cristo Mountains north of I-25.

westbound travelers at milepost 325. Much of the escarpment is heavily vegetated, but west of milepost 311, some of the formations can be distinguished. The Glorieta Mesa is capped with the limestones of the San Andres Formation. The white-yellow sandstone near the top is the Glorieta Sandstone and the red layers below the Glorieta are the Yeso Formation. Hard sandstone and limestone layers form ledges and cliffs, while soft mudstone and siltstone layers weather into slopes.

The red-brown color of the Yeso Formation comes from tiny grains of hematite, an iron oxide commonly thought of as rust. Iron oxides in different amounts impart many shades of pink and yellow to sedimentary rock. The iron was originally sourced from iron-bearing minerals like biotite and hornblende that were released when crystalline rocks eroded in the continental interior.

As the highway climbs toward Glorieta Pass, the rocky escarpment at the edge of the Glorieta Mesa curves with it. Near Pecos, these rock units form a backdrop for Pecos National Historic Park (exit 299), where ruins of a mission church and Native American dwellings are built of the rocks that surround them. Even thin slabs of clear gypsum were put to use—as windowpanes! Pecos Pueblo, well situated on a main trade route between the Plains and Rio Grande region, was one of the most important pueblos in this area when the Spanish missionaries built their mission next to the pueblo.

Flat-lying Pennsylvanian to Permian rocks exposed on Glorieta Mesa. Viewed from the frontage road south of Pecos National Historic Park.

Glorieta Pass (7,540 feet), a few miles west of Pecos, was crossed by plodding horses and footmen of Coronado's expedition in 1540–1541. After 1821, it became a part of the Santa Fe Trail. Later, the railroad and highway followed the same route.

West of the pass, the sedimentary formations dip southward off the Sangre de Cristo Mountains. Some steeply tilted layers can be seen in the roadcut at milepost 296. The Sangre de Cristo Formation composes the hills north of the highway; behind them rise Proterozoic rocks in the core of the uplifted fault block. The Proterozoic rocks extend to the highway west of milepost 292. Here, they are very hard, very resistant, greenish-black gneiss, patterned with veins and dikes. Between here and Santa

Fe and for some distance north of Santa Fe, a broad band of this gneiss stands as the west side of the range, rising well above the down-faulted Rio Grande rift.

To the north, in the Proterozoic rocks above Santa Fe, remnants of a meteor impact crater were discovered along a roadcut on NM 475 in 2005. The structures, called shatter cones, form only when rocks are subjected to the high, sudden pressures that occur with meteor impacts. The meteor is thought to have hit the Earth between 1.6 and 1.4 billion years ago.

PECOS CANYON STATE PARK

From Pecos, head north on NM 63 and follow the Pecos River upstream into Pecos Canyon State Park, which nestles in the southeastern flank of the Sangre de Cristo Mountains. The town of Pecos is built on Quaternary gravel terrace deposits. Brownish-red, gray, and purple mudstone and sandstone units of the Permian- and Pennsylvanian-age Sangre de Cristo Formation form the hillslopes around the town. A northeast-trending syncline tips the bedrock downward into a shallow cup. Several faults associated with Laramide tectonism cut the rocks.

The canyon walls become steep where the road enters the park. Landslides are common, especially on the east side of the canyon. Bedrock swiftly transitions into 1.7-billion-year-old metavolcanic basement rocks that originally formed in a volcanic arc. The lower portions of the canyon walls on both sides of the Pecos River are granites, gneisses, granodiorite, and tonalite. A dark-green-to-black amphibolite schist with visible garnet and chlorite is present on the eastern side of the road approximately 20 miles north of Pecos and at the park headquarters in Tererro.

The 1.7-billion-year-old Windy Bridge Tonalite, weathering to a warm orange-brown, forms striking vertical cliffs on west side of the road. This unit has large, approximately 1-inch-diameter quartz cIrystals, which form as much as 35 percent of the rock body.

Cave Creek Cave, in the Sangre de Cristo Mountains near the headwaters of the Pecos River, is eroded in La Pasada Formation, a sparsely fossiliferous limestone unit.

US 64
RATON—UTE PARK—TAOS
89 miles

About 8 miles south of Raton, US 64 leaves I-25 and crosses the western part of the Raton Basin before tackling the Sangre de Cristo Mountains, here divided into the Cimarron Range and the Taos Mountains. The highway follows the mountain branch of the old Santa Fe Trail between Raton and Cimarron.

The broad valley south of Raton is floored by Cretaceous Pierre Shale, a soft, dark-gray marine unit that in places contains beautifully preserved fossil shells. Look for the shale in gullies just west of I-25. Between Raton and Cimarron, the highway crosses three flat-topped, terrace-like pediments—erosion surfaces—cut into the slightly tilted Pierre Shale. Where US 64 crosses the Canadian River, just west of I-25, the tilting of this formation and beveling by erosion show up clearly.

Small, isolated buildings along this route are pumping stations that move water from Eagle Nest Lake, in the Sangre de Cristos, to Raton. Volcanic highlands visible to the east are more than matched by steep, high bluffs flanking Raton Mesa to the west. The bluffs are capped with resistant sandstone of the Raton Formation—rock deposited just at the end of Cretaceous time. Small landslide scars reveal dark- and brownish-gray shales that form parts of this unit. In this area, the sedimentary rock is fine-grained, reflecting its depositional environment in lagoons and estuaries, but

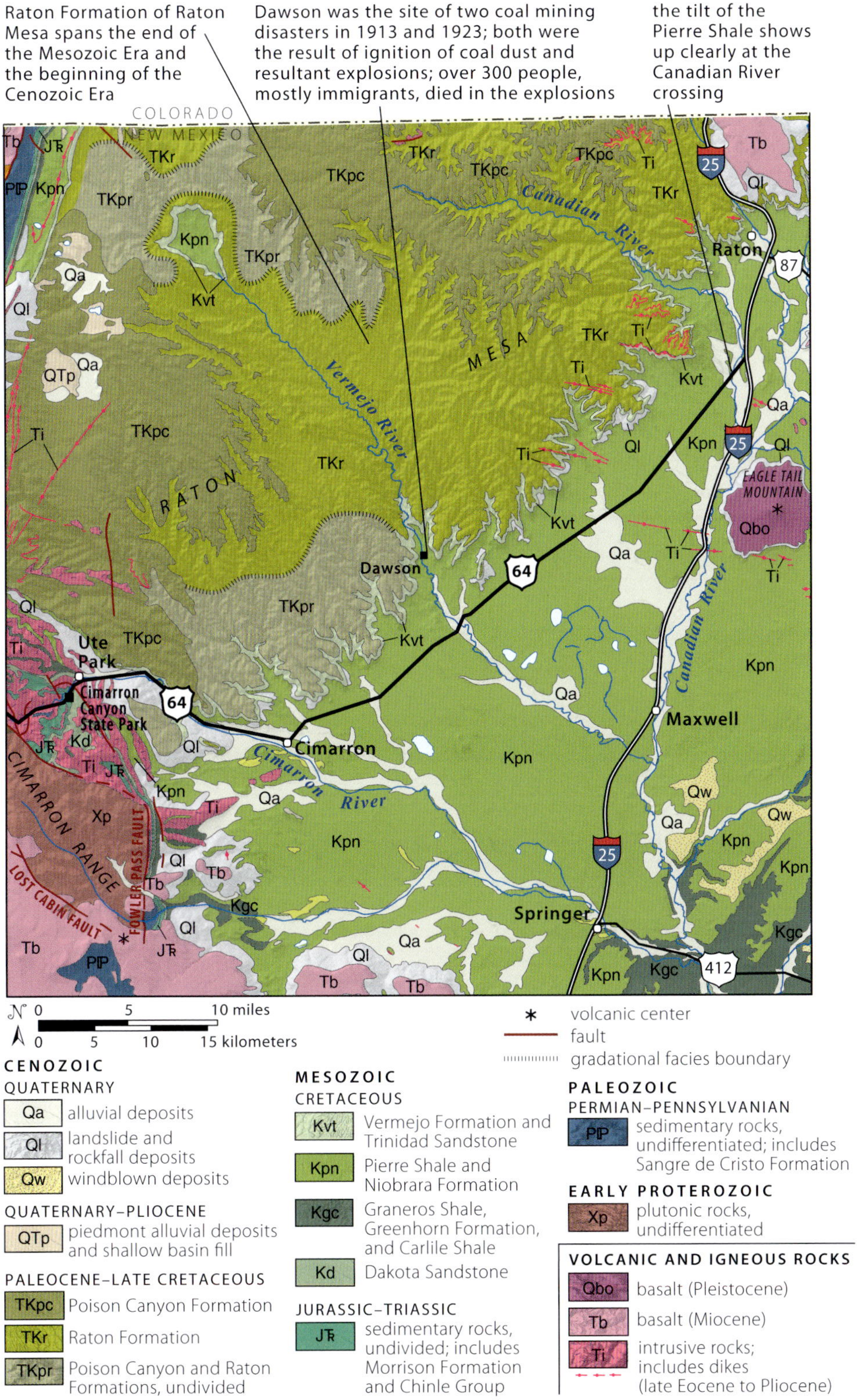
Raton Formation of Raton Mesa spans the end of the Mesozoic Era and the beginning of the Cenozoic Era
Dawson was the site of two coal mining disasters in 1913 and 1923; both were the result of ignition of coal dust and resultant explosions; over 300 people, mostly immigrants, died in the explosions
the tilt of the Pierre Shale shows up clearly at the Canadian River crossing
COLORADO
NEW MEXICO
RATON
MESA
Canadian River
Vermejo River
Cimarron River
Raton
Dawson
Ute Park
Cimarron Canyon State Park
Cimarron
Maxwell
Springer
EAGLE TAIL MOUNTAIN
CIMARRON RANGE
LOST CABIN FAULT
FOWLER PASS FAULT
25
87
64
412
0 5 10 miles
0 5 10 15 kilometers
volcanic center
fault
gradational facies boundary
CENOZOIC
QUATERNARY
Qa alluvial deposits
Ql landslide and rockfall deposits
Qw windblown deposits
QUATERNARY–PLIOCENE
QTp piedmont alluvial deposits and shallow basin fill
PALEOCENE–LATE CRETACEOUS
TKpc Poison Canyon Formation
TKr Raton Formation
TKpr Poison Canyon and Raton Formations, undivided
MESOZOIC
CRETACEOUS
Kvt Vermejo Formation and Trinidad Sandstone
Kpn Pierre Shale and Niobrara Formation
Kgc Graneros Shale, Greenhorn Formation, and Carlile Shale
Kd Dakota Sandstone
JURASSIC–TRIASSIC
JTR sedimentary rocks, undivided; includes Morrison Formation and Chinle Group
PALEOZOIC
PERMIAN–PENNSYLVANIAN
PP sedimentary rocks, undifferentiated; includes Sangre de Cristo Formation
EARLY PROTEROZOIC
Xp plutonic rocks, undifferentiated
VOLCANIC AND IGNEOUS ROCKS
Qbo basalt (Pleistocene)
Tb basalt (Miocene)
Ti intrusive rocks; includes dikes (late Eocene to Pliocene)

Geology along US 64 between Raton and Ute Park. See the map on page 115 for US 64 between Ute Park and Taos.

westward the sediment becomes coarser, reflecting the landward transition from coastal marine to river and alluvial systems. Here, shales predominate over sandstones; farther west, conglomerate and sandstone layers increase in number and thickness.

The bluffs along the edge of the Raton Mesa demonstrate an erosional pattern quite common in New Mexico and other arid regions: differential erosion of hard and soft rock layers. Resistant sandstones form cliffs and ledges; less resistant shale, siltstone, and coal form slopes.

There are a number of abandoned coal mines in the Raton Formation and overlying units, but most of them are out of sight, up canyons cut in the mesa flanks. The coal was used by local ranchers and the railroad or shipped south to Fort Union and Santa Fe. These mines closed after World War II.

Eagle Tail Mountain, the dark peak south of milepost 336 and east of I-25, is composed of Pleistocene basaltic and andesitic lava flows of the Raton-Clayton volcanic field. This mountain is an excellent example of inverted topography. The basalts here once flowed into and along the lowest parts of the landscape, but as the region rose and canyons eroded through the deposits, the resistant basalt that once filled the lowest portions of the ground formed the high points in this younger landscape that are seen now.

As the route drops into the valley of the Vermejo River, crossing it at milepost 321, the Cimarron Range is in view ahead; its highest peak, Baldy Mountain, reaches 12,441 feet. The Santa Fe Trail forded Vermejo River about 0.8 mile west of the US 64 bridge.

About 4 miles west of Cimarron, where the highway curves northward near milepost 305, the Raton Formation appears in cliffs to the northeast. Here, the formation

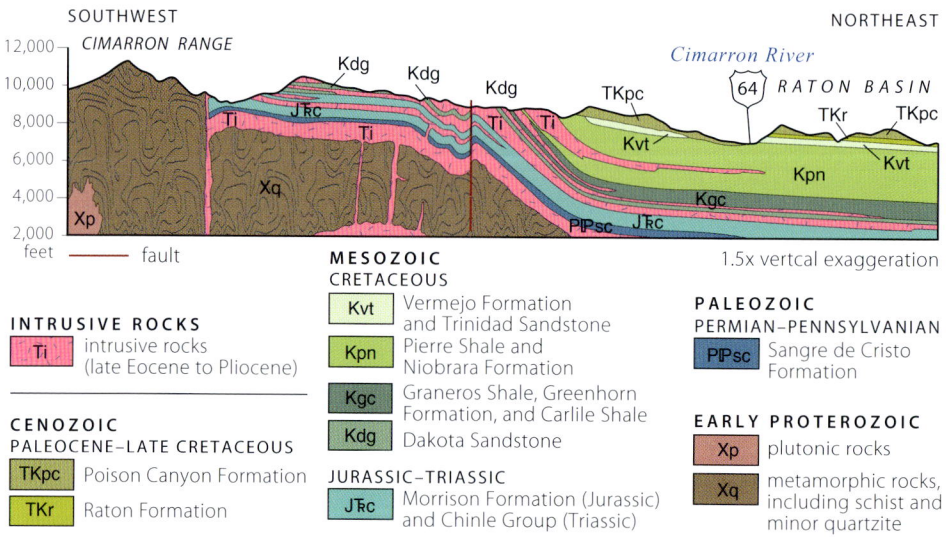

The basement core of the Sangre de Cristo Mountains warped and faulted the overlying sedimentary units, which bend down under the Raton Basin. This cross section extends from near the historic Philmont Boy Scout Camp in the Sangre de Cristos northeast toward Raton Mesa.
—Modified from Wanek and others, 1964

contains more of the cliff-forming sandstone and less of the slope-forming shale than it does farther east. Along with its grains becoming coarser to the west, it also cuts more deeply into formations below, suggesting the mountains had begun to rise and erode at the time it was deposited. The lowest Raton Formation unit, a coarse conglomerate, reflects the beginnings of mountain uplift near the end of Cretaceous time.

The highway climbs along the canyon of the Cimarron River, passing dark-gray Pierre Shale. This unit is weak and easily erodes, causing many slumps and small landslides. Look for slides and crumpled rock layers west of milepost 303. South of the river, the entire slope is landslide material and difficult to see due to vegetation cover.

The valley of the Cimarron River opens at Ute Park (milepost 297), which is floored with silty soil derived from the Pierre Shale. Copper and gold from the basal Raton Formation conglomerate were mined at the Mystic Lode, Aztec, and other mines. Rivers transported the metals from veins and bedrock lodes and deposited them with stream sediments, where they had settled into crevices between pebbles and cobbles, just as gold does in more modern placer gravels.

Visible on the mountain ridge between Baldy Mountain and the highway is a rock glacier, a mass of broken rock that moves slowly downhill, lubricated by ice between the rock pieces. The rock glacier formed during the Pleistocene glaciation.

Just west of the community of Ute Park are three sharp hogback ridges. Two are resistant sills—sheets of igneous rock intruded between sedimentary layers. The light-gray rock of the sills, called trachydacite, is speckled with white feldspar and quartz and black biotite and hornblende. It can also be seen at the Palisades Sill at Cimarron Canyon State Park. The sills are part of the Cimarron pluton that intruded this area 26 million years ago.

Cimarron Canyon State Park extends from Ute Park to Eagle Nest Lake (mileposts 297 to 288). The canyon, which runs along the Fowler Pass fault, separates Proterozoic basement rock (to the south) from Cenozoic rocks (to the north). Exposures of metamorphosed quartzite, amphibolite, and granite as well as dark sections of gabbro are visible from the road to the west. These units have been dated to about 1.6 billion years old.

The famous Palisades of the Cimarron Canyon State Park (milepost 294) are cut in a fine-grained, light-gray sill that intruded between the existing sedimentary and basement rocks about 26 million years ago. This sill cooled slowly at first, allowing large crystals to form; then cooling sped up, creating the fine-grained matrix as the remaining magma solidified. This protracted, variable cooling allowed joints to form within the sill as it cooled. Weathering, especially the dramatic effects of freezing and expanding water, enhanced the breakdown and removal of the rock along those fractures. As the Palisades were uplifted during mountain building, erosion accelerated, forming the stunning pillars we see today.

Upstream from the Palisades at milepost 294, watch for some very dark but glittering rock below the sill. This coarsely crystalline rock is gabbro, the intrusive equivalent of basalt. The gabbro is Proterozoic, 1.6 billion years old, as are the ancient volcanic rocks metamorphosed into greenstone and schist around it. This grouping of rocks may represent an ancient mid-ocean ridge. Two miles farther west the road comes into lighter-colored quartzite, also part of the mountain core and still Proterozoic in age.

The Palisades in Cimarron Canyon State Park are volcanic sill. The vertical fractures formed during cooling.

The magma of the Palisades sill crystallized slowly at first, and some of the minerals had time to crystallize out of the melt, then the magma cooled more rapidly so that the matrix appears gray and homogeneous.

At the mouths of several tributary streams, US 64 cuts through coarse, cobbly gravel of several terrace deposits. Rounded cobbles and pebbles are good evidence that the gravel is stream deposited rather than landslide deposited. Bouncing along in a stream inevitably rounds the corners of rock fragments.

Downstream from Eagle Nest Dam (milepost 289), the highway crosses three fault zones, the Fowler Pass fault, the Lost Cabin fault, and an unnamed fault. These faults were likely active as recently as Neogene time. Eagle Nest Dam, completed in 1918, impounds the Cimarron River and uses its water for irrigation and municipal use.

Eagle Nest Lake State Park (milepost 287) is in the Moreno Valley, a fault-bounded valley with Wheeler Peak to its west and Baldy Mountain to its east. Along the east side of the valley lies the west-dipping Moreno Valley fault, which was active in Pleistocene time. With ample sources of sand and gravel in the mountains around it, the valley is partly filled with Quaternary sediments. Note the broad alluvial fans that reach into the Moreno Valley from surrounding highlands. About 5 million years ago, volcanic activity south of the valley blocked the south-flowing drainage, forcing a diversion into what is now Cimarron Canyon. The upper Moreno Valley, north of the community of Eagle Nest, saw boom days back in 1867–1872, following a discovery of gold in 1866.

Due west of Eagle Nest is Wheeler Peak, the highest point in New Mexico (13,161 feet), in the Taos Mountains, which are part of the Sangre de Cristo Mountains. This peak and summits north of it are composed of contorted, 1.7-billion-year-old gneiss and amphibolite, thrust southeastward over younger rocks during the Laramide orogeny. South of here, the range was not lifted as high, and Pennsylvanian sedimentary rocks still cover the Proterozoic rocks. Ice of Pleistocene glaciation scoured cirques on the sides of the peak.

South of Eagle Nest Lake, US 64 curves southward for some miles, crossing alluvial fans and valley fill. The turnoff to NM 434 to the village of Angel Fire is at milepost 276. Coyote Creek State Park is 17 miles south of Angel Fire in the north end of Guadalupita Canyon. The 1.7-billion-year-old Proterozoic Vadito Group and Ortega Quartzite, including gneiss, amphibolite, schist, and quartzite, form the Rincon Mountains on the western skyline. La Mesa, to the east, poorly exposes marine sandstone, siltstone, and shale of the Permian Sandia Formation. These units dip 30 to 70 degrees to the east, having been tilted during the Laramide deformation. La Mesa is capped with basalt that erupted from vents near Black Lake, north of Coyote Creek State Park, about 4.5 to 4.0 million years ago. This capping basalt protects the underlying softer sedimentary rocks from erosion.

Turning west through hairpin turns of the Palo Flechado Pass (9,101 feet), US 64 crosses the southern part of the Taos Mountains, where Pennsylvanian limestones and sandstones cover the basement rocks. As the route drops into the canyon of Rio Fernando de Taos, outcrops reveal these rocks—dark siltstone and shale deposited in a shallow Pennsylvanian sea. Patches of Proterozoic basement and Neogene intrusive rocks are visible in places.

Ponderosa forests conceal much of the geology on the western side of the range. This difference in vegetation is due to the topographic controls on precipitation, called the rain shadow effect. When most precipitation comes from one direction (in this case, from the west) toward a high mountain range, precipitation is first dropped

Exposed in a steep hairpin curve near milepost 272 west of Palo Flechado Pass, fine layers of sandstone and shale steeply dip due to Laramide faulting. These rocks weather along planes of weaker shale and mudstone, while more resistant sandstone stands out in relief.

Brachiopod fossils erode out of a dark limestone outcrop by the side of the road, part of the late Pennsylvanian Madera Group.

as rain and snow on one side as air is forced upward and cools; once the air flows over the mountain crest, it carries (and deposits) less moisture on the other side. Soil forms more thickly under the abundant vegetation on the "wet" side of the mountains.

At the junction of NM 585 (Paseo del Cañon East), just east of milepost 257, is a large roadcut in the blocky, bedded Pennsylvanian to Permian rocks. Northwest of here, as US 64 emerges from the mountains, the highway crosses a series of faults that border the Rio Grande rift. These faults dropped the valley down to the west and uplifted the Taos Mountains. The Rio Grande rift is filled with young sediments derived from old rocks in the mountains that border it. Here at the mountain front, the Chamita and Tesuque Formations of the Santa Fe Group record Miocene-age basin filling with layers of clay, silt, sand, and cobbles overlying layers of the Servilleta Basalt. The Servilleta Basalt is well exposed at the surface, overlying these same sedimentary units farther west on the Taos Plateau. This shows the complicated relationship between deposition and rift-related faulting and subsidence. See the next road guide for more on the basalt.

Across the valley are the Tusas Mountains, composed of Proterozoic bedrock, lifted high along the faults that edge the rift on the west.

Blocky, dark-brown sandstones fill in scoured channels within more finely bedded shales near milepost 259. These Pennsylvanian to Permian units are approximately 300 million years old, equivalent to the Sandia Formation to the south.

Peaks of the Taos Mountains back Taos Pueblo. Pueblo Peak, the smaller peak on the right, is a Paleoproterozoic intrusion surrounded by Permian and Pennsylvanian sedimentary rocks.

US 64
Taos—Chama
93 miles

Taos sits on an alluvial apron, a swath of sediment wrapped around the base of the west edge of the Taos Mountains, all deposited by the Rio Pueblo de Taos and Rio Fernando de Taos. North of Taos, US 64 crosses wide, sage-covered alluvial fans that bury faults at the base of the mountains along the edge of the Rio Grande rift. Rare, low-level seismic activity continues in the Rio Grande rift, though with very small rates of motion. The slip rate on these faults is less than 0.02 millimeter per year, with recurrence intervals of 50,000 to 250,000 years. Most of the recorded earthquakes in the region occur on non-rift-related faults.

US 64 turns northwest just a few miles north of Taos, crossing lower parts of the alluvial fans. Cerro de la Olla, a rounded mountain to the north, and a number of other visible domes and cones are part of the Taos Plateau volcanic field. The field erupted more than 100 cubic miles of lava across 2,700 square miles. At least 35 vents, which erupted basaltic, andesitic, and rhyolitic lavas largely between about 4.8 and 2 million years ago, are mapped. The Taos Plateau volcanic field is famous for these sheet basalts, called the Servilleta Basalt, that are spectacularly exposed in the Rio Grande Gorge west of Taos. Mantle-derived, highly fluid basalt magma rose along faults to pour out of five central volcanic vents, depositing up to ten lava flows that are 3 to 40 feet thick and surface the Taos Plateau. These flows have been dated to 4.8 to 3.1 million years old. Sedimentary intervals up to 15 feet in thickness separate individual flows, documenting pauses in volcanism that allows for soil development and sedimentary deposition.

West of the Taos Regional Airport, the Rio Grande Gorge becomes visible. Perennial and ephemeral seasonal streams that deposited the complex alluvial fans of the Taos Mountains take a plunging drop into the Rio Grande Gorge. US 64 crosses the deep canyon on the Rio Grande Gorge Bridge between mileposts 243 and 242. This gorge formed as the river cut down through layered lava flows of the Taos Plateau starting between 700,000 and 400,000 years ago, the exact timing of which is an actively researched topic. The bridge, at 650 feet above the tumbling waters of the Rio Grande, is the second-highest bridge in the US highway system. The Rio Grande Gorge and surrounding lands are part of the Rio Grande del Norte National Monument. The monument also includes the lower section of the Red River, which drains the Taos Mountains and joins the Rio Grande upstream of the gorge bridge.

Exposed in the walls of the gorge are many basaltic layers of the Servilleta Basalt. These layers commonly form columnar joints, which are long, polygonal pillars that formed as the lava cooled and contracted. In between some of these lava flows are layers of orange and red sediments, indicating sediment deposition and perhaps soil development during periods of volcanic quiescence. Occasional plagioclase and olivine crystals can be seen in the basalt.

West of the Gorge Bridge, US 64 climbs through roadcuts in blocky basalt whitened by calcium carbonate, sometimes referred to as caliche. The basalt is dotted with small holes called vesicles that form when gases are trapped and expand within the blob of cooling lava but cannot escape the already hardened shell. Small volcanoes near the road are the sources of some of these flows.

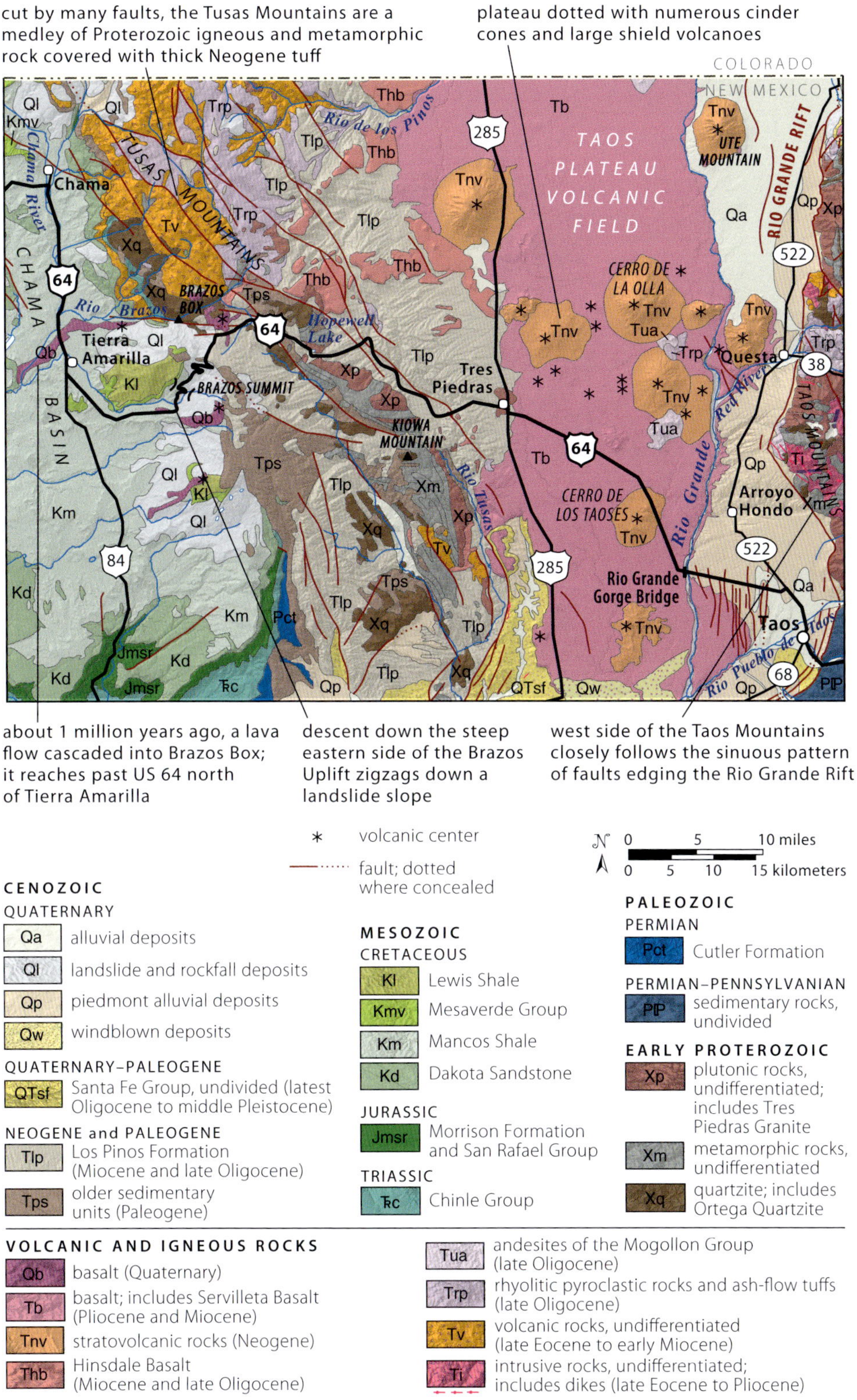

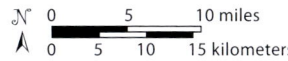

Geology along US 64 between Taos and Chama.

US 64 crosses the Rio Grande where it carved a vertical-walled chasm through the many basaltic lava flows of the Taos Plateau.

Hot lava burnt the soil underneath, oxidizing the iron in it. This photo was taken in the canyon to the north of the highway, near Arroyo Hondo, where a small road follows the Rio Hondo downstream to its confluence with the Rio Grande at the Dunn Bridge.

Views to the east of the Taos Mountains show the deep canyons of the Rio Grande's tributaries: the Rio Hondo and the Red River. Note the large, rounded outcrops of Proterozoic granite near Taos. From this distance, the difference between the high Proterozoic northern part of the range and the lower southern part, south of Taos, where Pennsylvanian sedimentary rocks and Cenozoic volcanics form the surface, can be seen.

The highway goes north around Cerro de las Taoses, a shield volcano composed of Neogene andesite and dacite, a somewhat stiffer lava than the surrounding basalt. The andesite contains visible olivine crystals, which are unusual in this kind of rock.

At milepost 223 is the village of Tres Piedras, which means "three rocks." The name comes from the massive granite boulders that stand dramatically above the sloping surface of the Los Pinos Formation of the Santa Fe Group, upon which the highway sits. The boulders are bedrock outcrops of Proterozoic granite that poke up through the lava and sediments of the Taos Plateau. The Tres Piedras Granite intruded about 1.65 billion years ago. This landmark has been noted on maps since the 1770s.

Near Tres Piedras (milepost 229), granite boulders slowly weather into rounded shapes. Rain, wind, and seasonal freeze-thaw cycles shape the boulders over time, creating small hollows that enlarge into cracks. Curved plates of rock eventually slough off, or spall.

West of Tres Piedras, US 64 climbs toward the Tusas Mountains, following the Rio Tusas through the Los Pinos Formation, a heterogeneous, conglomeratic unit composed of overlapping and coalescing Oligocene-age alluvial deposits. The Los Pinos Formation, mapped as part of the Santa Fe Group, is present all the way north into southern Colorado, collecting detritus shed from the San Juan and Sangre de Cristo Mountains. In the upper valley of the Rio Tusas, near the junction with Forest Road 133, US 64 runs along a fault zone that separates Paleoproterozoic metasedimentary and plutonic rocks to the south from much younger Paleogene sedimentary and volcaniclastic units to the north. The fault zone is as much as 3 miles wide, with several fault lines that trend generally northwest-southeast. Displacement totals about 700 feet, enough to make the valley lopsided.

Proterozoic rocks in the core of the Tusas Mountains are composed of three main types: metavolcanics and metasedimentary rocks of the 1.7-billion-year-old Vadito Group, vitreous quartzite and conglomerate, and numerous small stocks, all of which have undergone regional metamorphism. They may have formed in a volcanic back-arc setting along a subduction zone during the early formation of the North American continent. The rocks, among the oldest rocks in the southern Rocky Mountains, were intensely metamorphosed between 1.72 and 1.66 billion years ago and then intruded by the Tres Piedras Granite 1.65 billion years ago. The Tusas Mountain Granite, a single stock underlying the Tusas Mountains, dates to 1.5 to 1.4 billion years old.

The rosy color of the granites comes from the abundant potassium feldspar minerals that compose a large portion of the rock. Loose gravel and sand surround many of the boulder piles. Called grus, the gravel and sand are in fact disintegrated granite. The particles of angular rock fragments have broken down from larger granite boulders. This breakdown is a chemical and mechanical weathering process and is very common in arid to semiarid locations.

Above the Proterozoic rock, glacial gravels and volcanic tuff form much of the mountain surface. The tuff came from eruptions in the Jemez Mountains to the south and in the San Juan Mountains to the north and northwest. The soft, pinkish tuff appears in many roadcuts. Some of it is stratified, showing that water deposited it.

At milepost 204, the Continental Divide National Scenic Trail crosses US 64 near Hopewell Lake. High rolling land atop the Tusas, such as that in the vicinity of Hopewell Lake, bears the imprint of glaciation. In Pleistocene time, these rolling uplands were covered with an ice cap that reached long, frozen fingers down steep mountain canyons to a lower elevation of about 8,000 feet. Underfoot are glacial outwash gravels. These rounded boulders and cobbles were quarried from bedrock by flowing ice and smoothed during transport by running water from higher mountains to the north. Among them are fragments of Vadito Group metavolcanics and the white Ortega Quartzite.

Low outcrops near milepost 204 are Proterozoic quartzite. The quartzite in this photo was once a conglomerate but has been metamorphosed so that when it fractures, it breaks across what were once pebbles.

From the Brazos Overlook near milepost 196, the relatively level top of the Brazos Mountains and the dramatic 2,000-foot-deep Brazos Box, the cliff-walled canyon of the Rio Brazos, can be seen.

The impressive Brazos Box is three times as deep as the Rio Grande Gorge. As part of the Colorado River drainage basin, this canyon was probably carved in the last 5 to 4 million years. It is carved into the 1.7-billion-year-old Ortega Quartzite, a surprisingly white rock composed of nearly 98 percent quartz that is 7,200 feet thick in places. The north wall of the canyon, lifted along the Brazos fault, is higher than the overlook. A broad, grass-covered slope between the overlook and the highest cliff is a dip slope of east-dipping Cenozoic rock layers. The flat top is a peneplain, a regional erosional surface beveled onto the crystalline rocks uplifted during the Laramide orogeny. This area is just one small remnant of a once much larger peneplain.

At the Brazos Summit (milepost 193), where US 64 reaches its highest point at 10,507 feet, this flat surface is easily viewed. Beneath the Cenozoic cap, Mesozoic rocks overlie Proterozoic rocks, with no intervening Paleozoic sediments. There were Paleozoic layers here once, but they were eroded away during Late Pennsylvanian and Permian time, during the rise of the Ancestral Rocky Mountains. More Proterozoic granite and metamorphosed sedimentary rock are visible west of the summit.

Several small volcanic vents are visible at the summit to the northeast. From these vents erupted young basaltic to andesitic lavas that plunged down Brazos Canyon and out into the Chama Basin to the west. Farther north, Oligocene volcanic deposits from the San Juan volcanic field, which lies mostly in Colorado, reach south almost to Brazos Box.

To the west, the view from the summit takes in the Chama Basin, and beyond it the Gallina-Archuleta arch. This north-south-trending anticline is nearly 100 miles long, extending from the Nacimiento uplift into Colorado and separating the San Juan and Chama Basins. The Gallina-Archuleta arch is a Laramide structure, forming in response to the continental-scale compression that also uplifted the major Rocky Mountain ranges. To the south rise the volcanic Jemez Mountains.

Landslides play an important part in shaping the western slope of the Tusas Mountains. Rainfall is relatively heavy here, and wet rocks and soil slump and slide easily. In places, glacial gravels overlie Mancos Shale, a particularly weak Cretaceous rock unit that is famously sticky and slippery when wet. Highway builders chose what looked like an easy way down the mountain—a ramp-like slope—but it is almost entirely landslides! Even though highway cuts were beveled back at a shallow angle, they altered the natural angle of the slope. The roadcuts often initiate slides, some of which damage the highway today! As the highway zigzags down the west side of the Tusas Mountains, watch for other evidence of slides—hummocky topography, leaning trees, ponds, springs, marshy hollows, varied vegetation, roadcuts that show little scarps above bulging surfaces, and patched pavement. The road moves west onto Mancos Shale at about milepost 178. Roadcut exposures show that the weak Mancos Shale, along with man, is the culprit.

With thinner vegetation at lower elevations, rocks become more visible. Ledges and slopes of sandstone and shale are part of the Mesaverde Group, which forms the tawny hillsides. These units dip westward into the Chama Basin.

As the highway turns north, the Brazos uplift comes into view again near milepost 175, this time from below. Note the nearly horizontal surface at the top of the Proterozoic rocks, part of a regional erosion surface seen throughout much of the Rocky Mountains.

A Quaternary lava flow cascaded down Brazos Cliffs and out onto the plain just north of Tierra Amarilla at milepost 173.

A roadside shrine near milepost 173 has been built against the dark rock of a Quaternary lava flow. The dark, columnar basalt also shows up in roadcuts.

Brazos Cliffs, carved into pristine white Ortega Quartzite, follows an east-west fault with the down-dropped block to the south. Below the cliffs is the canyon, called Brazos Box, where the ephemeral Brazos Falls, the highest waterfall in New Mexico, cascades over the 1,300-foot cliff wall during springtime and monsoon season.

The Brazos Box shows up clearly from the junction of US 64 and US 84 and farther north in Tierra Amarilla. The town's name (which means "yellow ground" in Spanish) refers to yellowish soil derived from weathered Mancos Shale. There are other box canyons along the edge of the uplift, but none as large and spectacular as that carved by the Rio Brazos. The Rio Brazos joins the Rio Chama near milepost 171 north of the lava flow. The base of the cliffs can be reached by driving east on NM 512 on the north side of the Rio Brazos.

US 64 proceeds northward on the fertile floodplain of the Rio Chama and the terrace that borders it. Bluffs west of the town of Chama expose Mesaverde Group sandstone and capping Lewis Shale. Chama is a terminal for the Cumbres & Toltec narrow-gauge railway, built in 1880. It bridges the 64 miles between Antonito, Colorado, and Chama, New Mexico. The railroad was constructed as a component of the San Juan Extension to the Denver & Rio Grande railroad network that served to provide transportation between Denver and the San Juan Mountains mining camps. After the silver boom of the late 1800s ceased, the railway was a conduit for transporting timber, livestock, and agricultural products. Passenger service ended in 1951, and freight service ceased in 1968. The Cumbres & Toltec Scenic Railroad was added to the National Register of Historic Places in 2007 and designated a National Historic Landmark in 2012. The New Mexico Bureau of Geology and Mineral Resources has an excellent guide to the geology along the railway, which currently operates from May to October and during seasonal holidays.

US 285
Santa Fe—Colorado
106 miles

Santa Fe is built upon the Santa Fe Group, alluvial deposits derived from the Sangre de Cristo Mountains and deposited in the Rio Grande rift. Heading north out of Santa Fe, US 285 climbs to the top of alluvial fans that form the foothills along the west side of the mountains. Streams from the mountains deeply channel these Pleistocene deposits. Canyons and roadcuts reveal the complexly layered sands, silts, gravels, and ashes that compose the Santa Fe Group sediments. Benches dip slightly westward, away from the mountains and toward the major drainage of the Rio Grande.

The Santa Fe Opera (milepost 171), which overlooks the Rio Grande valley, is an open-air venue that holds world-class seasonal performances each summer. This architectural and acoustic wonder sits on the Tesuque unit of the Santa Fe Group. These sandstones, siltstones, and mudstones, collectively more than 2,000 feet thick, were deposited in the floodplains of rivers that drained the Sangre de Cristos during the Miocene Epoch. The Santa Fe Opera has an exceptional view across the Rio Grande rift, with the Jemez Mountains to the west. These mountains grew almost entirely from the voluminous outpourings of two massive caldera eruptions 1.6 and 1.2 million years ago in Pleistocene time.

US 285 follows the ephemeral Rio Tesuque north into the Española Basin of the Rio Grande rift. The basin here is nearly 2 miles deep and completely filled with sediments brought from bordering mountains and through-flowing streams. Deposits dip 5 to 7 degrees to the west here. Plants have difficulty growing in these easily eroded soils, especially in an arid climate, and wind and rain freely remove loosened clay and silt.

At milepost 175, Camel Rock stands to the west of the highway. A more resistant layer of the Santa Fe Group forms a caprock (the head of the camel) that protects a

Camel Rock, west of US 285 near milepost 175, is composed of the Santa Fe Group. The slope is littered with blocks of the resistant sandstone layer that forms the camel's head.

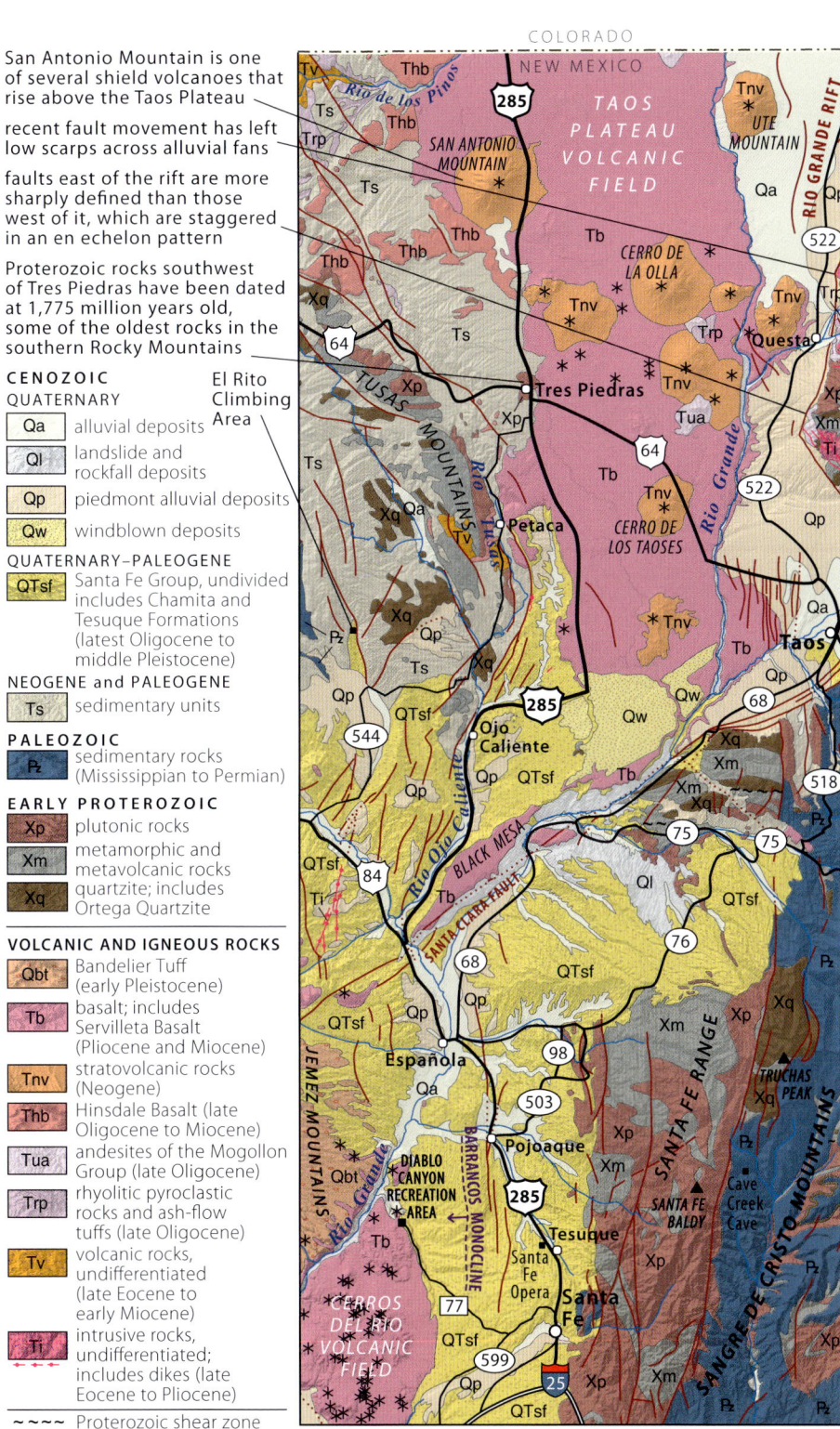

Geology along US 285 between Santa Fe and the Colorado border.

spire beneath it (the neck), which connects to its humped back. Though more resistant layers may exist, the Santa Fe Group is a fantastically weak, unconsolidated unit that has not solidified into rock. The "nose" of the camel recently broke off due to erosion. Wind is an important factor in shaping such isolated rock remnants. Just as water uses gravel to scour a riverbed, the wind uses blowing sand and dust particles to attack the exposed flanks of the Santa Fe Group. The power of erosion is visible in the carving of the badlands throughout the Rio Grande rift.

The silty, sandy deposits of Los Barrancos, a badlands bench west of milepost 182, record deposition in the Española Basin from 16 to 12 million years ago in Miocene time, based on age dates from white and gray ash layers erupted from volcanoes. Windblown and sheetflood deposits preserve the fossils of now-extinct mammals such as ancestral horses, deer, camels, and bears. Roadcuts just south of Española show the layered ash, pinkish sandstone, and caliche—calcium carbonate deposited by groundwater—of the barrancas, which means "ravines" in Spanish. Conglomeratic lenses deposited in alluvial fans or as channel-fill deposits are particularly well cemented with calcium carbonate and form ledges and the barranca tops. Muddy or silty sediments were deposited on a muddy floodplain every time the river overtopped its banks.

The highway drops down almost to river level at Española (milepost 188), on the fertile bottomlands of the Rio Grande. The valley is broader here than it is farther north, where the resistant lava flows of the Taos Plateau volcanic field hold the river in a narrow gorge. Fields along the river are irrigated directly from it or from tributary streams draining the Sangre de Cristo Mountains.

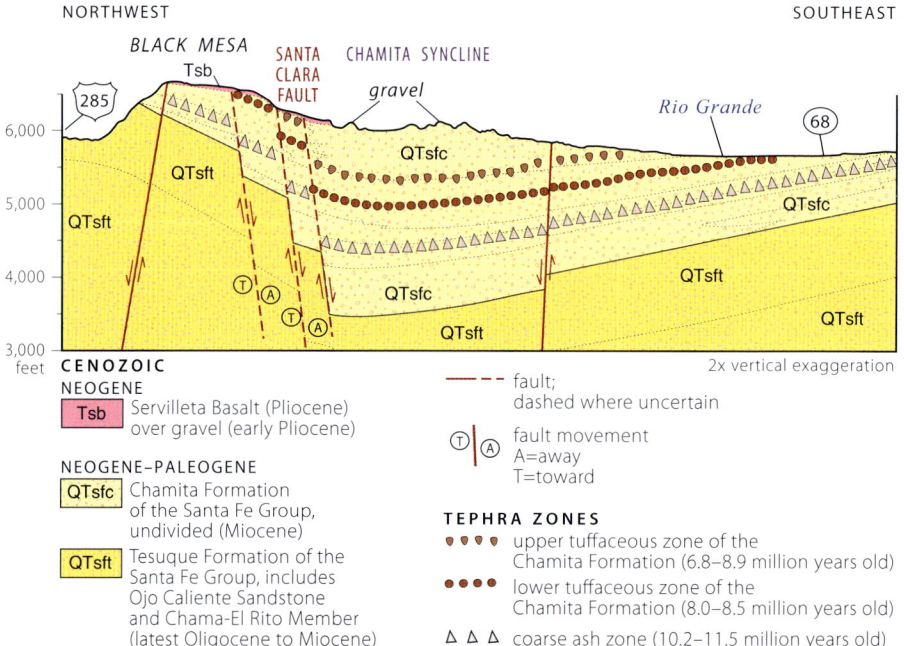

Cross section from US 285 southeast across Black Mesa and the Santa Clara fault to the Rio Grande. —Modified from Koning, McIntosh, and Dunbar, 2011

Crossing the Rio Grande at Española, US 285 heads north across the tip of an alluvial fan from the Jemez Mountains. North-bound travelers have their first good look at Black Mesa at milepost 196. Black Mesa consists of at least ten relatively thin, almost horizontal lava flows of Servilleta Basalt. These flows erupted 3.7 to 2.7 million years ago from volcanic centers within the Taos Plateau volcanic field, to the north.

Near the junction with US 84, US 285 crosses the Santa Clara fault, one of several in the Embudo fault system, which accommodates differences in downfaulting between the Española Basin to the south and the San Luis Basin to the north, both part of the Rio Grande rift.

North of its junction with US 84, US 285 rounds the northwest side of Black Mesa and milepost numbering changes. The highway follows the Rio Ojo Caliente north into the San Luis Basin. Extensive mass wasting and landslide deposits are visible on the flanks of Black Mesa, as well as Pliocene-age sandy gravels and alluvium directly below the basalt. These pale-brown and white, cross-stratified sands belong to the Ojo Caliente Member of the Tesuque Formation, part of the Santa Fe Group. The Ojo Caliente is thought to have originally been a dune field that was at least 1 mile wide. Near milepost 340, the Ojo Caliente sandstone layers have been eroded into mushroom rocks, hoodoos, and other unusual erosional forms. Their disintegration supplies sand for sand dunes.

The easily eroded Tesuque Formation of the badlands is responsible for the light-colored dust and sandy soils here. Thin, hard layers of silica and caliche deposits between sand and gravels represent times when the water table was static, allowing these resistant minerals to precipitate out of solution.

Hot mineral springs at Ojo Caliente, 1 mile west of milepost 353, come to the surface in a faulted area where Proterozoic granite and quartzite are lifted above the Santa Fe Group, defining the west side of the Rio Grande rift. The springs, with

Hot springs at Ojo Caliente rise to the surface along a fault. The dark-pink rock near the top of the hill is Proterozoic; the lower rock is the Santa Fe Group.

temperatures of 98 to 113 degrees Fahrenheit, contain minor amounts of iron, arsenic, soda, lithium, and other minerals commonly found in thermal waters.

North of Ojo Caliente, the highway is set into alluvial fans that contain cobbles of sedimentary rocks derived from the San Juan Mountains, volcanic highlands straddling the New Mexico–Colorado border to the northwest. In the Tusas Mountains between Ojo Caliente and Petaca on NM 519, about 15 miles to the north, Proterozoic granite contains bands of coarse pegmatite with large crystals of mica, beryl, garnet, feldspar, tourmaline, and other minerals. The Tusas Mountains are faulted on the east and tilted up like a trapdoor. In many parts of the range, their ancient rocks are covered with volcanic tuff and glacial outwash deposited by streams from Pleistocene glaciers. The same Proterozoic rocks underlie the Taos Plateau, where they have dropped many thousands of feet as part of the Rio Grande rift.

As the highway swings eastward and climbs onto the basalt surface of the Taos Plateau near milepost 362, views open of the towering Proterozoic edifice of the northern Taos Range to the northeast, and the lower southern part of the range, surfaced with Pennsylvanian sedimentary rocks, directly ahead. Wheeler Peak, in the northern part of the range, is New Mexico's highest point, at an elevation of 13,161 feet. The southern end of the Taos Range merges with the Truchas Peaks region, whose highest summits are also more than 13,000 feet. The Truchas Peaks end near Santa Fe and are the southern end of the great Rocky Mountain chain.

Tres Piedras (milepost 384) is named for the picturesque granite rockpiles that surround it—geologically part of the 1.7-billion-year-old bedrock of the Tusas Mountains. Many joints cut this rock, and weathering along them has gradually rounded outcrops into individual granite boulders that stand out above the sloping sediments of the valley floor.

North of Tres Piedras, US 285 passes volcanoes on the Taos Plateau. The field includes at least thirty-five vents, which erupted mostly between 4.8 and 2 million years ago. Some cinder cones are being quarried for cinder or perlite, a silicic volcanic glass with the same composition as rhyolite. When it is heated, perlite pops and expands. San Antonio Mountain, west of US 285 between mileposts 395 and 401, is

San Antonio Mountain is a shield volcano and is the highest peak of the Taos Plateau volcanic field, reaching 10,908 feet in elevation. The mountain erupted about 3 million years ago and consists of dacite flows that each reach up to 100 feet in thickness.

one of several shield volcanoes on the Taos Plateau. A breached crater is at its summit. Basaltic volcanism at San Antonio Mountain continued until 2.6 million years ago.

North of San Antonio Mountain (due west at milepost 399) are several cinder cones quarried for their very red cinder, also used for lightweight aggregate. Colorado's Sangre de Cristo Mountains rise to the northeast, behind Ute Mountain in the distance, another dome-shaped volcano on the Taos Plateau.

NM 68
Española—Taos
45 miles

NM 68 leads north from Española on the Rio Grande floodplain, following the river most of the way to Taos and crossing from the northern Española Basin into the San Luis Basin of the Rio Grande rift. Upstream and sidestream sources deposited pinkish, basin-filling sediments of the Santa Fe Group over millions of years, from Oligocene to Pleistocene time, as the rift opened. As climate shifted to colder Pleistocene ice ages, the increase in snow and ice pack also increased the ability of rivers to erode. The increased stream power, paired with the river becoming a through-flowing stream of the formerly interior-draining basins, resulted in more rapid river incision and the cutting of many of the canyons seen today. The Rio Grande rift shifted from a depositional to an erosional setting sometime between 700,000 and 400,000 years ago following the integration of the Rio Grande with the Gulf of Mexico. The exact timing of this event and many others in the evolution of the upper Rio Grande is hotly debated by geologists.

Instead of depositing sediments, the river now is cutting a path through its own sediments, revealing older river levels in the terraces that flank it. Several terrace levels—old river floodplains—can be seen on either side of the river. The Santa Fe Group, now eroding into complex badland topography, is visible across the river. A thin veneer of sands, gravels, and windblown deposits marks the modern deposition of sediments.

Alluvial fans from the Sangre de Cristo Mountains reach out toward the river. High mesas to the west are capped with orange-tan ash-flow tuffs from the Jemez Mountains, the location of two massive caldera eruptions at 1.6 and 1.2 million years ago.

The high plateau to the north and west from mileposts 6 to 7 is Black Mesa, capped with Servilleta Basalt that marks the southernmost flows of the Taos Plateau volcanic field. The basalts overlie Santa Fe Group sandstones and siltstones that are visibly tilted by continuing movement along Rio Grande rift faults and landslides. This basalt, which covered the lowest land surface when it erupted, illustrates how much erosion has occurred since then. The highway parallels its flank for some distance, crossing several dry arroyos that drain the Truchas Peak region of the Sangre de Cristo Mountains. Highest of three peaks in close vicinity, Truchas Peak reaches 13,108 feet in elevation, towering nearly 7,000 feet above the Rio Grande Valley.

North of Velarde (milepost 14), the Rio Grande's valley narrows to a canyon cut by the river, and NM 68 stays near the river bottom, providing great views. Once the Rio Grande started carving down into the resistant basalts, it became trapped,

Proterozoic gneiss appears in roadcuts as the highway climbs out of the gorge of the Rio Grande

from the rest stop, look over the vast Taos Plateau, cut by the twisting gorge of the Rio Grande; distant round-topped mountains are shield volcanoes

at least four terrace levels edge the Rio Grande in this area

partly because of their large component of volcanic ash, tilted Miocene and latest Oligocene deposits of the Santa Fe Group erode into barren badlands south of Española

CENOZOIC

QUATERNARY
- Qa — alluvial deposits
- Ql — landslide and rockfall deposits
- Qp — piedmont alluvial deposits
- Qw — windblown deposits

QUATERNARY–PALEOGENE
- QTsf — Santa Fe Group, undivided; includes Tesuque and Chamita Formations (latest Oligocene to middle Pleistocene)

NEOGENE and PALEOGENE
- Ts — sedimentary units; includes Picuris Formation

PALEOZOIC
- Pz — sedimentary rocks (Permian to Pennsylvanian); includes Alamito and Flechado Formations (Pennsylvanian)
- M — sedimentary rocks (Mississippian)

EARLY PROTEROZOIC
- Xp — plutonic rocks
- Xm — metamorphic rocks
- Xq — quartzite; includes Ortega Quartzite
- Xvm — metavolcanic rocks

- historic mine site
- volcanic center
- fault; dotted where concealed
- Proterozoic shear zone

VOLCANIC AND IGNEOUS ROCKS
- Tb — basalt; includes Servilleta Basalt (Pliocene)
- Tnv — stratovolcanic rocks (Neogene)

Geology along NM 68 between Española and Taos, including the High Road to Taos.

The Servilleta Basalt forms a resistant caprock of Black Mesa, sheltering the more erodible Santa Fe Group below.

and continued incision served to further entrench the river in its canyon. Tumbled black boulders of Servilleta Basalt cover slopes on either side of the river along with hummocky landslide deposits. Patches and occasional pinnacles of buff-tan Santa Fe Group sandstone and siltstone stand out.

The basalt is many-layered and in places shows vertical cooling joints that make it look like a log stockade. These regular columnar cooling joints, a pattern called colonnade, form as the lava cools and contracts. Famous examples include those of the Giant's Causeway (Ireland) and Devils Tower (Wyoming).

Embudo (milepost 18) is home of the Embudo Stream Gauging Station, a designated Historic Civil Engineering Landmark. The original stream gauge was established in 1888 as part of the US Geological Survey Irrigation Survey.

NM 75, which heads south from NM 68 at milepost 20, passes through the old mining town of Dixon. The Harding Pegmatite mine extracted lithium, tantalum, and beryllium from a pegmatite sill from 1918 to 1958. More than sixty minerals have been identified here. A few additional economically important minerals include almandine, spessartine, and pyrochlore. The pegmatite intruded approximately 1.3 billion years ago between two parts of the Vadito Group basement rocks. The coarse-grained pegmatite belt is about 2,500 feet long and up to 250 feet wide. Distinct layers of unique chemical and physical characteristics formed within the pegmatite during its protracted cooling. Some layers have abundant apatite and beryl, lower layers are muscovite rich, and another section contains bladelike spodumene crystals. The Harding Mine, extremely popular with mineral collectors and rock hounds, was donated to the University of New Mexico. Walking tours can be arranged through the university. The Harding Mine Historic Site in Dixon has further information on the mine.

North of milepost 20 on NM 68, roadcuts reveal old stream channels filled with basalt boulders. Descending again through the lava flows, the road travels below cliffs and highway cuts of Proterozoic metamorphic rocks—highly jointed gneiss, schist, and slate that break into angular blocks and skid downslope toward the Rio Grande. The river flows right along the Embudo fault zone between these ancient metamorphic rocks and the much younger Taos basalts to the northwest.

The Embudo fault zone is a northeast-trending transform fault, a type of fault that accommodates movement between two faulted areas. In this case, the faults are the east-tilted San Luis Basin to the north and the west-tilted Española Basin to the south, both basins of the Rio Grande rift. The Embudo fault lies along the Jemez lineament, an ancient zone of structural weakness in the underlying rock. The Jemez lineament contains many volcanoes and faults and stretches from Arizona to eastern New Mexico. The fault zone here takes up the differences in dip of the adjacent basins, as well as differences in rift spreading rates of the two basins, and the lateral motion that displaces the San Luis Basin to the east relative to the Española Basin to the south.

From Pilar, NM 570 continues north into the canyon about 6 miles to reach the south end of the Rio Grande Gorge, accessed only by trails. NM 68 climbs out of Rio Grande's narrow canyon through the Santa Fe Group and basalt flows to the

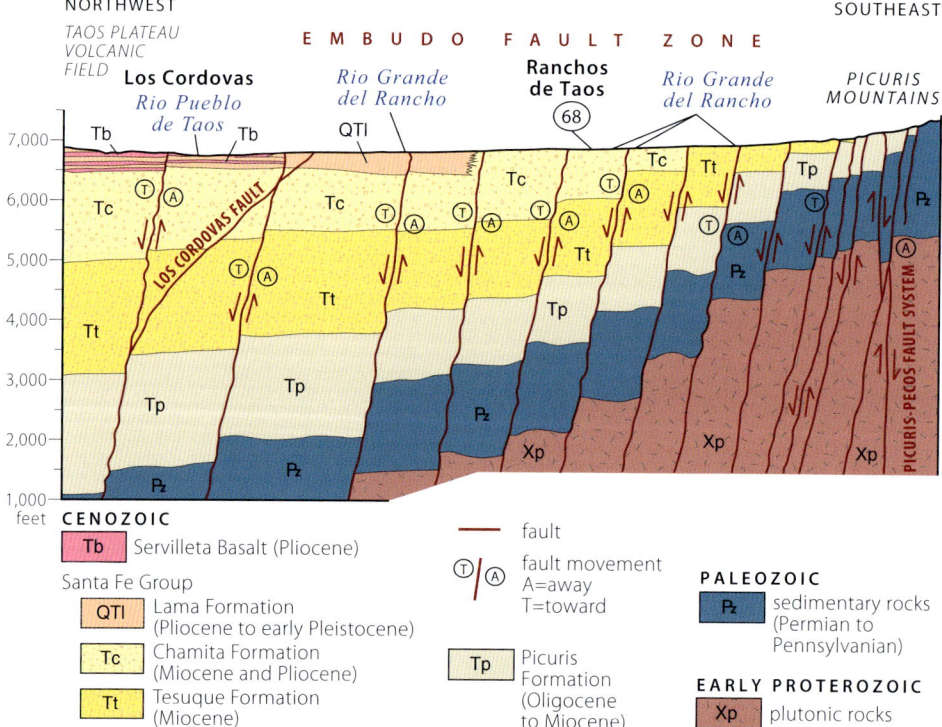

This cross section runs from the southern edge of the Taos Plateau volcanic field southeast through Ranchos de Taos to the Picuris-Pecos fault zone at the base of the Sangre de Cristo Mountains. It shows the highly faulted and en echelon nature of the Embudo fault zone at the edge of the Rio Grande rift. Note that the Embudo fault is strike-slip; the rocks on the left side of the faults slide toward (T) the foreground, and the rocks on the right side slide away (A) from the foreground. —Modified from Bauer and others, 2016

From the trailhead near milepost 34, the magnificent Rio Grande Gorge slices through lava layers of the Taos Plateau volcanic field.

surface of the Taos Plateau. Coarse gravel visible along the highway near milepost 29 is largely Pleistocene, washed from the Sangre de Cristo Mountains. Notice its well-rounded boulders and cobbles and the suggestion of stratification in pebble, cobble, and boulder layers.

At the trailhead near milepost 34 is a great view of the Taos Plateau, a flat surface of Miocene to Pliocene basaltic and andesitic lavas covered with sagebrush. To the west, not particularly prominent, is one of the volcanic centers from which the plateau lavas flowed. Several large, round dome volcanoes rise above the plain north of Taos, from which andesitic lava flowed.

East of the highway on the approach to Taos, the Sangre de Cristo Mountains are lower than elsewhere. Vertical uplift is not as great there, and the mountains are surfaced with Pennsylvanian sedimentary rocks, largely poorly exposed limestones, shales, conglomerates, sandstones, and siltstones of the Alamitos and Flechado Formations, among others. Some of these limestones are roughly equivalent to the Madera Group in the Sandia Mountains to the south. The Picuris-Pecos fault juxtaposes Proterozoic basement and these Pennsylvanian sedimentary units, while the sedimentary units form a slight topographic saddle between the high points of the Truchas Peaks to the south and the dramatic backdrop to the north from Taos that includes Wheeler Peak.

Taos (milepost 45), a global art hub, and the Taos Pueblo, a UNESCO World Heritage Site honoring more than one thousand years of inhabitation, are cultural gems. The iconic architectural style of Taos Pueblo has become immortalized by photographer Ansel Adams and painter Georgia O'Keefe. Today, the pueblo's original and reconstructed adobe buildings incorporate native rock, soil, and clay.

HIGH ROAD TO TAOS SCENIC BYWAY
Pojoaque—Ranchos de Taos
56 miles

See map on page 104.

The High Road to Taos is a scenic byway that goes through the high country between Pojoaque and Taos. From Pojoaque, head east on NM 503 toward the Sangre de Cristo Mountains. The Pueblo of Nambe has been inhabited since about 1300. The valley, hosting a sprawling population, is floored by Santa Fe Group, and NM 503 follows the modern alluvial deposits of Pojoaque Creek. These banks are abundantly covered in cottonwood and more invasive Russian olive and salt cedar trees.

Exiting the tree-filled river valley, the High Road to Taos rises slightly northeast of milepost 3 onto a rolling badland topography cut by water and wind into the Santa Fe Group sediments. Here, the diversity of the unit can be seen, with varying layers of buff-colored sand, brown silty lenses, elongated conglomeratic sections, and white rinds of ash, caliche, or both. More-resistant sections of the unit form caprocks for the many towering fins, pinnacles, and hoodoos that rise from the west-sloping main surface. The Tesuque Formation of the Santa Fe Group is the pinkish-to-brown unit forming the badlands. The unit consists of sandstone, siltstone, claystone, and conglomerate, some of which are visibly cross stratified. The unit was laid down largely by rivers as overbank and channel deposits between 16 and 14 million years ago, as determined by fossils and ashes in the many beds within the formation.

The Truchas Peaks to the east come into full view at milepost 6. These towering peaks of Proterozoic basement rock were first uplifted during the Laramide orogeny. Truchas Peak, the highest of a cluster of often-snow-capped peaks, reaches 13,108 feet in elevation, towering nearly 7,000 feet above the Rio Grande valley. Truchas Peak is composed of schist, quartzite, and amphibolite. To the north of this main peak, quartzite of the 1.7-billion-year-old Ortega Group dominates the skyline.

Badlands topography of the Santa Fe Group with a caliche caprock.

View from Juan Medina Road of the Proterozoic rocks rising high above Chimayo.

The High Road to Taos turns sharply to the left at Juan Medina Road at milepost 7.5 (before the Santa Clara Lake exit) and continues across a spectacular badland valley. A high pass drops the road down into the village of Chimayo (milepost 8 of NM 76), famous for the historical Santuario de Chimayo. Chimayo sits within the narrow valley of the Santa Cruz River.

East of Chimayo, the High Road (now NM 76) climbs back up several hundred feet of Santa Fe Group sediments past Cordova (milepost 11). The deeply eroded hillsides are testament to the erodibility of the soft, weakly cemented Tesuque Formation of the Santa Fe Group. The road reaches a high point at the village of Truchas (milepost 16.5), located at 8,040 feet elevation. The Truchas Peaks, a ridge of glacially carved basement rock, loom directly in front of northbound travelers. From south to north, they are Pecos Baldy (12,530 feet), Truchas Peak (13,108 feet, the second-highest peak in New Mexico), North Truchas Peak (13,025 feet), and Chimayosos Peak (12,841 feet). The peaks form a drainage divide, separating the Pecos River drainage to the east from the west-flowing tributaries to the Rio Grande. Several north-trending faults help shape this upthrown range, including the Picuris-Pecos, Rio Medio, and Pecos Baldy faults to the east of Truchas Peak and the Santa Barbara and Jicarilla faults on the eastern flanks.

After a sharp north turn just as the road enters Truchas, the route enters the gravel- and pediment-covered erosion surfaces of these upper portions of the Santa Fe Group. Landslide deposits and other poorly sorted debris become common as the route moves across the high terraces north of Truchas. These gravel-covered escarpments are visible in many roadcuts. To the west, across the rift, the Jemez Mountains are occasionally visible.

This road guide is the only one in this book that passes the Picuris Formation, a Cenozoic sedimentary unit on the west side of the Picuris Mountains. The upper

View from milepost 11 looking north over the valley of the Rio Quemado, an ephemeral stream, at badlands eroded in the Tesuque Formation in the distance. The flat surface in the middle ground is an erosional surface.

The Tesuque Formation, north of the road near milepost 16, incorporates layers and lenses of gravel containing Proterozoic clasts eroded from the mountains. The level surface at the top of the roadcut is the Truchas surface, one of many erosional surfaces that now provide a foundation for farming and upon which the village of Truchas sits.

volcaniclastic member of the Picuris Formation outcrops on the north side of the highway near milepost 18 of NM 76. This very friable, multicolored unit crumbles easily and contains mostly Cenozoic-age volcanic clasts but also some rounded Proterozoic basement quartzite gravels sourced from surrounding ranges. Age constraint on this unit is poor, but the upper unit is likely early Miocene.

The route continues through the villages of Ojo Sarco (milepost 22), Trampas (milepost 24), and Chamisal (milepost 28) before turning east at Peñasco on NM 75. Peñasco sits on the banks of the Rio Santa Barbara, one of the larger west-flowing tributaries of the Rio Grande. Here, the foothills to the east are in Pennsylvanian and Permian marine sediments, though they are covered in a blanket of Ponderosa pine forest.

North-trending rift faults are buried beneath the soils around milepost 18, including the Picuris-Pecos fault, a major fault on the west side of the Sangre de Cristos. It exists mainly in the basement rocks and is dropped down to the west.

One of Magdalena's daughters stands next to a white, fine-grained dike cutting the 1.68-billion-year-old granitic porphyry on the north side of NM 76 near milepost 25 northeast of Trampas.

At the T junction with NM 518 are fabulous exposures of Paleoproterozoic metasedimentary units. Less than a half mile east of the junction, NM 518 curves north into Borrego Canyon, bending around a large roadcut in the lower member of the Picuris Formation. The lower unit was probably deposited beginning about 34 million years ago. An overlook opposite the roadcut provides a place to view the rocks and valley below.

The T-junction of NM 75 and NM 518 displays fabulous exposures of Paleoproterozoic metasedimentary units.

White, tan, crumbling cliffs of the lower member of the Picuris Formation, an uncemented to moderately cemented conglomerate with minor pebbly sandstone lenses. Bouldery alluvium truncates the tilted layers at their top.

The first major exposure of Pennsylvanian to Permian bedrock is at milepost 60, and the road continues northward with these marine layers visible in cliffy roadcuts. On the north side of the road near milepost 63, well-cemented siltstone forms the blocky layers, while less-cohesive mudstone and shales form the crumbly interlayers. Several faults trend north-south in this area, tilting the bedrock. Glimpses of Wheeler Peak, the highest point in New Mexico, can be seen to the northeast. The road skirts bedrock to the west at milepost 64 before emerging into the wider canyon. Here, coarse-grained, feldspar-rich sandstone of the Pennsylvanian Alamitos Formation forms the hills to the west.

About halfway between mileposts 67 and 68, the road passes another large roadcut through the layered, conglomeratic lower member of the Picuris Formation, this one with greenish hues. It contains clasts of mainly well-rounded Proterozoic quartzite. This chaotic unit is interpreted to have been deposited as either debris flows or alluvial fan deposits at the beginning of Oligocene time.

The conglomeratic lower member of the Picuris Formation on the east side of NM 518 halfway between mileposts 67 and 68. Note the quartzite clasts in the close-up.

Downcanyon from milepost 68 on the east side of NM 518 is a lengthy roadcut in the Pennsylvanian Flechado Formation. Here, the unit is equivalent to the La Posada and Sandia Formations of the Madera Group. Numerous faults of the Embudo fault zone step units down to the west. Near milepost 70, the route opens into the Ranchos de Taos as it traverses broad alluvial fans of the Rio Pueblo de Taos and the Rio Fernando de Taos, major streams that drain the Sangre de Cristos.

Dark-brown and gray shales and siltstones contrast with coarse-grained sandstones in the Pennsylvanian Flechado Formation downcanyon from milepost 68.

NM 522 AND NM 38
ENCHANTED CIRCLE SCENIC BYWAY
TAOS—QUESTA—EAGLE NEST
83-mile loop

The Enchanted Circle Scenic Byway circumnavigates the mountainous area surrounding Wheeler Peak.

Taos sits in the Taos embayment, a subbasin of the main Rio Grande rift that forms a crescent-shaped scoop out of the Sangre de Cristo Mountains. Four major faults, hidden beneath the tumbled slopes of alluvium and other Quaternary cover, intersect here and record an eastward jump of faulting in the Sangres, disparate from the general north-south trend to the west. The Embudo fault zone accommodates the transition between the Española and San Luis rift basins that lie south and north of the embayment, respectively. The Sangre de Cristo fault, a west-dipping normal fault at the base of the mountains, marks the eastern edge of the rift basin. The Picuris-Pecos fault zone exists largely in basement rocks, with those rocks being deeply dropped to the west. The Los Cordovas fault zone—west-side-down, north-trending normal faults between Taos and the Rio Grande—record Neogene to Quaternary faulting.

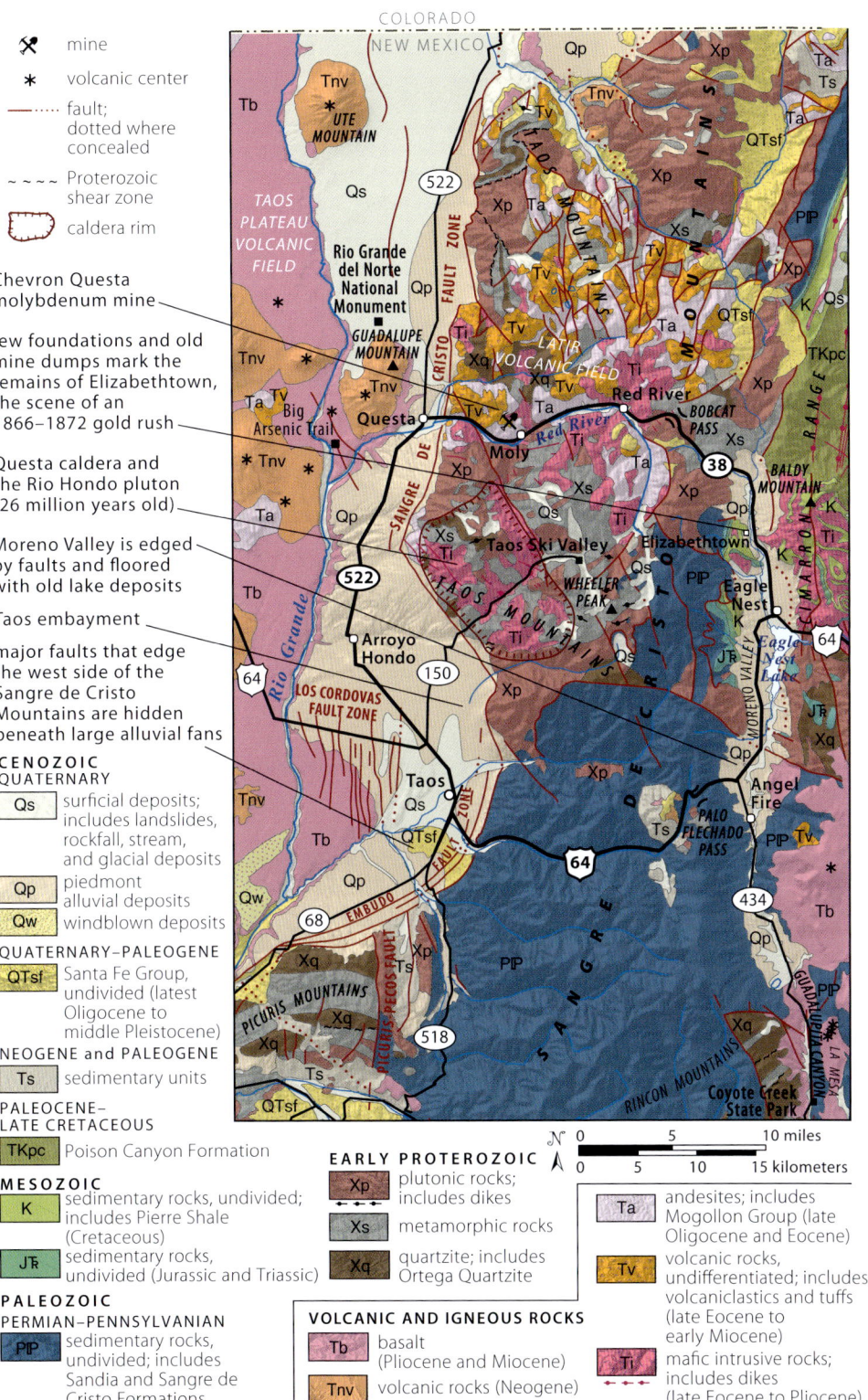

Geology along the Enchanted Circle Scenic Byway.

NORTHERN RIO GRANDE RIFT (522) (38)

Pick up NM 522 north of Taos from milepost 250.5 of US 64. NM 522 crosses the great alluvial apron at the base of the Sangre de Cristo Mountains, paralleling the dramatic mountain front. The Sangre de Cristo Mountains rise more than 5,600 feet above the Taos Plateau and include spectacular scenery as well as New Mexico's highest point, Wheeler Peak, at 13,167 feet. The Sangre de Cristos of New Mexico extend north into those of the same name in Colorado. These mountains have seen numerous mountain building events, preserved in their rock's metamorphic history. The mountains are composed of 1.7-billion-year-old Proterozoic metamorphic rocks of the Vadito Group, slightly younger (1.6-billion-year-old) Miranda granite, and Ortega Quartzite and Trampas Group metamorphics. Paleozoic-age sedimentary rocks exist and are often faulted against basement rock. The modern mountains got their start during the Laramide deformation, and rifting has continued to lift and attenuate these jagged ranges. The numerous alluvial fans that extend from their flanks catalog the uplift, erosion, basin filling, and movement in this region.

Taos Ski Valley is visible to the east near Arroyo Hondo (milepost 7). The range, while cored in crystalline basement rocks, is blanketed here with Oligocene volcanics. Extrusive entities include the 25-million-year-old Amalia Tuff, which originated from the Questa caldera, centered in the mountains northeast of Questa. Plutonic rocks include the white-to-orange, highly foliated granite of the Rio Hondo pluton that intruded 26 million years ago. These rocks formed during the transition from the Oligocene ignimbrite flare-up to the Rio Grande extension.

Wheeler Peak, the highest point in New Mexico, viewed to the northeast from near milepost 5 on NM 522. Glaciated in the Pleistocene, the mountain retains snow for much of the year.

At milepost 18, as the road drops into the Red River valley, Guadalupe Mountain to the northwest, one of a cluster of volcanoes rising above the lava expanse of the Taos Plateau, can be seen. These steep-sided volcanoes, all formed within the last 10 million years, erupted andesite and dacite (intermediate to silicic lava). Much less fluid than the basalt of the lava plateau on which they rest, these lavas pile up to form small stratovolcanoes. Most of the volcanoes have summit craters not visible from the highway. Ute Mountain, a pointed peak west of NM 522 near the Colorado border, is a volcanic peak that erupted rhyolite and dacite in Miocene to Pliocene time.

RIO GRANDE DEL NORTE NATIONAL MONUMENT

About 2.6 miles north of Questa on NM 522, turn west on NM 378 to reach Rio Grande del Norte National Monument. Over the next 10 miles or so, the road follows the canyon rim of the Rio Grande south as it cuts through the Taos Plateau. Numerous trails provide access to the river far below the rim. The overlook at La Junta Campground provides a view of the Red River confluence.

View to the west from the Big Arsenic Springs Trailhead of an orangish dacite intrusion incised by the Rio Grande. —Photo by Marisa Repasch

Although basaltic volcanism dominated the Taos Plateau volcanic field, numerous dacite and quartz latite intrusions were emplaced between 6 and 2.2 million years ago. In contrast to the basalt, which formed as lava rapidly cooled at the surface, these massive, red-orange rocks are considered subvolcanic because they cooled at shallow depths below the surface. These intrusions, seen on the east-facing wall of the Rio Grande Gorge from La Junta Point in Rio Grande del Norte National Monument, stand in contrast to the horizontal layers of basalt.

Basalt layers line the canyon walls of the Rio Grande at Rio Grande del Norte National Monument. —Photo by Marisa Repasch

At Questa, the Enchanted Circle National Scenic Byway heads east on NM 38, following the Red River upstream into the mountains. Immediately, the road plunges into a variety of extrusive and intrusive volcanics from the Latir volcanic field, all faulted together and lumped into the 30- to 24-million-year-old Mogollon Group. Rhyolite, ash-flow tuffs, and other rocks are exposed along the highway between mileposts 2 and 5, followed by more basaltic andesites and andesites (29- to 26-million-year-old) from mileposts 6 to 8. Rhyolitic tuffs cap high points.

Near the town of Moly at milepost 7 are the first basement exposures, 1.7-billion-year-old granodiorite, diorite, and gabbro exposed in the hillside directly west and east of the town. These ancient rocks are interpreted as the intrusive parts of a juvenile volcanic back arc, an ocean basin that opened on the continent side of a colliding volcanic island arc.

The Chevron Questa molybdenum mine, visible in the slopes above and to the north of Moly, opened in 1920 and operated as an open-pit mine until 1982, when mining moved underground. The mine closed permanently in 2014, and activity associated with it is now related to remediation activities of this Superfund site.

The high-grade molybdenite ore was mined from fissure veins within the Questa caldera that formed about 26 to 23 million years ago, part of the Latir volcanic field.

Calderas—circular collapse features walled by cliffs—form by inward collapse into their partly emptied magma chambers following a major eruption. The basin shape of this caldera is long since gone, but thanks to sixty years of molybdenum mining and unusually good exposures in the deep canyon incised by the Red River, its complex geology is well known.

Explosive eruptions showered the area with an ash-flow sheet, rhyolite, and other more mafic lavas. An incomplete ring dike is also present along the southern margin of the caldera. Hydrothermal activity associated with intrusions following the caldera collapse deposited the molybdenum ore in two layered, flat-lying zones approximately 500 feet thick. Veins of quartz and molybdenite filled cavities.

A large granite intrusion, the cooled and hardened contents of a resurgent mass of magma within the original magma chamber, is thought to be hidden under the mountains. Scars on the hillslopes may appear related to mining but are actually landslides that developed on the hydrothermally altered bedrock.

Red River, a quaint tourist town at milepost 12, sits on modern stream deposits brought in by several streams that converge in the narrow canyon. The town grew from the silver, gold, and copper boom of the late 1800s. Mining in the area was largely over by 1905, but the town continues to draw visitors who value the great outdoors.

NM 38 crosses back into basement rocks around milepost 14. Younger dikes cut the familiar 1.7- to 1.65-billion-year-old granitic gneiss and metasedimentary rocks

Tailings of the Questa mine rise high above a group of local residents near milepost 6 on NM 38.

of the Vadito Group. NM 38 crosses Bobcat Pass, the high point of the Enchanted Circle at 9,820 feet, between mileposts 16 and 17. South of the pass, the road descends toward the Moreno Valley, with pink-tinged Proterozoic granite forming angular blocks in roadcuts. Wheeler Peak, the highest point in New Mexico, is visible to the southwest. NM 38 crosses onto rolling Quaternary deposits at milepost 21 as the valley widens.

The highly colored and altered rocks east of the town of Red River form beautiful and treacherous cliffs to the north of the highway near milepost 14 on NM 38. The alteration occurred when magmatic fluids moved along fractures and faults.

Golden-colored Paleoproterozoic metasedimentary rocks erode on the north side of NM 38 near milepost 20. These extremely deformed and metamorphosed quartz muscovite schists and quartzites were once sand dunes or sandstones. The iron striping is a weathering feature.

Elizabethtown, between mileposts 24 and 25 (about 4 miles north of Eagle Nest), no longer exists, though a few crumbling walls and foundations mark the original townsite, and piles of bouldery gravel show the old diggings. Beginning in the 1860s, gold and copper were mined from quartz-rich veins, metamorphosed Cretaceous Pierre Shale, and related placer deposits in stream alluvium. Mineral-rich solutions from Oligocene intrusions cooled and interacted with the calcareous shale to precipitate the ore deposits. Dumps near Elizabethtown contain galena and pyrite as well as molybdenite, rhodochrosite, fluorite, and other minerals.

The crumbling Cretaceous Pierre Shale is occasionally visible at milepost 27 in foliage-covered slopes as the road heads toward Eagle Nest. Agua Fria Peak forms the high point at the south end of Moreno Valley south of Eagle Nest.

Complete the Enchanted Circle Byway loop by following the road guide for US 64: Raton—Taos, beginning on page 87.

JEMEZ MOUNTAINS

The Jemez Mountains are the remnants of the Valles Caldera, a huge caldera complex whose massive eruptions resulted in beds of basalt, thick layers of welded and unwelded tuff, and other volcanic deposits. The caldera is part of the Jemez volcanic field, one of three Quaternary continental caldera systems in the United States. The others are Yellowstone in Wyoming and Long Valley in California. The Jemez

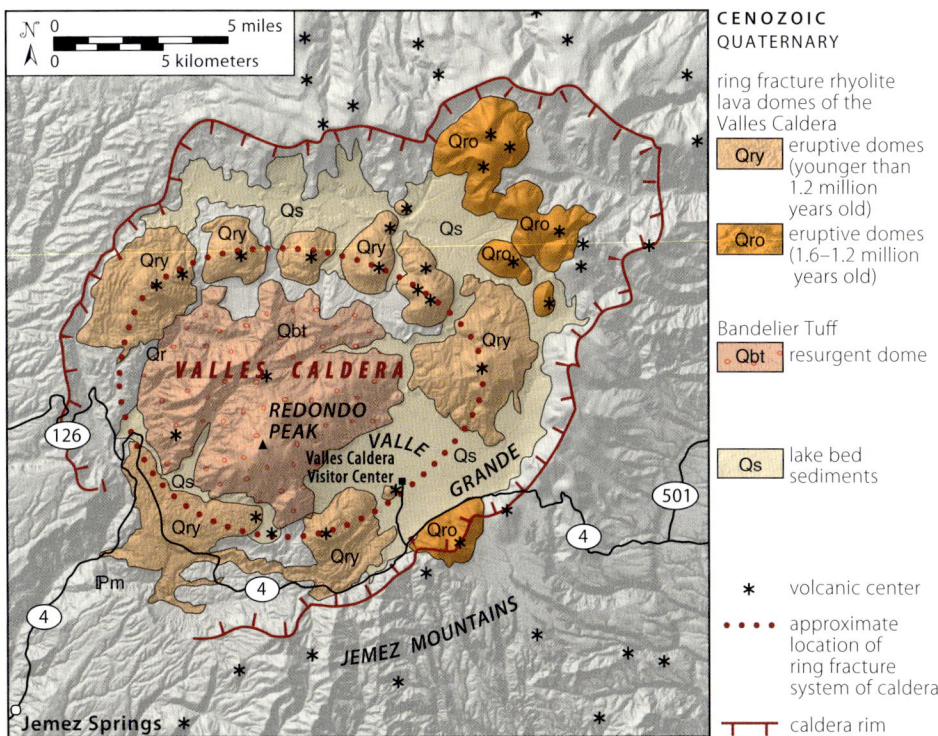

The Valles Caldera in the Jemez Mountains volcanic field. The caldera's collapsed centere is the Valle Grande. Resurgent domes that grew along ring fractures following caldera collapse punctuate the valley.

Mountains are located at the intersection of two lines of crustal weakness: the Jemez lineament and the Rio Grande rift.

The Jemez lineament is a series of volcanic centers that trend northeast from Arizona across New Mexico to Oklahoma, following an ancient Proterozoic scar in the Earth's crust. The scar formed where the Yavapai and Mazatzal provinces, two ancient pieces of crust, smashed together between 1.8 and 1.6 billion years ago. Though their movement is long over, younger volcanism has taken advantage of this crustal weakness and erupted along its length. The Jemez Mountains, the volcanic jewel in the chain of the Jemez lineament, lies on the western side of the Rio Grande rift, perhaps benefitting from the thin crust of the rift.

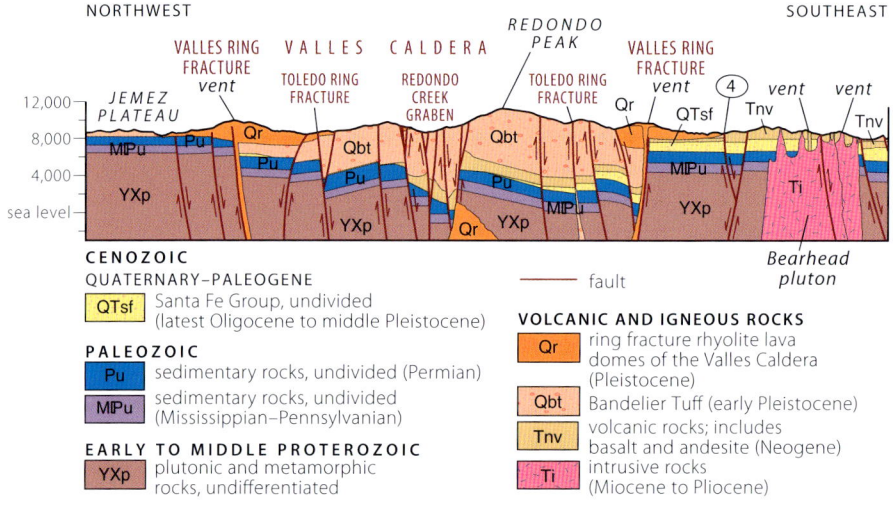

Cross section of the Valles Caldera. —Modified from Scholle, 2020∆

Following each major caldera eruption, the volcano collapsed into the empty magma cavity, forming a caldera. Each volcano erupted nearly on top of the previous one, so the majority of the caldera seen today is from the last eruption 1.25 million years ago. Parts of a caldera that formed 1.85 million years ago are also preserved. The central valley of the Jemez Mountains is the collapsed center of the Valles Caldera, the mountains that ring this valley are the broken slopes of the caldera walls, and the long plateaus that extend outward are the caldera's basalt- and tuff-covered flanks. The most recent outpouring of ash and rock covered many of the older volcanics.

These violent eruptions were on a huge scale; caldera eruptions alone deposited more than 145 cubic miles of lava, tuff, pumice, and ash. Ash was deposited as far east as the Mississippi River and as far north as Saskatchewan. The Mt. St. Helens 1980 eruption was tiny in comparison.

Magmas melted from both the mantle (basalts) and continental crust (rhyolite lavas and tuffs) erupted, along with various combinations of the two sources. The resultant volcanics have a variety of compositions and surface expressions, including towering cliffs of silicic welded tuff like those at Bandelier, basalt flows such as those in canyons below White Rock, pumice layers like those edging the highway in and near the Valles Caldera, and distinctive red layers and glassy obsidian like those in

Frijoles Canyon in Bandelier. The red layers and obsidian were produced when hot magma encountered groundwater, causing an explosive eruption known as a maar volcano.

The volcanic history of the Jemez Mountains begins with small volumes of Oligocene-age basalts that extruded during early extension of the Rio Grande rift, particularly the expansion of the Española Basin, which lies directly to the east of the current Jemez Mountains. Localized eruptions between 14 and 10 million years ago deposited lavas and tuffs of mafic to rhyolitic composition. These are largely buried by younger volcanics but are visible in Santa Clara Canyon west of Española. These relatively small eruptions led to more voluminous volcanic activity starting around 10 million years ago.

Most of the 10- to 9-million-year-old volcanics are basalts, trachyandesite, andesite, or rhyolite in composition. The oldest series of large basalt flows forms the modern Chamisa and Borrego Mesas in the southern Jemez Mountains, erupting 10 million years ago. At the same time, broad shield volcanoes in the northern Jemez volcanic field erupted basalts and crystal-rich dacites of the Lobato Formation. Andesite and dacitic lavas were erupted between 9 and 7 million years ago from multiple volcanic centers, many of which are controlled by the presence of the major Cañones fault zone on the western edge of the Rio Grande rift to the north of the Jemez Mountains.

The basalt and andesites that cap the iconic Cerro Pedernal (made famous by the paintings of Georgia O'Keefe) and other nearby mountains erupted between 7.9 and 7.7 million years ago and are part of the Lobato Formation. Volcanism intensified with eruptions from at least twenty vents depositing rhyolitic domes, flows, and pyroclastic deposits between 7 and 6 million years ago in the southern Jemez Mountains.

Caldera volcanism began with eruptions of the Tschicoma Formation between 6 and 2.7 million years ago. These viscous, voluminous flows (dacites) form many of the notable peaks in the northeastern Jemez Mountains. They were sourced from vents that were later obliterated by at least three massive caldera eruptions in Pleistocene time.

The collection of tuff and pumice erupted from the Pleistocene calderas is collectively called the Bandelier Tuff. It has three main tuff members. The 260-foot-thick La Cueva Member is the oldest, erupting at 1.85 million years ago. This unwelded unit is known for many pumice deposits. The La Cueva Member is very visible as the road passes Jemez Springs in San Diego Canyon. The 1.6-million-year-old, 390-foot-thick Otowi Member of the Bandelier Tuff is a massive, unwelded, rhyolitic ash flow. It contains abundant ice-blue sanidine (a high-temperature form of potassium feldspar) and quartz crystals. This member is visible all over the Pajarito Plateau and can be seen forming the mesas upon which the town of Los Alamos is built. These ignimbrites erupted from the Toledo Caldera, which the later Valles Caldera eruptions mostly obliterated.

The massive Tshirege Member of the Bandelier Tuff erupted 1.26 million years ago from the Valles Caldera. It deposited more than 3,000 feet of rhyolitic ash-flow tuffs that range from unwelded to densely welded. This orange-buff-colored, welded tuff forms the steep walls of the Jemez Mountains. The tuff shows zoning within the unit as magma composition shifted throughout deposition: lower flows that erupted first from the upper parts of the magma chamber are more silicic, while upper portions of the deposits, which were erupted later from the lower, hotter portions of the magma chamber, are more mafic. The material erupted as ignimbrites, which became welded tuff as they cooled. This requires volcanic ash and material to have been erupted at over 600 degrees Celsius. During welding, the hot pieces of ash, glass, and lava flattened into pancake shapes and adhered together to form an extremely hard, resistant deposit. The Tshirege Member has variable welding. More strongly welded tuff is of darker orange coloring than the underlying tan-buff cliffs, a feature that is nicely exposed in Frijoles Canyon in Bandelier.

The Tshirege Member also has abundant sanidine crystals. The very bottom of the Tshirege, sandwiched between the upper Tshirege and lower Otowi Member, is the Tsankawi Pumice Bed, a surge deposit of violently erupted material. A careful viewer can see the icy blue flecks of glassy sanidine within a hand sample.

After eruption of the Tshirege Member, the surrounding walls and overlying rocks collapsed into the emptied magma chamber. This collapse formed a circular pattern of faults and cracks known as ring fractures. While major eruption ceased, minor volcanism continued as the underlying magma chamber refilled. This refilling and eruption of young volcanics formed resurgent domes within the caldera; Redondo Peak is one such dome. The Valles Caldera is a classic example of a resurgent dome caldera, a geologic feature that was first studied here in the 1960s due to the iconic preservation of the caldera and the presence of Los Alamos National Laboratory on its flanks.

The collapsed caldera has been filled with a series of lakes throughout its history. Geologists know about the lakes from studying lake sediments that fill the valley center. Plant pollen that settled to the lake bottoms has been analyzed from drill cores, providing information on vegetation and climate changes in the Southwest. The lake level gradually rose in the caldera until the water breached through San Diego Canyon about 500,000 years ago, on the southwest side of the Valles Caldera. Once the rim was breached, lake waters flowed out with increasing rapidity, triggering a catastrophic flood that scoured the deep San Diego Canyon that exists today.

Valles Caldera has been an important place to the people of New Mexico for at least 11,000 years. Native Americans mined obsidian, a volcanic glass used for knives and points, from the basaltic outcrops and traded it across the Southwest. The sprawling Baca Ranch, established by land grant in 1876, encompassed the Valles Caldera proper. The area was used for grazing animals and timber mining until the 1970s. The Valles Caldera Preservation Act of 2020 formalized the process for a transfer from private to public stewardship. Today, the Valles Caldera National Preserve is open to the public.

NM 502 AND NM 4
POJOAQUE—WHITE ROCK—SAN YSIDRO
79 miles

From Pojoaque, NM 502 heads west, following the Pojoaque River toward the Rio Grande. Views to the north show Los Barrancos, badlands etched in soft, tan, easily eroded Tesuque Formation of the Santa Fe Group. These weakly to well-consolidated muddy siltstones, sandstones, and conglomerates were laid down in alluvial fans and river channels. The Barranca tops are formed of rock layers that are well cemented with calcium carbonate, called caliche. Muddy or silty deposits represent the lower

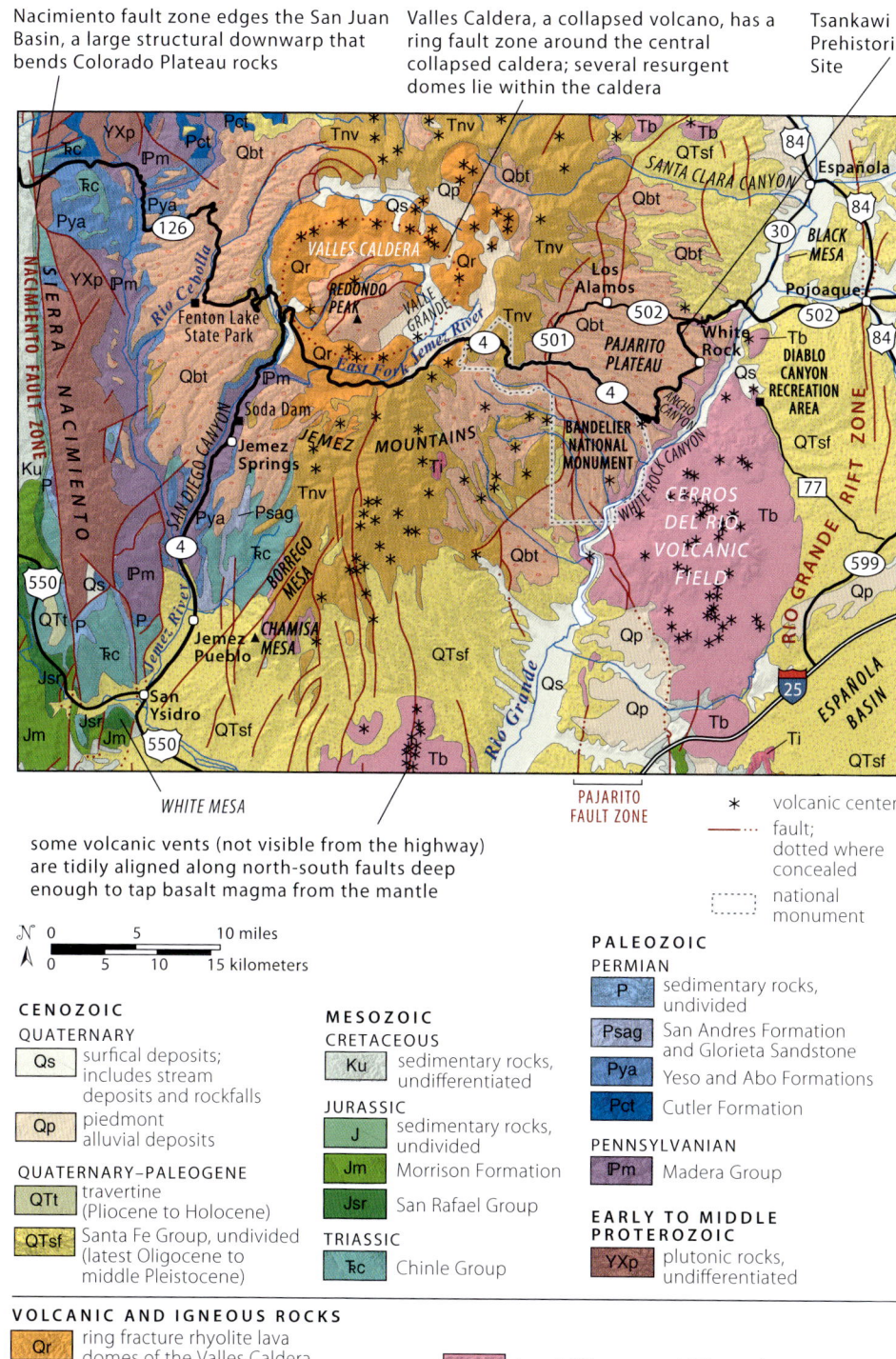

Geology of the Jemez Mountains.

energy deposition of a muddy floodplain or overbank deposits. White ash layers and other volcaniclastic detritus derived from local eruptions stripe these deposits and provide radiometric dating, which indicates these erupted 14 to 13 million years ago.

These rock layers were tilted into a monocline by faulting associated with the formation of the Española Basin, a major basin of the Rio Grande rift. The basin here is nearly 2 miles deep and filled with sediments brought from bordering mountains and through-flowing streams. Deposits dip 5 to 7 degrees to the west here, clearly visible where mesa tops slope toward the Rio Grande. These dry, poorly consolidated, easily eroded rock units crumble and blow or wash away before soils can form in this arid landscape. Plants have difficulty growing in such soil-poor conditions, as evidenced by the sparce vegetation. The silty, sandy deposits of the Tesuque Formation contain fossils of now-extinct mammals such as ancestral horses, deer, camels, and bears that thrived here in Miocene time.

At milepost 12, views to the north include Black Mesa, not to be confused with the long, narrow Black Mesa north of Española. This smaller black basalt edifice is a cinder cone that built around a volcanic neck approximately 4.4 million years ago, older than the nearby caldera volcanism of the Jemez volcanic field. This cinder cone is perhaps related to the Cerros del Rio volcanics that form the volcanic topography south of Santa Fe.

After crossing the Rio Grande on the modern Otowi Bridge (milepost 11), the route climbs through tan-buff, 15- to 12-million-year-old Tesuque Formation (Santa Fe Group) sandstone, mudstones, and fanglomerates. The historic Otowi Bridge, built in 1924 and on the National Register of Historic Places, lies just to the south of the modern traffic bridge.

Columnar jointing in basalt high on the southern slope is striking. The basalt is the Cuerbio Basalt, a flow of the Cerros del Rio volcanic field. These basalts have slumped and slid, visible in the many angles at which the columnar basalts (originally vertical) now careen. The basalts are overlain by the tawny-pink Bandelier Tuff, which composes the majority of the Pajarito Plateau, the flat surface atop the tuff. A pumice

View to the west of slumped basalt columns on the hillside west of Otowi Bridge near milepost 11 on NM 502.

DIABLO CANYON RECREATION AREA

To reach Diablo Canyon Recreation Area from near Santa Fe, take exit Camino La Tierra from the Santa Fe Bypass (NM 599) and head northwest about 4.5 miles on Camino La Tierra (Santa Fe County 77). Turn right onto Old Buckman Road, a well-graded dirt road, and go another 7.5 miles. The recreation area, popular for climbing and hiking, is known for its vertical walls composed of columnar basalt. The Rio Grande drove downcutting, resulting in the rapid carving of this ephemeral tributary canyon.

Diablo Canyon is located at the northern end of the Cerros del Rio volcanic field. Volcanic activity occurred here between 3 and 1 million years ago, with the majority of the basaltic to andesitic lavas erupting about 2.5 million years ago. The source of the Diablo Canyon basalt was a maar volcano. Basaltic magma began rising through the sediments of the Tesuque Formation then encountered the water table, resulting in an explosion. The blast formed a 1-mile-diameter ring of tuff. More eruptions produced cinder, scoria, and bombs as well as small lava flows, filling in much of the tuff ring. Finally, basalt was erupted both extrusively and intrusively, squeezing into cracks and cooling in large sills, dikes, and plugs. A lava lake pooled around the central scoria cone and a moat of earlier erupted lavas prevented it from escaping.

The columnar basalts in the canyon walls are a cross section of the solidified lava lake. More than 300 feet of spectacularly jointed, vertical to slightly curved basalt columns are visible. Columns form as the magma cools and contracts slowly over hundreds of years or more. Fractures first form at the top and bottom surfaces of a flow or intrusion, where the hot magma interacts with cooler host rock or the atmosphere, then propagate into the rock perpendicular to these cooling surfaces.

A climber utilizes cracks that separate vertical basalt columns at Diablo Canyon. —Photo by Timothy Elliot

The cliffs above the road near milepost 6 are Bandelier Tuff. The exposed section is the Tshirege Member of the Bandelier Tuff, which sits above the Otowi Member of the Bandelier Tuff. The angular fractures of the tuff formed as the ignimbrite cooled and contracted after eruption.

Tafoni in the Bandelier Tuff at Ancho Canyon.

layer at the base of the Bandelier Tuff, called the Guaje Pumice, is mined about 5 miles north of Tsankawi Prehistoric Site.

At milepost 6, head south toward White Rock and Bandelier National Monument on NM 4. Tsankawi Prehistoric Site is just south of the junction between mileposts 66 and 67. This section of Bandelier National Monument provides a short hiking tour that includes petroglyphs and excavated ruins abandoned in the late sixteenth century. The trail is exposed and requires comfort with ladders and high ledges. Cliff dwellings and fading petroglyphs are carved into the soft and erodible Bandelier Tuff.

The town of White Rock (milepost 63), named for the nearby White Rock Canyon of the Rio Grande, is built largely upon basalt flows. The orange-tan cliffs of the Bandelier Tuff form the hillslopes to the north.

Hairpin turns descend sharply into Ancho Canyon at milepost 58. Look for holes in the rock in the sheer walls of Bandelier Tuff. Known as tafoni, these weathering features form in sandstone, granite, limestone, and tuff, often in arid or semiarid environments, due to fluctuations in moisture, temperature, and wind. Cavities begin to form as salt crystals are deposited on a rock surface when water in cracks, fractures, or holes evaporates. Once a tiny hole or crack has formed, mechanical weathering continues to attack the rock surface, enlarging the holes over time with freeze-thaw, wind, and rain. The high desert setting of the Pajarito Plateau is ideal for the creation of pockmarked cliffs.

BANDELIER NATIONAL MONUMENT

To visit Bandelier National Monument, exit NM 4 as directed by highway signs at milepost 55.5. The canyons of the Pajarito Plateau were home to several Ancestral Puebloan communities. The canyons where the biggest ruins lie are deep and precipitous and provided warmth and shelter for farming and safety from other tribes. Tuff, which is cohesive and yet soft as rocks go, was an ideal building material for the extensive cliff dwellings and pueblos. In Frijoles Canyon, where the visitor center and main trails are, the upper Tshirege Member of the Bandelier Tuff forms the upper rim of the canyon.

Ruins are composed of quarried blocks of tuff.

Ancestral Puebloans hollowed out caves in the cliffs of Bandelier Tuff.

Many features of the tuff can be seen near the ruins. The rock is composed of bubbly fragments of volcanic ash, hard shards of volcanic glass, and irregular lumps of other volcanic ejecta. Elsewhere, the tuff is studded with volcanic bombs, blobs of magma that formed a chilled crust as they spun through the air. This chilled crust then cracked as the hot gases within continued to expand. Scattered through the tuff are natural cavities called tafoni. In Frijoles Canyon, ancient Puebloans enlarged some of the larger cavities to form living and storage areas.

As welded tuff cools and contracts, it forms generally vertical fractures known as cooling joints. These are similar to the vertical columnar jointing seen in basaltic flows near Los Alamos and generally form large polygonal (five- or six-sided) columns. These vertical fractures affect the hydrology and weathering of the tuff. Weathering has followed vertical joints to the extent that conical rock forms locally known as hoodoos or tent rocks have developed, as visible along the main Ruins Trail.

Downstream from the ruins, the canyon walls are composed of dark lava flows that predate the Bandelier Tuff. Here, the Rio de los Frijoles has cut a narrower canyon. Greenish olivine crystals are present in the basalt. Olivine basalt is sourced from the Earth's mantle deep below the solid crust, indicating that there was a deep connection between the faults that edge the Rio Grande rift and the lower crust and upper mantle. Farther down the canyon is evidence of other older eruptions. Cinder cones and reddish maar volcano deposits can be seen on the trail to the river. Here, the magma met groundwater associated with the Rio Grande, and the resulting steam blasted out a maar volcano.

BANDELIER TO SAN YSIDRO

West of the Bandelier turnoff, NM 4 climbs gradually across the Bandelier Tuff and the ring of older volcanic rock that forms the base of the Jemez Mountains. The strongly welded unit of Bandelier Tuff forms distinctive, nearly vertical walls, while the less-welded, upper layer is not as cohesive and forms slopes. The nearly treeless mesas to the south at milepost 52 were burned in the 1977 La Mesa wildfire. The formerly ponderosa-covered mesas never regained full forest cover. Higher on the horizon, black tree skeletons that stand above cliffs of blocky Bandelier Tuff are the result of the 2011 Las Conchas Fire, which burned more than 156,000 acres. The 2000 Cerro Grande Fire also burned in the Jemez Mountains, including Bandelier National Monument and even parts of the city of Los Alamos. This fire scar is visible to the north at milepost 52.

At milepost 50, a turnoff to NM 501 leads to Los Alamos, home to one location of the multisite Manhattan Project National Historic Park. Other locations are in Hanford, Washington, and Oak Ridge, Tennessee. Los Alamos Scientific Laboratory was critical to the development of nuclear weapons during World War II. Living in the then secret city of Los Alamos, some of the world's leading scientists developed the theoretical and experimental tests that built the foundations of atomic weapons. Los Alamos National Laboratory is still an important pillar of scientific research today. The Bradbury Science Museum hosts exhibits to explain the history and progress of work conducted at the national laboratory.

NM 4 climbs steeply west of the turnoff to Los Alamos, crossing many splays of the Pajarito fault zone. This major rift-bounding fault system drops rock layers down to the east and forms the western margin of the Española Basin. The fault zone is still active; occasionally, residents feel small tremors. As the route proceeds up the fault scarp on hairpin turns, the Bandelier Tuff is exposed at milepost 49 on the uphill side of the road. The flat-topped mesas of the Pajarito Plateau extend to the east.

At milepost 45, the upper reach of Frijoles Canyon, which hosts Bandelier National Monument, opens to the south. Look down the canyon for a long view to the Sandia Mountains near Albuquerque.

West of the ridge at milepost 43, NM 4 emerges into the Valle Grande, the inner valley of the Valles Caldera. The caldera is nearly 14 miles across from rim to rim, though the view to the western rim is blocked by the 11,258-foot-high Redondo Peak, the largest resurgent lava dome in the caldera. Located entirely within the caldera, this

View looking northwest into the broad valley of the Valles Caldera at the small resurgent dome called Cerro la Jara, the smallest of the resurgent domes. It rises from the valley floor to the right. The dome to the left is South Mountain. The treeless hills that form the skyline are the northern flank of Redondo Peak.

mountain formed approximately 1 million years ago when viscous rhyolite magma filled the chamber below, pushing up and deforming the surface rock and erupting onto the caldera floor. Around it are several smaller resurgent domes that erupted between 1 and 0.13 million years ago along a circular ring of fractures within the caldera. These are known as moat domes.

Warm springs within the caldera are visible from the highway as spots of greenery during summer and as locations bare of snow during winter. A meandering river, formed from the convergence of small tributaries and many springs, snakes its way through the caldera and becomes the East Fork of the Jemez River.

A young lake filled the caldera 60,000 to 50,000 years ago when the El Cajete pyroclastic unit dammed the southern outlet of the East Fork of the Jemez River. This 9 mile-long, 3-mile-wide lake was likely very short lived. The former geometry of this lake is evident in wave-cut benches, spits, and shorelines visible from the highway. Old wooden houses, which have been used as props in numerous western movies, are built on these features. The small rhyolite resurgent dome that stands out in the valley near the preserve entrance and administration buildings (milepost 39) is called Cerro la Jara.

NM 4 exits the caldera at milepost 38 and descends gradually through more Bandelier Tuff exposed in large boulders and rugged pinnacles among the trees. The distinction between tightly welded and unwelded tuff can be seen in roadcuts. Black volcanic glass (obsidian) directly overlies some of the tuff. Obsidian develops where silica-rich lava cools rapidly, with minimal time for crystal growth. The lava that forms the obsidian is similar to the rhyolite that forms the surrounding mountains; the difference is the rapid cooling of the glassy obsidian.

At the Las Conchas Trailhead, between mileposts 37 and 36, a wall of Bandelier Tuff towers to the north. This popular hiking and climbing area lies along the East Fork of the Jemez River, a National Wild and Scenic River that is the main outlet of the Valles Caldera. The innocuous-looking creek that now exists was once a conduit for the torrential draining of lakes held within the caldera proper.

Roadcuts south of milepost 36 contain several spectacular pumice, basalt, and tuff exposures of the crumbly, snow-white El Cajete pyroclastic unit. Look within it for the occasional volcanic bomb, a piece of rhyolite that has fallen through the air and chilled to form a crumbly, bread loaf–shaped clast. The El Cajete pyroclastic beds and the Battleship Ignimbrite (seen later at milepost 23) both erupted at around 74,400 years ago. Ash blown from this eruption spread through the Jemez Mountains and even reached across the rift to settle near Santa Fe and farther east across New Mexico. The even younger (68,000 years old) Banco Bonito Flow caps this ash.

Redondo Peak, the second-highest peak in the Jemez Mountains, rises to the north and east of the highway. This resurgent dome reaches 11,258 feet in elevation. Near milepost 32, NM 4 crosses the East Fork of the Jemez River. The turnoff to Jemez Falls is on the south, to the west of the bridge. A short hike winds through a slot canyon to a lovely cascade on the East Fork of the Jemez River. The river cuts canyon walls in the South Mountain Member and capping East Fork Member (specifically the El Cajete Pyroclastic Flow) of the Valles Caldera rhyolite.

At the Jemez Canyon Overlook, just west of milepost 29, a short nature walk provides views down into San Diego Canyon, the slot that drains the Valles Caldera. Before this canyon outlet was cut, the caldera held a series of lakes, similar to how

As NM 4 dives into the canyon of the East Fork of the Jemez River, a spectacular roadcut is visible on the north side of the road near milepost 32. Here, the layered beds of the El Cajete Pumice are overlain with the black, blocky, glassy beds of the Banco Bonito Flow.

The El Cajete Pumice on the north side of the road at a roadcut near milepost 35 on NM 4. The pumice and ash settled out of the air and is well layered. A large lithic fragment in the center of the deposit has weathered to yellow. An ashy layer is riddled with holes made by modern wildlife.

Crater Lake, Oregon, fills a modern caldera. Obsidian, flecked with white amygdules (gas-bubble holes filled in with quartz, feldspar, or other minerals), is exposed just across the highway from the overlook.

The turnoff to Seven Springs, also called Sulfur Springs, is at milepost 27. Located on the western flank of Redondo Peak, this geothermal area is known for acid-sulfate hot springs, fumaroles, and mud pots. The area was initially explored for geothermal resources in 1959. While determined to be too small a resource for commercial development, interest has continued. Two scientific wells were completed in the late 1980s, and a third was completed in 2004. These wells have provided abundant information on caldera stratigraphy and structure, hydrothermal alteration, and Pleistocene climate change.

NM 4 descends San Diego Canyon, initially occupied by San Antonio Creek, passing more exposures of Bandelier Tuff on the way. Sedimentary rocks are exposed in the canyon: gray and tan Pennsylvanian Madera Group limestone and shale become visible; bright-red overlying beds of the Permian Abo Formation are visually striking. A large cave worn into the softer tuff at milepost 26 is home to bats. Battleship Rock, a massive exposure of glassy welded tuff and obsidian, is visible to the east at milepost 23, at the confluence of San Antonio Creek and the East Fork of the Jemez River. The combined streams form the Jemez River.

Soda Dam is located just north of Jemez Springs at milepost 19. This massive travertine deposit, a carbonate rock, historically blocked the canyon and the Jemez River. While Soda Dam no longer dams the canyon because highway construction disrupted the blockage, mineral-rich waters are still depositing travertine here. The gray and pearly white draped folds of rock that form the dam developed where mineral-saturated water emerged from hot springs in the canyon and cooled. Calcium carbonate precipitates out of the oversaturated waters, forming the travertine. Hot springs are numerous in the Jemez Mountains, a result of the high heat flow associated with their volcanic origin. Groundwater, heated by hot rocks heated in turn by shallow magma, rises through the many cracks and fractures to the surface. This depressurized, cooling water bubbles out into hot springs such as those at Soda Dam. Active travertine deposition can be seen on the north side of the highway and from other springs in the dam and river. Visitors can see bright algae growths on both sides of the road during all seasons.

FENTON LAKE STATE PARK

Fenton Lake State Park, a short drive north on NM 126 from near milepost 26, is loved for its broad, grassy fields and the picturesque rhyolite cliffs of 1.6-million-year-old Otowi Member and 1.2-million-year-old Tshirege Member of the Bandelier Tuff. The valley houses the Rio Cebolla, which carved the valley and joins the East Fork of the Jemez River. The oldest rocks visible in the park are red beds of the Early Permian Abo Formation, as well as other sandstone and siltstone deposited in basins as the Ancestral Rocky Mountains eroded. Cliffy walls collect landslide deposits, and the modern stream deposits of the Rio Cebolla are the youngest units in the park. Mountainsides here have been severely burned in recent wildfires.

Soda Dam in San Diego Canyon is formed of travertine, calcium carbonate deposited by hot springs near the road. The Jemez River has tunneled a cavern passage through the natural dam. While icicles form during winter on the deposits, the springs heat the river year-round.

The Great Unconformity is present here. To the east of Soda Dam, Proterozoic gneiss outcrops in the steep canyon wall, overlain by the Pennsylvanian-age Sandia Formation. This gneiss is the only outcropping of basement rock in the Jemez Mountains. The gneiss lies south of the Jemez fault zone, visible as a crumbly, steeply dipping zone just south of the cement barricade on the western side of the road at Soda Dam.

The Bandelier Tuff forms high cliffs east and west of Jemez Springs. Its three members (La Cueva, Otowi, and Tshirege) can be easily distinguished in many places, with conical hoodoos, or tent rocks, developing prominently in the middle layer. Note that the ash flows were deposited on an irregular surface full of valleys and ridges. The base of the tuff gives information about the pre-eruption local topography. The incredible thickness of the tuff, as seen from the bottom of the canyon, gives a feeling for the truly enormous magnitude of these caldera eruptions.

On the east side of Jemez Springs, Jemez Historic Site (milepost 18) contains the remains of a seventeenth-century Spanish mission church and an older Native American pueblo. The walls of the church are built primarily from blocks of Madera limestone and Bandelier Tuff. The Madera Group limestones and shales contain marine fossils such as brachiopods and crinoids. The pueblo, likely established in the fifteenth century, is estimated to have consisted of nearly two hundred rooms, built of similar limestone and volcanic materials. The buildings sit on the floodplain of the Jemez River. Snowpack and springs in the Jemez Mountains feed this river, making it prone to flooding, which is combatted today by the visible roadworks along NM 4.

Hoodoos, or tent rocks, in the Bandelier Tuff decorate the high walls of San Diego Canyon. Each spire, or tent, is topped with a resistant capstone that protects the fragile volcanic materials beneath from erosion, much like an umbrella protects the bearer below. As the soft volcanic material is removed from around the protecting caprock, the tent or tower grows in height. These spires are ephemeral; all eventually collapse when the spire becomes unable to support its capstone.

The walls of the San José de los Jemez Mission Church at Jemez Historic Site are composed of blocks of the nearby Madera Group Limestone and Tshirige Member of the Bandelier Tuff. In the distance is the Bandelier Tuff in the northern wall of San Diego Canyon.

The river has a high concentration of dissolved minerals in its water due to the inputs from the numerous upstream hot springs.

South of Jemez Springs, brilliant red mudstone and sandstone beds of the Permian Abo Formation are exposed. Look for great roadcuts on the east side of the road between mileposts 16 and 15. The low band of red rock right next to the road from mileposts 15 to 14 is also the Abo Formation.

North of San Ysidro, the Triassic sandstones and shales of the Chinle Group gradually take over the scenery. These rocks dip southward off the Laramide-age Sierra Nacimiento uplift west of Jemez Pueblo. Cross-bedding can be seen in some of the large sandstone exposures.

Red Mesa, northwest of San Ysidro, is striped with Triassic red beds and older sedimentary rocks, some of them standing on edge. The ridge is a fault-edged syncline.

View to the west near milepost 14 of layered red mudstone and sandstone of the Abo Formation (lower half) overlain by the lower, thin San Diego Tuff and, above it, the Bandelier Tuff. The white layer is in the Bandelier Tuff, which forms the greater part of the sheer cliff face. A landslide that also covers the Abo Formation truncates the white layer on the right.

Triassic, Jurassic, and Early Cretaceous sedimentary rocks surround San Ysidro. The Chinle Group cuestas south of town are capped with the Jurassic Todilto Formation, formed mostly of shales and limestones. Its upper unit, however, is gypsum, deposited approximately 159 million years ago in a quiet, oxygen-poor body of saltwater isolated from the main Jurassic-age ocean basin. Gradual evaporation resulted first in the precipitation of thinly varved, or layered, limestone, followed by the precipitation of thick beds of gypsum. The gypsum is visible at the top of White Mesa southwest of San Ysidro. This gypsum is actively mined and used for cement, plaster, dry wall, and other building materials. The White Ridge bike trails, southwest of White Mesa, are an excellent access point for exploring the gypsum-covered mesas. See *New Mexico Rocks!* by Nathalie Brandes for more information.

NM 96

CUBA—ABIQUIU
59 miles

North of Cuba, near milepost 68, take NM 96 northward to circle through the northern Jemez Mountains. The scenery changes again as the Paleogene sandstones, shales, and conglomerates of the Ojo Alamo, Nacimiento Formation, and San Jose Formation form tan, buff, and brown hills and buttes of badlands. Rounding the north end of the Nacimiento uplift at about milepost 14, NM 96 crosses the Gallina-Archuleta arch that separates the Chama Basin to the northeast from the San Juan Basin to the west. The arch is a north-south-trending anticline nearly 100 miles in length, extending from the Nacimento uplift into Colorado. The Gallina-Archuleta arch formed in response to the continental-scale compression of the Laramide orogeny.

North of this feature, NM 96 passes through gaps in steeply tilted hogbacks of Mesozoic sedimentary rocks that get older to the east. Jurassic rocks form the whitish

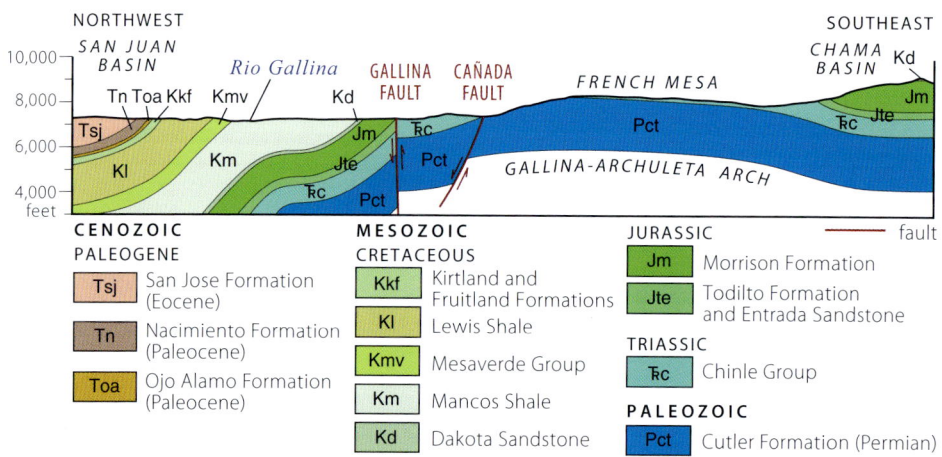

Cross section between the Rio Gallina and French Mesa, showing the arching of rock layers that occurred during the Laramide orogeny. —Crouse and others, 1992

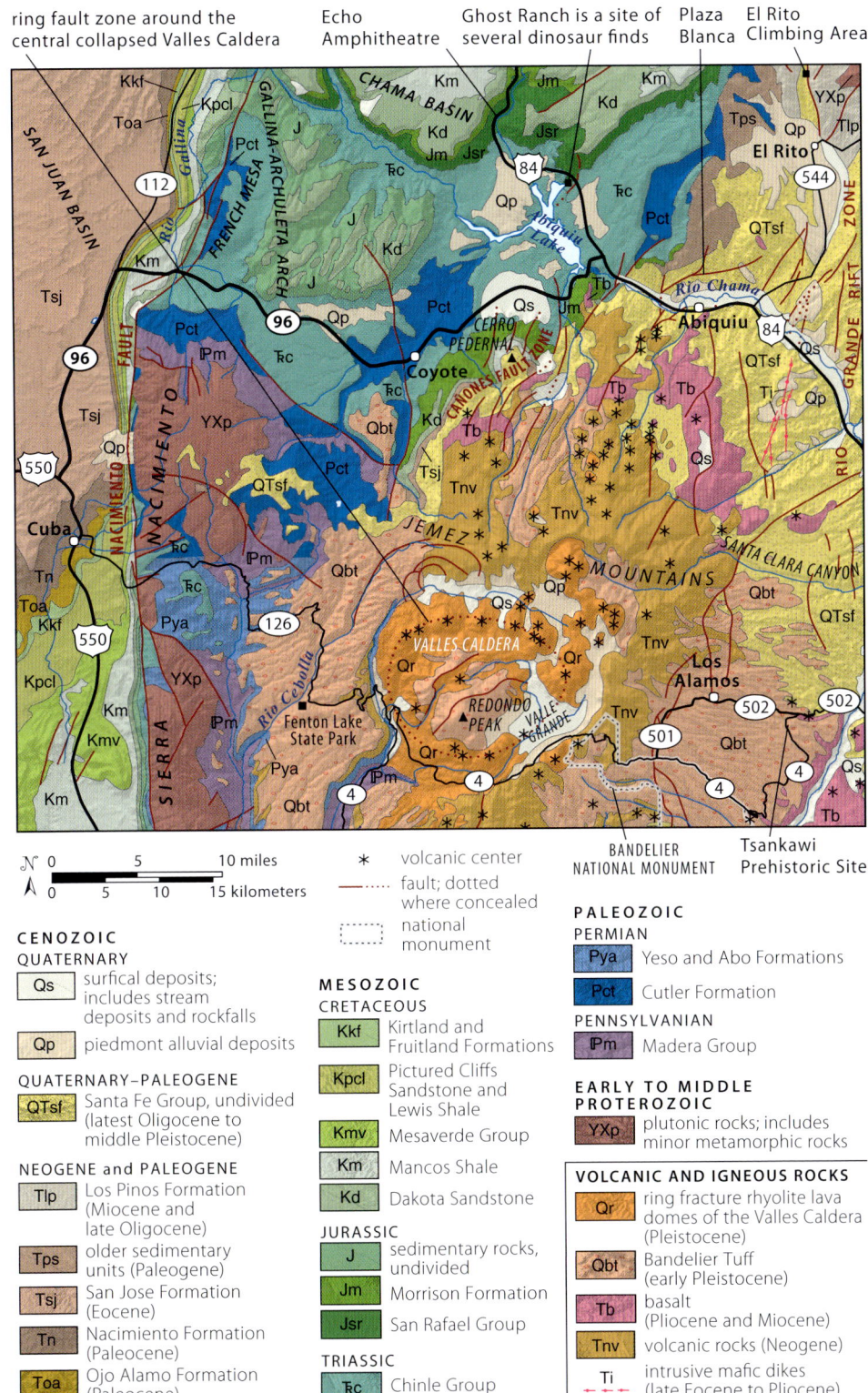

Geology along NM 96 between Cuba and Abiquiu.

cliffs between mileposts 17 and 18. Forested mountain slopes are surfaced with Paleozoic rocks, particularly Pennsylvanian and Permian limestones. There are no Mississippian or older Paleozoic units in this region; these were eroded away in Pennsylvanian time when this area was lifted to form one of the ranges of the Ancestral Rocky Mountains. Gypsum in the ridge to the northeast of milepost 17, is common in Jurassic rocks at this end of the mountains, too. Thick gray-white beds of it are sandwiched between other Jurassic rock layers. This soft rock is used for carving sculptures because it can be whittled with a pocketknife. Very fine-grained, compact gypsum is alabaster.

To the south, visible east of milepost 18, is the north slope of the Jemez Mountains, the location of the Valles Caldera.

Near Coyote, the highway descends into the Rio Puerco valley (yet another Rio Puerco) and passes through the Triassic Chinle Group into the Permian Cutler Formation, then returns to Chinle Group just before the end of the valley. Cerro Pedernal, east of Coyote, is a flat-topped landmark in the Jemez Mountains, made famous by Geogia O'Keefe's many paintings. It is capped with at least five andesite and basalt flows erupted in the late Miocene.

Cerro Pedernal literally means "flint hill," and the Pedernal Chert for which the mountain is named is a massive, cryptocrystalline, limey chert interlayered with beds of sandstone and conglomerate of the 27- to 25-million-year-old Abiquiu Formation. This chert-bearing unit forms a steep cliff, where the black, white, gray, and blue multicolored chert is exposed. Native Americans historically quarried chert and obsidian from Cerro Pedernal and other sources in and near the Jemez Mountains to make arrowheads, spear points, and scrapers.

Northeast of milepost 41, a grand panorama opens up. Abiquiu Lake is backed with colorful Triassic red beds and high cliffs of Jurassic sandstone. The reservoir sits within the Abiquiu embayment, a shallow extensional basin that forms the western edge of the Rio Grande rift here. The major Cañones fault zone separates the Abiquiu

Sunlit mesas near Abiquiu Lake expose Triassic red beds beneath a cap of Jurassic sandstone.

embayment from the Colorado Plateau. The El Rito Formation, Abiquiu Formation, and Santa Fe Group record Cenozoic basin-filling. All are capped with volcanic rocks of the Polvadera Group, including the 15- to 9-million-year-old Lobato Formation, the 5- to 3-million-year-old Tschicoma Formation (a dacite), and the 3-million-year-old El Alto Basalt. These basalts stand high above the modern Chama River, which supplies the Abiquiu Reservoir with water. Past river levels or tributaries deposited river gravels that are now buried beneath the basalt flows.

The El Rito Conglomerate, deposited in Eocene time, is spectacularly exposed at the El Rito Sport Climbing Area, northeast of Abiquiu off NM 554. Popular with outdoor enthusiasts, the El Rito climbing area is known for the sheer walls of this orange-to-red conglomeratic unit. The red color comes from the hematite (iron-rich) cement that binds the highly variable conglomerate together. The unit, deposited by a vigorous river, includes grain sizes ranging from fine to medium boulders, and rock types ranging from basement rock to local volcanics, including pumice, basalt, and the Amalia Tuff.

A short side trip north on US 84 is Ghost Ranch, where the upper Chinle Group holds the state fossil of New Mexico: the *Coelophysis*. Living 220 to 195 million years ago, this slender, bipedal carnivore was about 10 feet from head to tail. The first fossil of this dinosaur was found in a quarry in 1881 a few miles south of Ghost Ranch.

A climber ascends a towering cliff face of the El Rito Conglomerate at the El Rito climbing area. —Photo by Julia Martinez

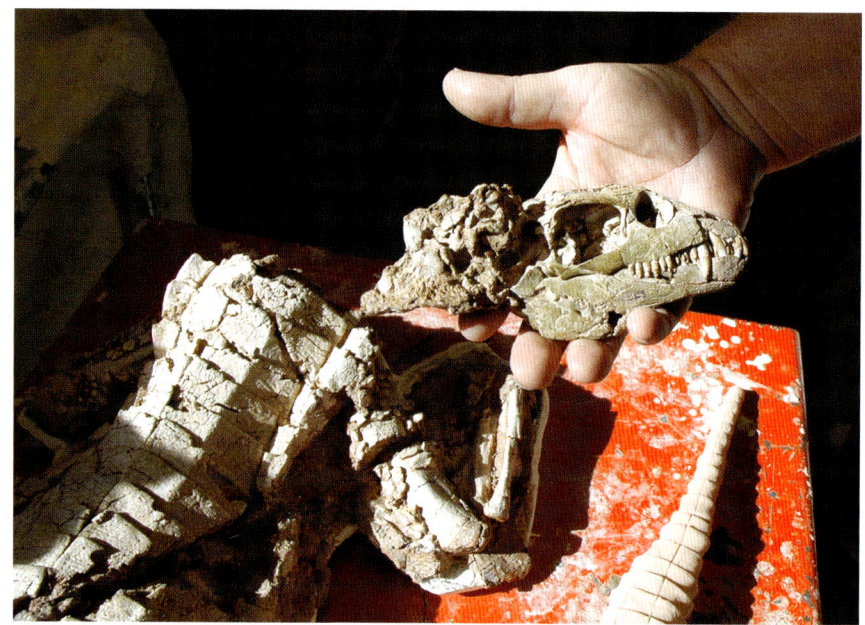

This small Coelophysis *ran swiftly on strong hind legs and sported a frightening array of sharp teeth. At least one thousand specimens have been unearthed near Ghost Ranch.*

Chimney Rock towers to the northwest of the road into Ghost Ranch off US 84 and is accessible via a hiking trail. The red colors in the foreground are iron-stained sediments of the Chinle Formation. The spires are 200 feet of Jurassic Entrada Sandstone. Cretaceous rocks top the mesa above. When undermined, the massive cliffs of the Entrada tend to break away along sometimes-arching joints, forming tall, arched coves, as they also do 9 miles to the north in Echo Amphitheater.

East of milepost 217, the highway cuts through the lower Chinle Group. —Photo by Kent Budge

The dinosaur lived and died in a floodplain with distinct wet/dry seasons in a warm, equatorial setting. These remains and other displays can be viewed at the Ruth Hall Museum of Paleontology at Ghost Ranch.

The loop continues southeast on US 84 to Abiquiu. East of milepost 217, the highway passes beneath vertical roadcuts of the lower Chinle Group. Plaza Blanca, accessed via County Road 155 just west of Abiquiu, was made famous as the "White Place" of artist Georgia O'Keefe. It spectacularly exposes the Abiquiu Formation, a 19-million-year-old volcaniclastic unit that includes material from the Latir volcanic field. The unit records the time when the Rio Grande rift began opening. Above the white cliffs, 4.8-million-year-old basalt caps the Sierra Negra, the small, faulted mountain that houses cliffs. Plaza Blanca is privately owned.

The Rio Chama meanders scenically through its valley near Abiquiu, as low-gradient rivers do. The Bandelier Tuff is visible on hills to the south. US 84 follows the river to the southeast, and river terraces along the Rio Grande come into view for southbound travelers, along with the glacially carved Truchas Peaks 35 miles away across the Rio Grande rift. Black Mesa, covered in flows of the Servilleta Basalt, is north of the highway at the US 84 and US 285 junction. This long, sinuous lava flow displays columnar jointing that formed during cooling. These Servilleta Basalt lava flows erupted from 4.5 to 3.6 million years ago, a span of nearly a million years, from five shield volcanoes in the Taos Plateau volcanic field.

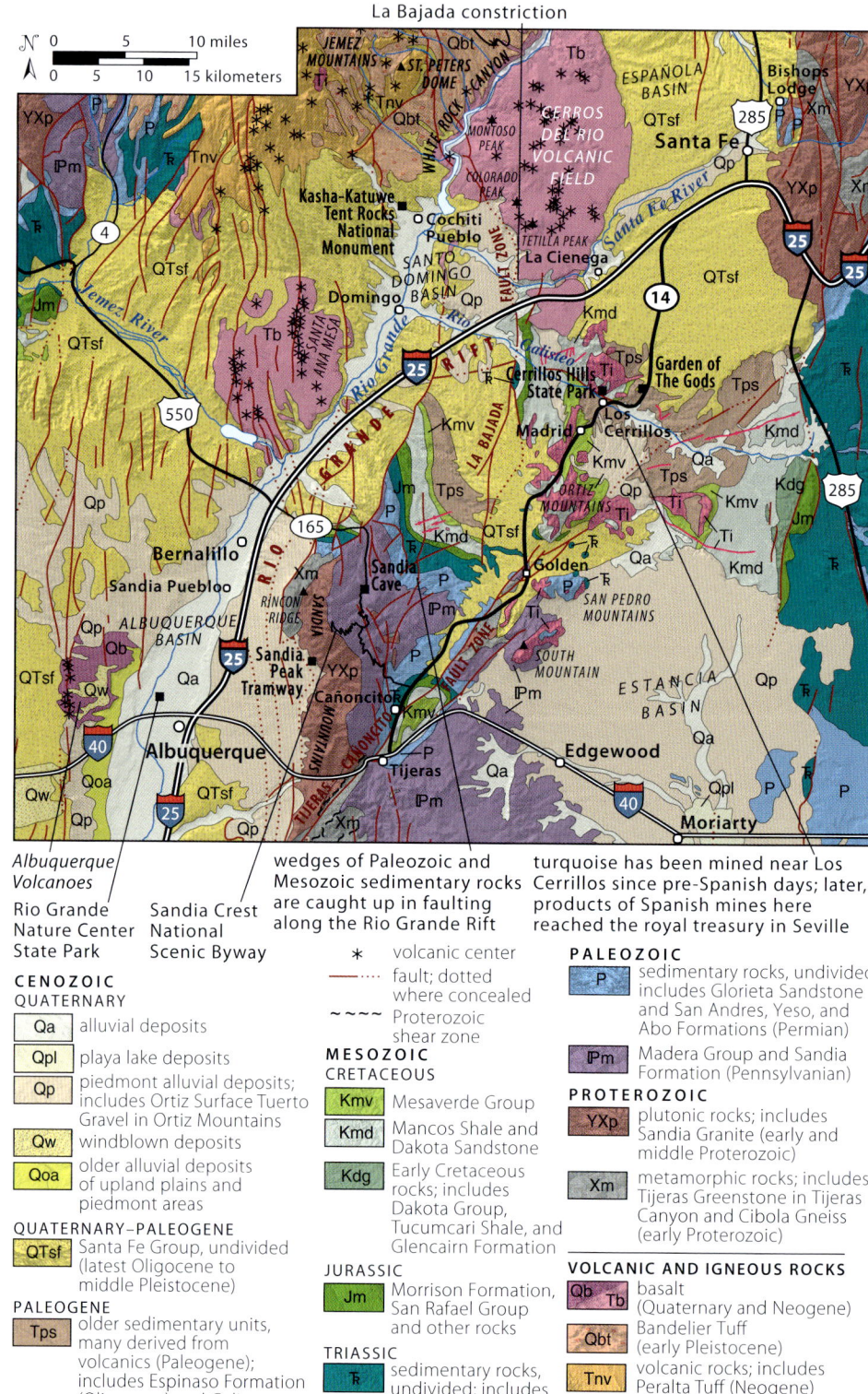

Geology along I-25 between Santa Fe and Albuquerque.

Southern Rift

I-25
Santa Fe—Albuquerque
56 miles

South of Santa Fe on I-25 at milepost 277, the tapering ends of the Sangre de Cristo Mountains are visible as low foothills to the east. The Paleoproterozoic granites exposed in these foothills are distinctly pink, especially at sunset. Strongly foliated amphibolite schists crosscut the granite in many places. The mafic schists weather more easily than the surrounding granites. Behind these foothills, and directly behind southbound travelers, the treeless summit of Santa Fe Baldy rises above the crenulated foothills to crown the mountain range above Santa Fe. At 12,632 feet, Santa Fe Baldy is the fifth-tallest mountain in New Mexico.

Between mileposts 258 and 262, I-25 crosses a broad, gentle slope eroded into the receding mountain base of the Ortiz Mountains to the south. The Ortiz surface, about 400 feet above the present Rio Grande, is a gravel-covered, late Miocene erosional surface. This surface rings the Ortiz Mountains and slopes away from the higher topography in all directions. The Tuerto Gravel covers this erosional surface. The Ortiz Mining District, designated in 1861 following the discovery of placer and lode gold in 1828, is one of the oldest in the United States.

The yellow-brown-tan silts and sands of the upper Santa Fe Group here are known as the Ancha Formation. This young unit of the Santa Fe Group is composed of basin fill deposited during the late Pliocene and Pleistocene, from 3 to 1 million years ago. The Ancha and the underlying Tesuque Formation of the Santa Fe Group are important aquifers for residents south of Santa Fe.

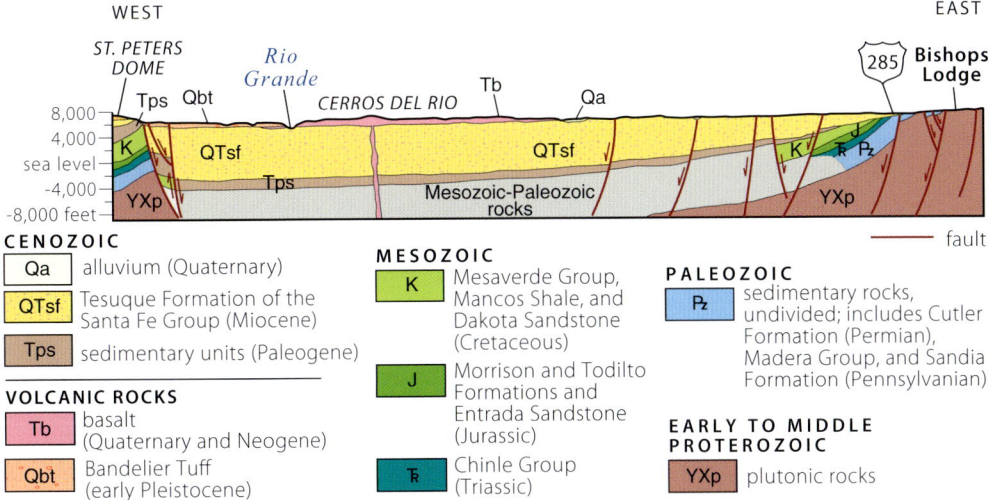

Section across the Cerros del Rio volcanic field in the Española Basin from Santa Fe to St. Peters Dome. —Modified from Kelley, 1978

At La Cienega (exit 271), vents and low, scrub-covered hills to the west are the volcanic remnants of the Cerros del Rio volcanic field. The Cerrillos Hills appear to the southeast, where mines are visible from here on a clear day. See the road guide for NM 14 (Turquoise Trail National Scenic Byway) on page 156 for more information about the mines and the Ortiz Mountains.

At milepost 268, south of the Sandoval County line at milepost 270, the highway rises from the Ortiz surface onto a basalt surface studded with small, low, somewhat obscure volcanoes. These are part of the Cerros del Rio volcanic field (also known as the Caja del Rio) that forms low hills to the west of the highway in front of the Jemez Mountains on the skyline. The Rio Grande flows mostly, between the two volcanic areas, out of sight from the highway in White Rock Canyon, a broad, deep gorge.

The Cerros del Rio volcanic field contains at least sixty volcanic vents, which erupted largely basaltic and andesitic lavas about 3 to 2 million years ago. The volcanic field lies at the junction of the Rio Grande rift and the Jemez lineament, a juxtaposition that resulted in the eruption of basaltic magma from deep beneath the continent crust. Eruptions occurred in three main phases. The oldest was the eruption of basalts from vents to form Montoso Peak, Colorado Peak, and Tetilla Peak, visible to the northwest at La Cienega. The middle eruptions were most voluminous, while the youngest, small eruptions were of tholeitic basalt flows in White Rock Canyon just over 1 million years ago. This type of basalt is very similar to that erupted at mid-ocean ridges, attesting to the deep source of these magmas. When eruptions mixed with water, steam blasted out maars. Cinder cones associated with the Cerros del Rio have been quarried for more than one hundred years. See Diablo Canyon Recreation Area on page 127 for more information.

At milepost 266, the interstate steps down the lava-walled rim of the Rio Grande Valley at La Bajada Hill. La Bajada, meaning "the Descent" in Spanish, is a 600-foot-tall escarpment capped with basalts of the Cerros del Rio. This blocky, black edifice is formed of multiple thin (13- to 50-foot-thick) trachybasalt flows that contain abundant olivine phenocrysts. The basalts are locally coated with white caliche crusts, which form as the basalt weathers in the arid climate.

It is not only the resistant nature of the basalt flows that forms such a startling escarpment; tectonic controls are also at play. The highway steps down through what is known as La Bajada constriction, a region of complex faulting where the Rio Grande rift is offset 18 miles to the right. La Bajada constriction, a complex system of faults and folds, separates the Española Basin of the rift (and the Cerros del Rio volcanic field) to the northeast from the Santo Domingo Basin to the southwest. To the north of the highway, the canyon of the Santa Fe River cuts this escarpment.

Below the basalt at the side of the highway, reddish and pinkish sandstone and mudstone of the the Santa Fe Group are well exposed between mileposts 264 and 262. Exit 264 to Cochiti Pueblo is notable for the striking red Morrison Formation that outcrops at the roadside. A small sliver of Jurassic to Cretaceous sedimentary units is preserved here, exposing the sequence of Morrison Formation–Dakota Sandstone–Mancos Shale.

At milepost 263, where the highway crosses the Rio Galisteo, a gypsum mine can be seen upstream to the east. Another is upstream from milepost 253. The gypsum, which occurs in the Jurassic-age Todilto Formation, is used in manufacturing cement, wallboard, and plaster. The Todilto Formation is composed of evaporite, minerals left

To the east of I-25 at milepost 264, red sandstones and siltstones of the Chinle Group and Entrada Sandstone are exposed in the uplifted side of the La Bajada fault zone. Here, normal faults cut these Mesozoic sedimentary rocks into clean angles. These units are overlain with basalt (not visible in this photo).

KASHA-KATUWE TENT ROCKS NATIONAL MONUMENT

Kasha-Katuwe Tent Rocks National Monument is accessed via exit 259. The monument is nestled into the flanks of the Jemez Mountains and highlights some of the earliest volcanism in this mountain range. Stunning layers of pearly white and red ash, gravels, and tuff are exposed in narrow slot canyons and towering hoodoos, landforms carved by torrential summer monsoon precipitation and strong seasonal winds.

At Kasha-Katuwe Tent Rocks National Monument, spires are carved in varying shapes dependent on the hardness of the tuff, lava, and ash in which they are carved.

Caprocks protect the spires from erosion.

Beautifully layered deposits at Kasha-Katuwe Tent Rocks National Monument with the Pajarito Plateau in the distance to the north. The Sangre de Cristo Mountains form the distant, right-hand skyline.

The monument has two main volcanic units on display: the white-to-orange Peralta Tuff Member of the Bearhead Rhyolite and the younger, gray Cochiti Formation. The Peralta Tuff was erupted from numerous volcanic vents about 7 million years ago, long before the massive eruptions that resulted in the cliffs of tuff at Bandelier National Monument. These early eruptions deposited the ash fall, ash flow, and pyroclastic flow deposits that are displayed at Tent Rocks. The Peralta Tuff also records a long period of deposition and reworking, evidenced by fine, uniform ash falls, tumbled ash flows, and areas where streams have reworked and deposited cross-bedded volcaniclastic sediments. In places, rift faults, which drop the eastern side of the monument down toward the main rift, cut these beautifully layered deposits.

Apache tears are abundant at the monument, but collecting them is prohibited by federal law. These rounded pebbles of obsidian form when groundwater hydrates the glassy volcanic deposits, transforming black obsidian into gray perlite. The perlite is friable and weak, easily removed from a harder obsidian core. These resistant glass tears weather out of the tuff and collect in ravines and canyon bottoms.

Seasonal rainfall carved vertical canyons in the Peralta Tuff. The Slot Canyon Trail is a short but steep climb through one such canyon to a viewpoint of both the Kasha-Katuwe Tent Rocks National Monument and the Rio Grande rift.

over from the evaporation of a highly saline body of coastal water that nearby seas occasionally flooded.

Southwest of the Ortiz and San Pedro Mountains, beyond a low saddle, is the triangular edifice of the Sandia Mountains. Proterozoic metasediments, metavolcanics, and intrusive granitic rock core this east-tilted fault block. A fault along the block's west side uplifted it some 11,000 feet above sea level, and 21,000 feet (4 miles!) above the corresponding granite in the down-dropped block buried beneath the Rio Grande rift. The relatively smooth, gentle eastern slope of the Sandia Mountains is surfaced with several hundred feet of Pennsylvanian sedimentary rocks, mostly limestone layers of the Madera Group. The steep, craggy western slope, different from the eastern slope, is the faulted side of the tilted block, worn back and ornately sculptured by erosion.

Look west from milepost 245 to see several of the Albuquerque Volcanoes poke up above the Rio Grande. The river removed the soft basin fill of the Santa Fe Group, exposing the small volcanoes, their flows, and their former central conduits, which stand above the surface as volcanic necks. In this general area, quite a few small volcanoes—the sources of lava flows—line up along north-south fissures. These five volcanoes erupted 210,000 to 156,000 years ago.

ALBUQUERQUE

During its early history, the Rio Grande rift consisted of a series of closed, non-draining basins, many of them occasionally occupied by lakes. Gradually, the basins filled with sediments washed in from adjoining ranges as well as with lava flows and volcanic ash. These deposits are now known collectively as the Santa Fe Group. For much of the history of the basins, there was no through-flowing river; the Rio Grande established itself as a through-flowing stream only about 2.5 million years ago. The largest of the former basins, now filled with several vertical miles of gravel, lava, and ash, is the Albuquerque Basin, with Albuquerque lying at its eastern edge.

At the north edge of Albuquerque, look west across the Rio Grande's inner valley to see the Albuquerque Volcanoes, a string of small cones lined up along a north-south fissure. These erupted in Pleistocene time.

A number of sources supply Albuquerque's water. In addition to surface water from the Colorado River via interbasin transfer, wells tap the Middle Rio Grande Aquifer, groundwater in the porous, permeable gravels of the valley fill. The aquifer acquires water from mountain-front recharge and underground flow from the Rio Grande and its tributaries, as well as from unrenewable fossil water trapped in these sediments when they were deposited. The water levels in the wells have fallen over time as humans continue to pump, sparking concern that the aquifer, once thought to be unlimited, could be depleted.

To maintain and support growing water demand, the San Juan–Chama Water Project, which brings water from southern Colorado via a network of diversions, tunnels, and reservoirs, was created. This interbasin water transfer project diverts water from the San Juan River watershed (which would normally flow into the Colorado River) to the Rio Grande watershed for irrigation and municipal supply. Heron Lake, a manmade reservoir near Tierra Amarillo in northern New Mexico, is part of the system. Since Albuquerque started using the diverted water from Colorado, the aquifer in the sediments below the city has started to recharge faster than it is being used in some places.

The City of Albuquerque prioritizes the use of the San Juan–Chama surface water over the use of groundwater to maintain sustainable resource usage.

Manmade levees along the mountain front protect the city from sudden floods originating from cloudburst-style storms over the Sandias. Some of the water caught by the levees sinks into the porous ground, adding to the general supply, rather than flooding through the city and onto the low bottomlands that border the river's present channel and natural levees. Located about 50 miles upstream from Albuquerque, the large, earthen-fill Cochiti Dam was constructed in the mid 1970s to control the seasonally changing waters of the Rio Grande as well as to regulate irrigation and to adhere to interstate and international water compact.

A major flood in 1874 prompted the taming of the Rio Grande through the Albuquerque area with small dams and levees that slowed the water. While protecting the community from flooding, these water control features had the unintentional result of increasing deposition of silt in the area. The riverbed was aggrading, or building up, over the valley floor. In tandem with this aggradation, groundwater levels were raised, creating a progressively swampy and waterlogged landscape that killed orchards and alfalfa fields and damaged homes. In the 1930s, the Middle Rio Grande Conservancy District was created as a means of managing fluctuating water, development, and agriculture. River infrastructure—including the construction of several upstream dams on the Rio Grande, additional drainage and irrigation canals, and river levees and jetties—was added to control water flow. The conservation district continues to work to restore and manage a more natural river bosque ecosystem in the more than 30,000 acres it owns. The Rio Grande through Albuquerque and other protected habitat restoration areas in the region (e.g., Valle del Oro, Bosque del Apache, and Servilleta National Wildlife Refuges) are key riparian habitats for migrating birds. Rio Grande Nature Center State Park has tens of miles of curated paths for walking and bicycling as well as an interpretive center.

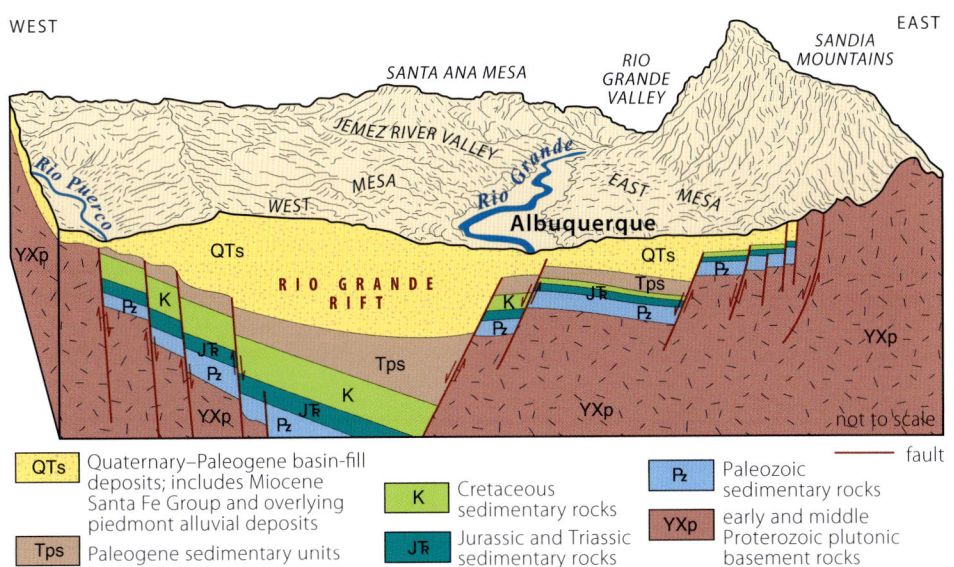

Cross section of the Sandia Mountains and the Albuquerque Basin. —Modified from Connell, 2008

Albuquerque stretches from the imposing cliffs of the Sandia Mountains westward onto the badland barrancas sheltered by the basaltic flows of the Albuquerque Volcanoes. The city sits on a broad, sloping alluvial apron, sometimes called the Llano de Albuquerque. These massive coalescing alluvial fans extend more than 10 miles from mountain front to river. The city is also built on river terraces on the banks of the Rio Grande. Beneath the recent gravels that coat them, the terraces are composed of Miocene and Pliocene sediments of the Santa Fe group.

The New Mexico Museum of Natural History and Science, located in Albuquerque, presents an excellent review of the geologic history of the state. Exhibits include bones and reconstructions of dinosaurs and other Mesozoic reptiles, dioramas showing fossil mammals that once roamed this region, a reconstructed volcano, and rock and mineral specimens.

Sandia Cave, a well-known Paleo-Indian site, lies in Las Huertas Canyon and requires a day use fee. The cave is formed in the Madera Group limestone. Excavated between 1937 and 1941, the cave apparently had multiple periods of occupation. Pueblo III pottery (AD 1150 to AD 1350) and Folsom points (10,800 to 10,200 years before present) are among the items found. Nineteen Sandia Point arrowheads, a variety of single-shouldered point, were reputedly found in association with the bones of extinct horse, bison, camel, mastodon, and mammoths from 35,000 to 17,000 years ago. The dates and the direct association of these points (and thus, human habitation) with the animal bones of that age are not considered valid by modern archaeologists. Later radiocarbon dating of wood fragments at the site gave dates of 3,447 years before present, a far cry from the original over-20,000-year-old date. Other researchers postulate that perhaps the cave originally contained remains from an earlier occupation, but the poorly documented excavations, along with the packrat digging and tourist impact, cleared the cave of anything useful.

Dinosaur sculptures welcome you to the New Mexico Museum of Natural History and Science in Albuquerque.

Sandia Mountains and the Tramway

The Sandia Mountains, which rise boldly behind the city, are an east-tilted fault block cored by the Sandia Granite, a 1.45-billion-year-old intrusive rock that cooled about 3 to 6 miles below the surface of the Earth. The range is famous for its pink hues (in Spanish, "sandia" means watermelon), which come from the abundant large potassium feldspar grains within the granite. The approximately 1,650-billion-year-old Cibola Gneiss is also present in the southern part of the Sandia Mountains, exposed along the roadcuts of I-40. The Cibola Gneiss was emplaced within the crust as a granite pluton during the Mazatzal orogeny, one of the stages of formation of the continent, when island arcs were converging to form what is now North America.

Sandia Mountains from the Rio Grande near Montaño Road.

A gray band along the crest of the Sandias is 300-million-year-old Pennsylvanian and Permian sedimentary rock that overlies the pinkish granite. The subsidence of the Rio Grande rift emphasizes the height of the mountains. The offset along the Sandia fault, the major fault that drops the Albuquerque Basin down to the west at Albuquerque, has as much as 21,000 vertical feet of offset. The marine sedimentary units of the Madera Group presently tower about 6,000 feet above the city of Albuquerque, resulting in a cumulative fault movement of close to 30,000 feet, about 6 miles. The fault zone along which all this shifting took place lies hidden beneath the alluvial fans that border the west side of the range.

Rincon Ridge, a distinctive line of foothills in northern Albuquerque, stands above the Sandia Pueblo and is visible from milepost 236 of I-25. The ridge is a slice of Proterozoic metamorphic schist and quartzite that is visibly intruded by dikes. Because city development does not cover them, the long alluvial fans at the western front of the Sandia Mountains are visible here.

There are two fantastic ways to get a closer view of the Sandia Mountains, which reach 10,678 feet in elevation: The Sandia Peak Tramway offers a view of the mountain front, and the stunning Sandia Crest National Scenic Byway zigzags up the east side of the range to the highest point on the mountains, taking advantage of the sloping Pennsylvanian units that ramp down eastward. Both transport you from basement rock to sedimentary units and right through the Great Unconformity, the contact between the Proterozoic crystalline rock and the overlying Pennsylvanian

As granite weathers along joints, it becomes more and more rounded, eventually forming residual (weathered in place) boulders like these in the foothills of the southern Sandia Mountains along the Eye of the Sandias Trail.

sedimentary rock. More than 1.1 billion years of Earth history is missing along this contact.

The Sandia Tram is an engineering marvel, having only two support towers in its 2.7-mile line length. The aerial tramway, built in 1966, goes from a base elevation of 6,559 feet to 10,378 feet at the top. The tram is the world's third-longest single-span and was the world's longest aerial tram until 2010. The tram covers this distance in approximately 15 minutes.

At the base of the mountains, large boulders of Proterozoic Sandia Granite weather in place. Loose gravel and sand piles that surround the larger boulders are composed of broken-down granite called grus. Many joints and cracks in the massive granite pluton have weathered into canyons, fins, and crags. Large feldspar minerals, blocky in shape and vaguely iridescent, dominate the granite matrix. Biotite mica shows shiny and black, while quartz is smoky and gray.

Above the second tower of the tram, a careful observer can see the scar of a massive landslide that occurred in 1936. Look for a transition from rounded, weathering granite boulders to jagged, angular rock the slide exposed. The tram converges with sedimentary rocks that cap the range. Look for four distinctive major cliff zones of Madera Group limestone separated by more easily erodible sandstone and shale that present as bands of vegetation between the gray cliffs of limestone. The limestone and shale layers contain fossil brachiopods, bryozoans, and corals, as well as button-like sections of crinoid (sea lily) stems.

From either Sandia Peak or Sandia Crest, the city of Albuquerque can be seen spread out below, its eastern edge crowding the great alluvial fans at the mouths of mountain canyons. Good eyes can pick out the flood-control embankments and storage reservoirs east of the city. Beyond the city proper, near the far side of the inner valley, the Rio Grande swings lazily in its channel below the floodplains that border the valley on the other side.

View to the north of Sandia Peak from South Sandia Peak. In the immediate foreground are gray, wrinkled, broken edges of the Madera Group limestone, which is also visible forming the smooth ramp of the flank of Sandia Peak. The next ledges and tumbled foothills of the Sandia Mountains are the Sandia Granite. The city of Albuquerque is visible in the left section of the photo. Rincon Ridge is the gray, jagged, mostly treeless ridge in the middle. The Jemez Mountains are visible in the far background.

To the west, across the Rio Grande Valley, the Albuquerque Volcanoes are visible on the horizon. These little cones and basalt lava flows erupted 210,000 to 156,000 years ago. The volcanoes line up along a north-south fault that acted as a conduit for magma to ascend to the surface. The basalt flows of the Albuquerque Volcanoes form the black, cliffy western edge of Albuquerque.

Far beyond them, silhouetted against the western skyline, is Mt. Taylor, a composite stratovolcano built of successive lava and ash-flow tuffs starting 3.7 million years ago. The Jemez Mountains, home to the Valles Caldera, are visible to the north of Mt. Taylor. The Sangre de Cristo Mountains and the city of Santa Fe are visible to the northeast. The Proterozoic-cored Sangre de Cristos are the southern tip of the Rocky Mountains.

NM 14 (TURQUOISE TRAIL NATIONAL SCENIC BYWAY)
Tijeras—I-25 near Santa Fe
45 miles

See map on page 144.

The Turquoise Trail National Scenic Byway heads through the San Pedro and Ortiz Mountains between Tijeras and Santa Fe. North of I-40, the tilted fault block of the Sandia Mountains is clearly visible. The east-dipping, tree-covered mountainside is floored with Pennsylvanian Madera Group and the underlying Sandia Formation. These units form a relatively smooth surface that dips to the east, having been tilted up along a fault to the west when Rio Grande rifting severed the landscape.

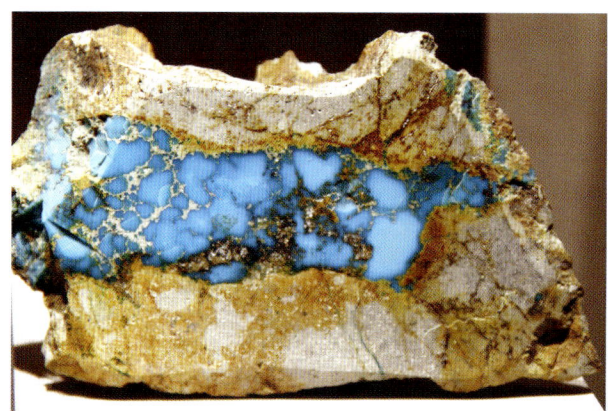

Turquoise from the Cerrillos Hills in the mineral collection at the Smithsonian Museum of Natural History in Washington, DC. —Courtesy of the Smithsonian Museum of Natural History, Creative Commons 2.0

The Turquoise Trail is so named because it passes through the Cerrillos Hills, a major historical source of turquoise, a blue-to-green hydrous phosphate mineral containing aluminum and copper. The Cerrillos Hills are in the Ortiz porphyry belt, which also includes (from north to south) the Ortiz Mountains, San Pedro Mountains, and South Mountain. Hot, mineral-rich fluids of Late Eocene to Oligocene intrusions transferred gold, tungsten, turquoise, and other base metal minerals through breccias, fractures, and veins, while also altering the surrounding rocks. One of the more interesting products of this alteration is turquoise, which has been mined here since pre-Spanish days, perhaps as early as AD 900 to AD 1100. Valued by prehistoric inhabitants of the Rio Grande Valley for both ornament and ceremony, turquoise from this area was also a major item of trade. Some of it reached Teotihuacan, the Aztec capital that became Mexico City, and after the Spanish Conquest found its way by treasure galleon to Spain, where it took its place among the crown jewels.

In the early 1820s, placer gold was discovered in Dolores Gulch in the Ortiz Mountains. Lode deposits began to be developed in the 1830s. A small gold rush to the region preceded the more famous California gold rush by more than a decade, but lack of available water limited production. Later, lead, silver, and hard anthracite coal, which is rare in the West, were mined in these volcanic hills. Gold occurs here in fractures in pyrite or in association with other minerals in breccia pipes and skarns.

The Ortiz Mountains, as well as the San Pedro Mountains farther south, consist of clusters of small intrusions: laccoliths, stocks, and sills. An initial group of intrusions pushed up between 36 and 33 million years ago, coinciding with the ignimbrite flare-up, and another group intruded between 31 and 27 million years ago as the Rio Grande rift began to extend the crust. These intrusions lie along the Tijeras-Cañoncito fault system, which forms the eastern edge of the Rio Grande rift between Albuquerque and Santa Fe. Movement within this system has been recurrent, but most activity is grouped into four major periods: Proterozoic movement, late Paleozoic movement, more notable movement on high-angle reverse faults associated with Laramide deformation, and normal faulting associated with rift-related crustal extension during Oligocene to Quaternary time. While most of the Ortiz porphyry belt is intrusive, slivers of Triassic and Jurassic San Rafael Group and Cretaceous Dakota, Mesaverde, Mancos Shale, and other sedimentary rocks are present. Numerous dikes and some volcaniclastic units are also present here.

As NM 14 leaves Tijeras, it rides on a valley floored by gray Mancos Shale. Around milepost 2, it crosses the Tijeras fault, east of which is a small syncline, where the road crosses a few miles of valley floored by Chinle Group shales. The red hills and white bluffs that are exposed as the road goes through the village of Cañoncito at milepost 4 and farther north are the Chinle Group. Then, in the next few miles, the highway crosses slivers of the San Andres, Yeso, and Abo Formations. The sloping Sandias are visible to the north. South Mountain is to the northeast at 2 o'clock. This Oligocene laccolith lies at the southern end of the San Pedro–Ortiz porphyry belt. The first views of the Cerrillos Hills appear just north of the turnoff to NM 536 near milepost 6. NM 536 leads to the Sandia Crest, a turning, twisting side trip to the top of the Sandias (see Albuquerque, page 153).

At milepost 11, horizontal bedding of the Madera Group limestone is visible in the long ridges to the east. These hills are backed by the treeless summits of the intrusions of the San Pedro–Ortiz porphyry belt. Valley floors are built of Quaternary alluvium and colluvium.

The settlement of Golden (milepost 17) was established in 1879, long after placer gold was discovered near Tuerto Spring in 1825. The town enjoyed a short heyday and is now mostly a ghost town. The general store that exists today has been in operation since 1918. Like many projects in New Mexico, the lack of water hampered efficient gold mining.

At milepost 20, the road crests the hill to take in sweeping views of central New Mexico as seen across the Rio Galisteo drainage. The shell-shaped arc to the west is the Sandia Mountains, and the slopes of the Sandia Peak Ski Area are clearly visible. Far on the horizon to the northwest, Cabezon Peak, a volcanic neck in the Rio Puerco drainage, forms a steep-sided mountain across the rift. The Jemez Mountains are farther to the north across the valley.

At milepost 23, cliffy ridges are visible to the north-northeast. To the east, landslides are visible on the steep slopes. Deformed laminar beds of the black Mancos Shale are visible under blocky, light-colored sandstones of the Point Lookout Sandstone (part of the Mesaverde Group). The valley ahead is floored by tumbled detritus known as the Tuerto Gravel, the unit in which the original placer gold was found.

Coal mined near Madrid (milepost 27) was used for small-scale local smelting; it also supplied the growing city of Santa Fe. The crumbling, black- to red-brown edifices of mine dumps can be seen near the town, which is set in the Mesaverde Group. A few old miners' cabins are also visible. Mines at Madrid produced both coking and noncoking bituminous coal and anthracite. The coal was originally mined in 1835 to fuel a mill at the Dolores mining camp, and commercial-scale mining began in the 1880s. The need for anthracite coal drove the development of a railroad spur to the town. The town has had a modern resurgence and is now a popular artists' town and tourist destination. Mining here ceased in the 1950s.

The coal is present in five principal, steeply dipping, thin seams within the Mesaverde Group, primarily in the Point Lookout Sandstone and Menefee Formation. These coal-bearing seams are faulted and crosscut by dikes that create difficult mining conditions. The coal is high quality due to the injection of the Madrid laccolith, an igneous body related to the 36- to 33-million-year-old Ortiz porphyry belt intrusions. Heat from the intrusion metamorphosed the coal to anthracite, creating the highly productive White Ash and Ortiz Arroyo coal seams. North of Madrid, the Mancos Shale is green.

CERRILLOS HILLS STATE PARK

Cerillos Hills State Park encompasses the rounded hills near Los Cerillos, where the park interpretive center is located. For centuries, the Cerrillos Hills have been mined for turquoise and precious metals. The hills are the eroded remains of an Oligocene-age laccolith, one of twelve laccoliths that intruded between 36 and 33 million years ago and form the Ortiz porpyrhy belt. At the Devils Throne, the shimmery, black metamorphosed Mancos Shale, into which the andesite porphyry sill intruded, can be seen. Both the intrusion and the bedding of the host shale are nearly vertical.

Native Americans mined turquoise at Mount Chalchihuitl near the modern village of Los Cerrillos. These mining practices began approximately AD 900, and the semiprecious stone was a major ornamentation and trade object. Turquoise in the Cerrillos mining district occurs in two

A colored hillside in the Cerrillos Hills indicates alteration and mineralization.

Devils Throne in Cerrillos Hills State Park. All of Devils Throne is an andesite sill, part of the Cerrillos Hill laccolith.

main deposits: Mount Chalchihuitl and Turquoise Hill. Color ranges from light to dark, bright blue to blue-green to dark green, and is present in veinlets and nodules in a matrix of monzonite and latite. The turquoise is often stained with limonite (a brown iron oxide mineral) and occasionally occurs with pyrite inclusions. Turquoise mining here peaked in the 1890s, though it continued until 1925.

Turquoise is a cryptocrystalline, blue-green aluminum phosphate mineral with small quantities of copper. It is the inclusion of copper that gives the mineral its blue hue. Turquoise forms after the deposition and weathering of copper-bearing host minerals. When rainwaters percolate through the copper porphyry body, dissolved oxygen in the water oxidizes the copper sulfides. The acidic, copper-enriched water solution moves along veins and sediments and reacts with aluminum and potassium in the host rock. This reaction triggers the precipitation of turquoise as bumpy nodules or thin crusts. It does not form individual crystals.

At milepost 30, the Cerrillos Hills are directly to the north. At milepost 31, black cliffy exposures weather to white and are cut by Waldo Gulch, an ephemeral stream. Cerrillos Hills State Park, less than 1 mile north of the town of Los Cerrillos, provides an unparalleled view into an intrusion.

At milepost 33, a regional, angular unconformity lies to the west. Red siltstones and mudstones of the Eocene-age Galisteo Formation form the mesa top. Beneath lie steeply tilted white beds of the Paleocene Diamondtail Formation, a sandstone with feldspar grains. It is locally conglomeratic. In some places, these tilted units were turned upside down due to deformation during the emplacement of the Cerrillos Hills igneous complex.

New Mexico's Garden of the Gods is located at milepost 34. The steeply dipping sandstone, conglomerate, and mudstone beds of the Galisteo Formation are visible on both sides of the highway, and a scenic pullout allows for close-up viewing,

At New Mexico's Garden of the Gods at milepost 34, near-vertically tilted bedding in the Galisteo Formation creates walls. The pockmarked weathering visible on the bedding surface is due to the removal and erosion of larger pebbles or cobbles, followed by the enlargement of the hole by rain and wind.

although the cliffs themselves are privately owned. The Galisteo Formation was deposited 40 million years ago in energetic river systems in a continental basin. In places, it contains petrified wood and fossilized mammal bones, including a type of rhinoceros known as the titanotheres. The Galisteo Formation now stands in near-vertical (or even overturned) fins, along with nearby and overlying rock layers, all deformed during the intrusion of the Cerrillos laccolith about 30 million years ago. This intrusion pushed its way through overlying sedimentary units, severely tilting and overturning the beds, which have weathered into spectacular features.

The Oligocene-age Espinaso Formation, a light-purplish-gray volcaniclastic and pyroclastic flow unit, is visible to the east. Several small peaks of volcanic rock, as well as a dike about a half mile west of the road, show that the volcanics of the Ortiz porphyry belt extended north here. The Ortiz Mountains and Cerrillos Hills are the most prominent of these features.

Between mileposts 39 and 44, NM 14 crosses a relatively flat surface that surrounds the Ortiz Mountains. This partly eroded, partly stream-deposited surface, known as the Ortiz surface, is about 400 feet higher than the Rio Grande. It is capped with the Pliocene to Pleistocene Tuerto Gravel, made notable for the placer gold deposits found within its tumbled gravels.

I-25
Albuquerque—Socorro
77 miles

The east-tilted block of the Sandia Mountains, approximately 20 miles long, dominates the geologic scenery for some distance south of Albuquerque. Tijeras Canyon, the major break in the uplifted range, cuts obliquely east-west. It formed as erosion removed fractured rock along the Tijeras fault zone. I-40 lies within this canyon, joining the eastern plains to Albuquerque. South of the intersection with I-40, I-25 crosses Tijeras Arroyo's huge, 3-mile-wide alluvial fan.

To the west, across the Rio Grande Valley, the Albuquerque Volcanoes are visible on the horizon. These little cones and basalt lava flows, forming the black, cliffy western edge of the city, erupted 210,000 to 156,000 years ago along a north-south fault.

South of the Sandia Mountains and forming a continuous uplift with it are the Manzano Mountains, another tilted block of Proterozoic rock wearing a cap of east-dipping Pennsylvanian limestone. Although the range records Laramide deformation, it was uplifted most recently 20 to 15 million years ago during Miocene time by faults on the east side of the Rio Grande rift. Manzano Peak, elevation 10,098 feet at the south end of the range, is dark with Proterozoic schists and quartzites that were originally volcanic and plutonic rocks accreted onto the continent of North America about 1.65 billion years ago.

The Manzano Mountain exposures have been important in the interpretation of Proterozoic mountain building at the edge of the supercontinent Rodinia. The rocks record two episodes of orogeny: one when the Mazatzal microcontinent collided with the main continent 1.65 billion years ago, and a later episode, the Picuris orogeny, 1.4 billion years ago. Between the two orogenies, a period of quiescence resulted in deposition of sediments in large intermontane basins. These sediments were faulted and

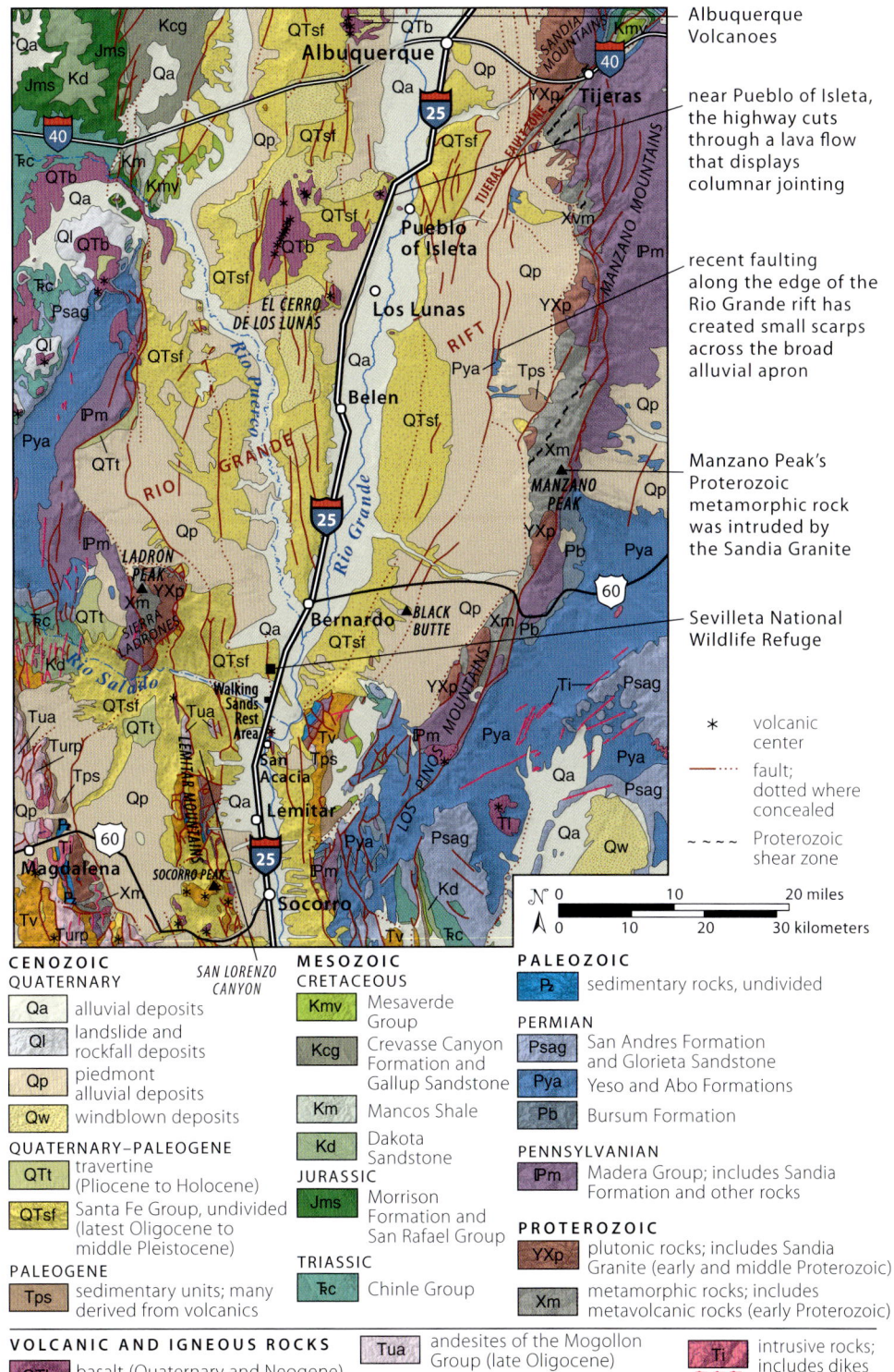

Geology along I-25 between Albuquerque and Socorro.

folded into two large thrust belts during the Picuris orogeny, one of which is recorded in the Manzano Mountains, where gigantic folds indicate the degree to which the rocks were compressed during the mountain building. This collection of rocks was not uplifted to anywhere near its modern heights until more than 1 billion years later.

Here, in the southern half of the Albuquerque Basin, the Rio Grande rift is much deeper than it looks. The basement rock is buried up to 2 to 3 miles below Albuquerque. This down-dropping occurred gradually over millions of years, and as the basin dropped, it filled with thousands of feet of alluvial fan, lake, and playa deposits. The majority of these sediments predate the through-drainage of the Rio Grande and most of them compose the ubiquitous Santa Fe Group. They were brought into the rift by alluvial fans and small streams draining the mountains that edge the rift.

After the Rio Grande became a major through-flowing river and emptied into the Gulf of Mexico about 800,000 years ago, it began cutting into these sedimentary deposits. The incision rate has increased over the last 600,000 years, meaning the river is cutting down faster now than it was back then. It is hypothesized that runoff from the southern Rocky Mountains has increased over this period, enhancing the Rio Grande's ability to carve through these underlying sediments. Between the bordering ranges and the river are five terrace levels that represent cyclic episodes of erosion (valley cutting) and deposition (valley filling). Over the last half million years, the river has carved an inner valley, where it built a floodplain now dotted with fields and orchards.

The highway crosses the Rio Grande north of Pueblo of Isleta between mileposts 215 and 214. Just prior to crossing the river proper, the road reaches a floodway contained by manmade levees. In addition to ferrying water to local agriculture and residents, these water channels protect buildings and farms from overbank flooding. The river valley here is about 22 miles wide and is part of the Albuquerque Basin, the largest of the rift basins. At mile 213, drivers pass through a basalt flow, visible on the west side of the highway. The ancestral Rio Grande eroded through this flow. A narrow belt of trees lines the modern Rio Grande, and its 3-mile-wide floodplain is used largely for cropland.

This lava flow in a roadcut near milepost 213 baked the underlying sediments, oxidizing the iron to red.

Farther south near Pueblo of Isleta, a small shield volcano seen to the west is the source of five basaltic to andesitic flows. The 3.5- to 1.5-million-year-old lava flows are interbedded with the Santa Fe Group deposits. The lava flows are exposed in spectacular roadcuts at mileposts 213 and 211.

Isleta reminds visitors that this broad basin, with its fertile floodplain and abundant water, has provided a bountiful living for its inhabitants for many thousands of years. The original pueblo, built on a lava flow that rises like a little island (*isleta* in Spanish) from the Rio Grande floodplain, was an established habitation when Francisco Coronado arrived in 1540. Climatic changes have strongly influenced vegetation change in the five hundred years following early Spanish settlement. Especially notable are droughts during the 1920s and 1930s that culminated in the Dustbowl. Overgrazing by cattle, horses, and sheep damage fragile high-desert ecosystems, resulting in the overgrowth of the now ubiquitous sagebrush and creosote bush and the introduction of invasive species. The native acacia bushes thrive and are highly visible along this route. Pecan orchards are visible from mileposts 190 to 186.

The broad, sloping plain below the Manzano Mountains, visible to the southeast from milepost 209, is broken about halfway up the slope by a fault scarp, part of the Manzano fault system that edges the west side of the Manzano Mountains. Movement on the fault postdates development of the sloping plain and probably last shifted about 130,000 years ago in the late Pleistocene. Sloping sediments on the flanks of the mountains usually bury the fault lines.

The Rio Grande rift is still active, and recent movement along the Manzano and other faults is noted in large GPS studies. Offset of sedimentary layers across these faults can be observed in young geologic deposits such as river terraces and travertine deposits. The area between Albuquerque and Socorro is one of the most seismically active in New Mexico. A handful of magnitude 4 earthquakes have been recorded in the last fifty years, and a magnitude 7 earthquake shook Socorro (and much of New Mexico, Texas, and Oklahoma) in November 1906.

Between mileposts 204 and 201, the highway passes close to the small volcano of El Cerro de Los Lunas to the west. Farther south and west, watch for exposures of the eroded edge of the uppermost terrace and its flat layers of reddish sand, clay, and volcanic ash, part of the Santa Fe Group.

West of the northern junction with US 60 is the dome of Ladron Peak, an upfaulted wedge of Proterozoic rock that dominates the skyline. Ladron Peak marks the western edge of the Albuquerque Basin and the beginning of the Colorado Plateau behind it. The granite and metamorphic rocks that form the mountain's body are bounded by normal faults of the rift. Dark volcanic mountains behind it mark the northeast corner of the Datil-Mogollon volcanic highlands of west-central New Mexico.

Black Butte, standing out in the midst of the rift east of the highway at milepost 175, is an Oligocene basalt flow that erupted 29 to 26 million years ago. Los Pinos Mountains are visible farther to the east. These mountains are cored by Proterozoic rocks, unconformably overlain with Pennsylvanian and Permian rock layers in another example of the Great Unconformity. Early Paleozoic units are wholly missing. They may never have been deposited here, or any units once deposited may have been eroded and removed during mountain building. The Sandia and Madera units in the Los Pinos Mountains were deposited in and adjacent to a shallow sea overlying the continental shelf.

View to the south from milepost 179 at the small, black hummock of Black Butte, an Oligocene volcanic vent, standing in the middle of the Rio Grande Valley. Whiteface Mountain forms the high point on the northern end of the range in the background. The verdant cottonwoods of the Rio Grande bosque are visible in midground.

The Bursum, Abo, and Yeso Formations show an input of sediments derived from the Ancestral Rocky Mountains, which were uplifted in Pennsylvanian time.

The interstate crosses the Rio Puerco at milepost 174. This ephemeral river, the headwaters of which lie north of Grants, is often dry and largely bare of vegetation. It consists of dry interleaves of a braided stream, though here it has incised deeply into Quaternary sediments and the Santa Fe Group. In some places, the streambed lies more than 30 cliffy, near-vertical feet below the overlying land surface. When the Rio Puerco does flow, it carries a heavy sediment load and is a major contributor of sediment to the Rio Grande. Poorly consolidated Quaternary gravel terraces edge the Rio Puerco drainage, evidence of past deposition. The valley has been extensively studied for information about how climate change affects arroyo filling and cutting in arid environments.

A drying climate and overgrazing within the Rio Puerco drainage basin (and much of the western United States) resulted in a dramatic increase of sediment carried by the river, much of which is deposited on its floodplain when the river slows and stops flowing. The Rio Puerco is considered an example of extreme sediment supply. It supplies nearly 80 percent of the sediment load of the Rio Grande, though it has only 25 percent of the Rio Grande drainage basin area. With so much sediment, the arroyos fill with their own load and then are rapidly incised, forming a complex and dynamic landscape that is highly responsive to seasonal and climatic change.

South of the Rio Puerco, the Rio Grande floodplain is broad, marshy, and home to many wetlands. The Sevilleta National Wildlife Refuge provides habitat to many species of migrating waterbirds following a flyway between winter feeding grounds and summer nesting sites. The refuge protects the land adjacent to the highway and extends 30 miles from one side of the rift valley to the other. A visitor center, accessed from exit 169, offers information on the refuge species and local geology.

Mountains on both sides of the Rio Grande gradually converge at the south end of the refuge, almost closing the southern end of the Albuquerque Basin as it transitions into the Socorro Basin of the Rio Grande rift. The transfer zone between the basins is known as the Socorro constriction. This area is underlain by the Socorro magma body, a pancake-shaped body of molten magma 12 miles below the surface. While the body is only about 400 feet thick (vertically), it spreads beneath an area of about 1,300 square miles and is the second-largest magma body on Earth. The extension, or pulling apart, of the crust in the Rio Grande rift allows the mantle's molten rock to rise, melt, and spread out into the crust. The magma body has caused the land surface to rise about 6 inches since the early 1900s, enough to upset railroad tracks that cross above it. The land surface continues to rise at about one-eighth inch per year. The Socorro magma body has inflated at least twice in the last 26,000 years. The area is also prone to small earthquakes that will continue as the magma body continues to be injected with magma.

From the Walking Sands Rest Area, between mileposts 167 and 166, sand dunes can be seen along the edges of nearby terraces. In dry climates, sands shift and blow about too much to support vegetation, forming dunes when and where they are deposited. Here, the sand is derived from the bed of the Rio Salado, a normally dry desert wash, which, when flowing, brings a supply of new sand from mountains to the west. Braided patterns common to desert rivers often show up on the streambed. Even a quick glimpse of the dry wash from the bridge (south of the rest area) shows the sorting of sand, pebbles, and cobbles, one of the clues to water deposition. The rest area is home to the Rio Salado Sand Dunes Historical Marker.

Near Socorro, the Lemitar and Chupadera Mountains just to the south form the western edge of the Rio Grande rift but are not part of the Socorro caldera. The ranges are cored by complex Proterozoic units that include metasedimentary schists, mafic dikes and intrusives, and granites. Paleozoic- and Cenozoic-age sedimentary and volcanic layers overlie the basement units. Numerous carbonatites have intruded

An angular unconformity shows up well in a butte in San Lorenzo Canyon, south of San Acacia and within the Sevilleta National Wildlife Refuge. Here, two levels of the Santa Fe Group (Oligocene to Pleistocene) tilt at different angles—the lower one was affected by Rio Grande rift faulting before the overlying layer was deposited.

these. Carbonatites are unique, carbonate-rich intrusions of variable mineral assemblages and textures that are of volcanic origin, likely the fractionalization of magmas from the upper mantle. Carbonatites are valued for iron ore, vermiculite, barite, fluorite, phosphate, and economic concentrations of rare-earth elements including uranium, thorium, and others. Here, these occur as a swarm of more than one hundred dikes emplaced around 450 million years ago, ranging up to more than 3 feet thick and 2,000 feet long.

The mountains directly west of Socorro are primarily volcanic. Socorro Peak—the one with the "M" on its flank—is part of the 32-million-year-old Socorro caldera and is the northeasternmost exposure of the Mogollon-Datil volcanic field. Perlite, a volcanic siliceous glass with the same composition as rhyolite, is mined here from one of the deepest and largest perlite mines in the world. The mine, which extracts perlite from nearly 400 feet belowground, is visible to the south of Socorro Peak. The rock is removed, crushed, and heated until it expands and is distributed for use in horticulture and construction and as a filter aid and filler.

Socorro is home to the New Mexico Institute of Mining and Technology—hence the "M" on Socorro Peak. The New Mexico Bureau of Geology and Mineral Resources, headquartered on the campus, hosts a spectacular mineral museum open to the public and is a good source for maps, scenic trips, and field conference guidebooks, as well as further information about New Mexico geology. The university also runs the Magdalena Ridge Observatory, which sits just over the mountaintop of the Magdalena Mountains. The Magdalena Ridge Observatory Interferometer (MROI) is home to two instrument facilities. One houses a large telescope used for observing the solar system, artificial Earth satellites and space vehicles, and defense purposes. The MROI is a ten-element imaging interferometer. The goal of the interferometer is to produce images of faint and complex astronomical features at resolutions up to one hundred times as clear as those taken by the Hubble Telescope. An interferometer works by combining several small telescopes to create a clearer picture than a single instrument can. The interferometer will help scientists understand star and planet formation, how stars gain and lose mass, and how galaxies transform throughout their lifetimes.

Socorro Peak, west of Socorro, is part of a west-tilting, fault-bounded block that borders the Rio Grande rift. Oligocene volcanic tuffs and lavas of the Mogollon-Datil volcanic field give the mountains a reddish tint.

I-25
Socorro—Truth or Consequences
74 miles

Between Socorro and Truth or Consequences, I-25 crosses numerous dry arroyos channeling storm runoff from volcanic mountains to the west, part of the vast Mogollon-Datil volcanic field that extends westward to Arizona. From Socorro southward, the faults along the west edge of the Rio Grande rift are staggered, so successive mountain ranges come into view: first the south end of the Chupadera Mountains, then the Magdalena, San Mateo, and Mimbres Mountains. South of Socorro, the rift widens dramatically and is composed of multiple side-by-side basins. The interstate remains west of the river on terraces and alluvial fans that border the Rio Grande Valley. Both terraces and fans are veneered with sediments. Arroyos and highway cuts reveal the rocks beneath the veneer.

The Bosque del Apache National Wildlife Refuge, just south of Socorro near milepost 132, is a renowned bird sanctuary. The Rio Grande flows across young sediments hosting abundant cottonwoods, salt cedars, and brushy undergrowth. These flood deposits are less than 10,000 years old. Paleozoic sedimentary rock layers, heavily faulted during rifting, are forming the Little San Pascual Mountains, visible to the southeast. Oligocene calderas of the Mogollon-Datil volcanic field erupted enormous ash-flow tuffs, one of which caps Chupadera Peak to the northwest of the Bosque. It erupted 27.9 million years ago. Eocene and Pliocene to Pleistocene sediments are exposed in the mountain flanks.

The different surfaces within the Rio Grande rift have been carefully studied, defined, and named according to their elevation above the Rio Grande. The youngest of these surfaces reflect the dynamic nature of the Rio Grande rift and river. At least six major surfaces of young terraces, all less than 1 million years old and deposited since the end of basin filling, have been mapped in the Socorro Basin. These are, from

A low roadcut in the Santa Fe Group near milepost 131 shows the variability in the alluvial fan sedimentation, with cobble conglomerates interspersed with sandy layers. Pulses of runoff from the highlands interleaved sediments across braided channels.

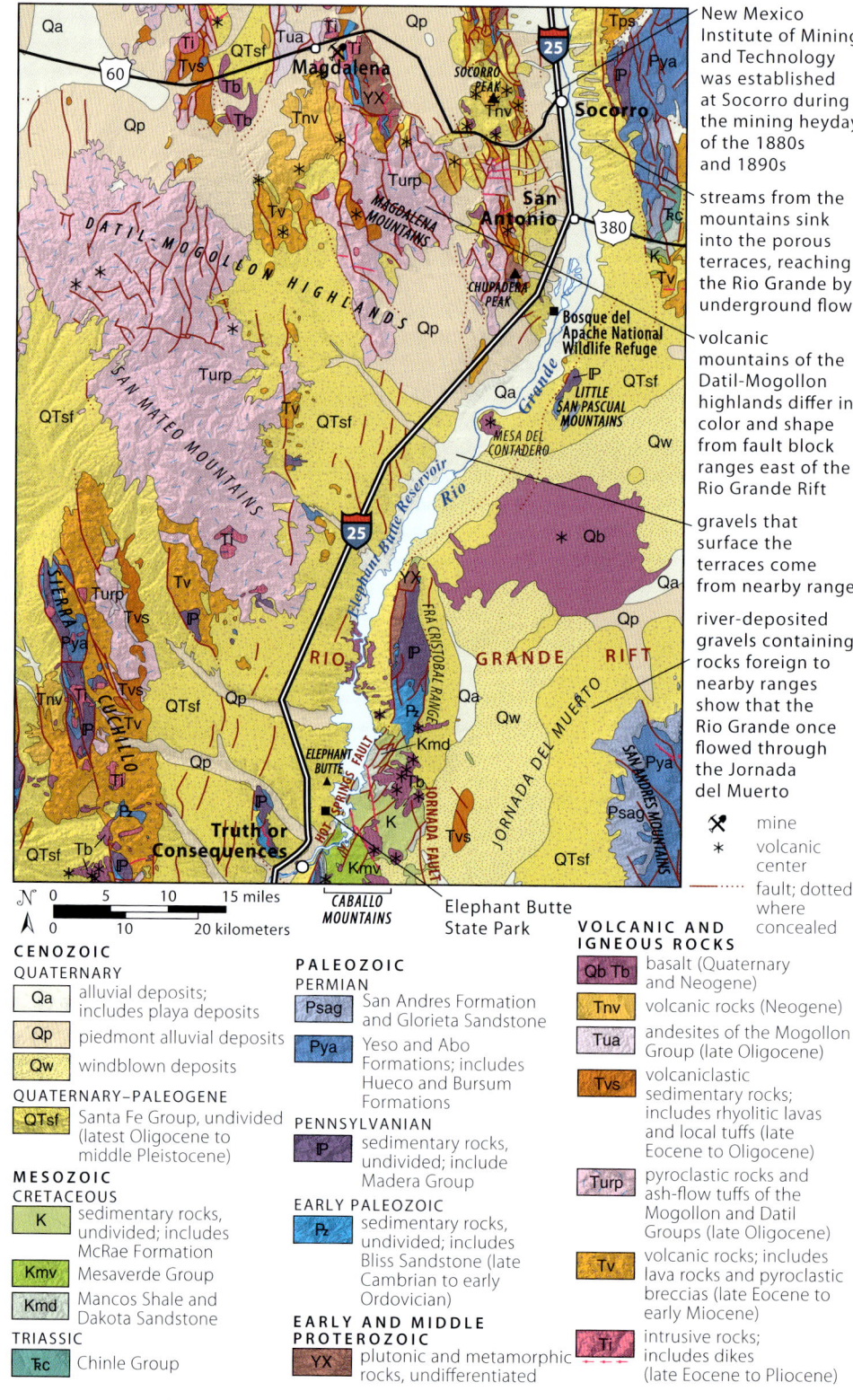

Geology along I-25 between Socorro and Truth or Consequences.

river level to higher surfaces, the Chamizal, Cañada Mariana, Loma Parda, Valle de La Parida, Tio Bartolo, and Las Cañas surfaces.

The Jornada del Muerto, a broad valley within the rift province, lies east of the Rio Grande Valley. The basin, named by Spanish conquistadors, is infamous for its nearly 100 miles of waterless travel between Las Cruces to the south and Socorro, to the north. Despite its brutal conditions, it was shorter than following the Rio Grande to the west. The Jornada del Muerto is bordered by the San Andres and Organ Mountains on the east and the Jornada fault and the volcanic Doña Ana Mountains to the southwest. The basin bedrock, downfaulted in a graben, tilts to the west, though the valley floor today is largely flat. An excess of 3,000 feet of Santa Fe Group sediments fills the basin and is covered with a gravelly veneer of more recent stream and alluvial deposits.

The Jornada del Muerto is a closed basin, meaning no river drains it. The Rio Grande flowed through the valley at one time, but its course shifted to its more western route in the early Pleistocene Epoch. During the wet climate of the Pleistocene ice age, Lake Jornada flooded the basin. This intermittent playa lake was shallow, ranging from 5 to 45 feet deep depending on location and time. It existed approximately 1 to 0.3 million years ago, stretching perhaps about 25 miles north to south. Runoff from the bordering limestone-rich mountains filled the basin and left gypsum-rich evaporites behind when it evaporated. Selenite crystals can be found in these deposits. The Santa Fe Group sediments within the valley are a discontinuous, highly faulted series of units. Groundwater levels associated with the ancestral Rio Grande that once supported the lake would have dropped as the Rio Grande migrated west and entrenched down into its own valley. Today, brackish groundwater is found hundreds of feet below the valley floor.

Black Mesa, or Mesa del Contadero, is visible from the rest stop at milepost 114.

Black Mesa, or Mesa del Contadero, visible across the Rio Grande from the rest stop at milepost 114, is topped with 9 to 20 feet of 820,000-year-old basalt. This caprock of resistant basalt shelters underlying river deposits. The difference in elevation from the river to these perched gravels that lie 600 feet above the modern river helps geologists better understand how fast the Rio Grande system is incising into the landscape.

South of milepost 105, Elephant Butte Reservoir becomes increasingly visible. The reservoir, the largest body of water in New Mexico, was built as part of the Bureau of Reclamation's Rio Grande Project. The dam was constructed in tilted, deformed bedrock, requiring engineering finesse. Construction on the 1,674-foot-long concrete dam began in 1911, and reservoir filling began in 1915. The dam was the largest in the world at the time and is the first structure designed for the international distribution of water. It is now a National Historic Civil Engineering Landmark.

Tilted dark shales and fine-grained sandstones of the Mesaverde Group, viewed on the south side of Elephant Butte Dam, form the walls of the Elephant Butte Reservoir.

The reservoir can store more than 2 million acre-feet of water, which is critical for the local agriculture and compliance with the Rio Grande Compact. White rings that surround the present lake show evidence of higher water levels in the past. Wetlands near and north of this reservoir are famous for their abundant and varied birds, including pelicans.

The reservoir, located in the Rio Grande rift, is surrounded by mountain ranges including the Fra Cristobal Mountains (east), Caballo Mountains (south), San Mateo Mountains (northwest), and Cuchillo and Black Ranges (west). The Fra Cristobal Mountains host Proterozoic metamorphic schists and plutonic rocks as much as 1.6 billion years old. The late Cambrian to early Ordovician Bliss Sandstone lies unconformably on the Proterozoic rocks. It was deposited in beach and nearshore environments and forms a dark band visible on the lower flanks of the Fra Cristobal

Elephant Butte, visible from between mileposts 80 and 88, is located within the reservoir named after the iconic hill. The Fra Cristobal Range forms the background. The butte is a basaltic intrusion approximately 3 million years old. Note the dark coloring in the butte, as well as on surrounding mountains and mesas.

Mountains. The Caballo Mountains expose a very complete section starting with units deposited in a shallow sea from about 500 to 340 million years ago, forming the Fusselman Dolomite, Percha Shale, and Lake Valley Formation. Above them, the Permian Abo, Yeso, Glorieta, and San Andres Formations record the retreat and return of shallow seas.

A few miles east of the interstate, the colorful resort community of Truth or Consequences nestles among gray travertine hills, a product of thermal springs that gave it its original name: Hot Springs. Several private spas allow visitors to bask in the hot water. Heated by contact with hot crustal rocks and relatively shallow magma, the hot water escapes to the surface along small faults. The crust, stretched thin here in the rift, allows magma to exist at relatively shallow depths. The crust here is only about 22 miles thick—much thinner than that of the Great Plains and the Colorado Plateau that border the rift, which has crust ranging from 28 to 31 miles in thickness. The young volcanic features along the rift valley are evidence of the persistence of heat and magmatism along the rift.

I-25
TRUTH OR CONSEQUENCES—LAS CRUCES
80 miles

South of Truth or Consequences, the Rio Grande rift continues to widen and its eastern edge is 60 miles from the highway in the Sacramento Mountains east of Alamogordo. The north-south-trending normal faults that dominate the northern sections of the narrow rift shift to being more northwest-southeast in the southern sections of the rift. These faults likely follow older faults that have existed within the crustal structures since the Proterozoic and have remained zones of weakness ever since. The rift bends as it continues southward into Texas and Mexico. The southern rift basins have extended approximately 2.5 times that of the northern rift basins.

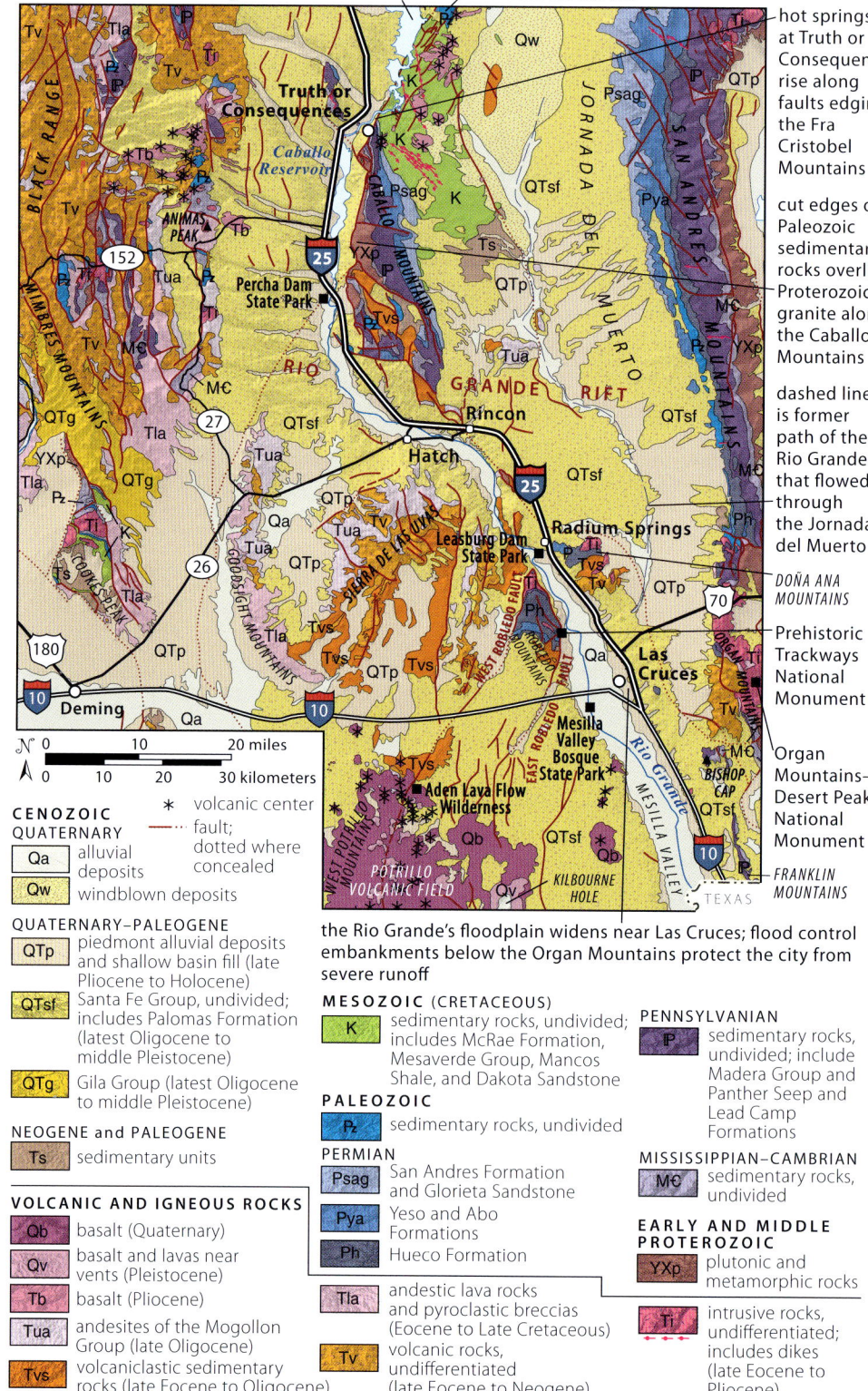

Geology along I-25 between Truth or Consequences and Las Cruces.

Sedimentary rocks in general dip east on both sides of the Rio Grande Valley. The tilt shows up particularly well on Animas Peak west of the highway. In the rugged, up-faulted west face of the Caballo Mountains, east of Caballo Reservoir (milepost 63) and Percha Dam State Park, contorted, Paleozoic sedimentary rocks overlie 1.4-billion-year-old granite. Oligocene-age volcaniclastics are also present in the southern parts of the Caballo Mountains.

View of Santa Fe Group near milepost 75, just south of Truth or Consequences. In places, the Santa Fe Group erodes into badlands with colorful pinkish stripes, indicating higher amounts of iron in the sediments.

Striking, southeast-dipping Paleozoic rock layers cap the Caballo Mountains, seen here across Caballo Lake State Park at milepost 63. The reddish rocks at their base are 1.4-billion-year-old granites with large crystals set in a matrix of smaller crystals. Between the lake and the red granite is the rolling surface of Santa Fe Group sediments, which alluvial deposits break.

The highway skirts the inner valley of the Rio Grande as far as the south end of Caballo Reservoir, dipping and rising over an irregular surface of coarse-grained, easily eroded gravels of the Santa Fe Group. This unit, though often treated as a single entity, is an accumulation of many smaller units that interfingered and overlapped to fill the topographic low of the Rio Grande rift. Here, the Palomas Formation of the Santa Fe Group was deposited from about 5 to 1 million years ago by east-flowing streams originating in the Black Range and Mimbres Mountains to the west of the highway. Within the Palomas Formation, many episodes of basin filling are recorded in channel-fill deposits, gravel horizons, paleosols, and silty deposits, some with cross-stratification, laminations, or imbrication of larger sediments. The deposition of this single formation within the Santa Fe Group was complex and existed in an evolving depositional setting over millions of years.

Younger terraces show the transition from aggradation and deposition that deposited the Palomas Formation to the incisional setting seen today, with the Rio Grande cutting down into its own deposits. Here, terraces show the progressive cutting that has happened over the last million years. These young terraces are often covered in stream deposits of pebbles, gravel, and cobbles. Roadcuts expose these sediments, some of them derived from sources far to the north, up the Rio Grande.

South of Caballo Reservoir, I-25 drops down toward the floodplain, then swings eastward with the river to go around Oligocene volcanic rocks of the Sierra de las Uvas, which lie directly to the south from milepost 42. These 29- to 26-million-year-old andesites, basaltic andesites, and volcaniclastic rocks are part of the Mogollon Group. Volcaniclastic sedimentary aprons that extend south of the Sierra de las Uvas formed at the same time as nearby eruptions. Roadcuts near mileposts 52 and 53 show yellowish and reddish Paleozoic, mostly Pennsylvanian, rock layers. The San Andres Mountains lie to the east across a broad valley.

The relatively broad gap between the Caballo Mountains to the north and the San Andres Mountains to the southeast is the southern end of the Jornada del Muerto, a nearly waterless corridor between Las Cruces to the south and Socorro to the north. Despite the brutally dry conditions, it took early travelers less time than following the Rio Grande north. During the Pliocene, the paleo Rio Grande flowed almost straight south through the Jornada, crossed its present course south of Rincon (milepost 35), and continued straight south toward what is now the Mexican border. Tectonic and volcanic activity prompted the westward migration of the paleo Rio Grande, and the river abandoned the Jornada del Muerto. Today, the Jornada has no through-flowing drainage. With the westward shift of the river during the early Pleistocene, the basin flooded, creating a shallow playa lake known as Lake Jornada, which has since evaporated.

Near Hatch, I-25 sits on the modern Rio Grande floodplain composed of young sediment as much as 120 feet thick. In the distance, many flat-topped and beveled hills record the incision of the Rio Grande. Light-pink and tan sandstones and conglomerates on both sides of the highway are Quaternary deposits of the ancestral Rio Grande. Many of these hills are capped by a resistant conglomerate, which somewhat protects the more erodible units below. To the far northeast, the peaks are 28- to 24-million-year-old basaltic andesite from the nearby Uvas volcanic field.

From exit 41 at Hatch or exit 35 at Rincon, travelers can access NM 26, a cutoff connecting I-25 with I-10. This cutoff is a narrow, mostly two-lane highway, which

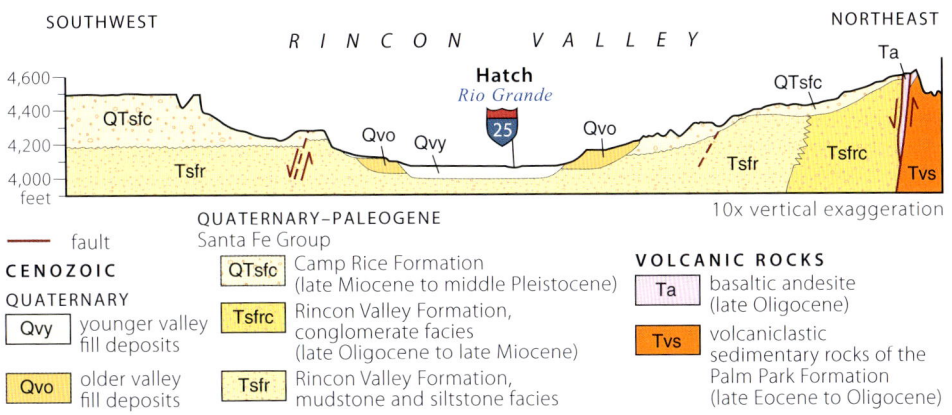

Southwest-northeast section across the Rio Grande Valley at milepost 43 east of Hatch.

offers stunning views of the Cookes Range, a fault-bound mountain range cored with a late Eocene intrusion of the Mogollon-Datil volcanic field. For more on this range, see I-10: Deming—Las Cruces.

The San Andres Mountains, east of the Jornada del Muerto, are composed of Proterozoic igneous and metamorphic rocks overlain by Paleozoic sedimentary rocks lifted along faults. But even they are not the edge of Rio Grande rift, which lies still farther east across the Tularosa Valley.

Rincon (milepost 35), whose name means "corner" or "angle" in Spanish, lies where the Rio Grande angles southward once more. From the top of the terraces near Rincon, a good view looks northwest up the river and north into the Jornada del Muerto. Wind is a dominant erosional force here; southwest spring winds whip across the Rio Grande Valley and up the Jornada del Muerto, blowing away sand and silt from natural surfaces and the plowed fields of the Rio Grande floodplain. At times, these can cause blinding dust storms, and signage throughout this section of road attests to their danger.

Take exit 19 at Radium Springs to reach Leasburg Dam State Park, which offers camping, hiking, and fishing. The dam, built in 1908, serves not to retain water but rather to divert flow into a network of agricultural channels in the upper Mesilla Valley, the local name for the Rio Grande Valley here. Radium Springs is named for hot springs that rise through a fractured Oligocene-age rhyolite flow and contain measurable radium. The fractures allow groundwater to infiltrate through hot buried rocks. Fort Seldon National Monument in Radium Springs gives the history of the fort that once stood at this entrance point to the Jornada del Muerto. Adobe walls still stand. Exposed Hueco and Abo Formations along with underlying Paleozoic rock layers are bent-over Oligocene intrusions of the Robledo Mountains to the southwest, all of which are faulted by the rifting. The East Robledo fault drops the Mesilla Valley down to the east, creating a stark mountain front. The volcanic rocks and ash-flow tuffs of the Doña Ana Mountains form prominent points to the north-northwest.

South of milepost 19, the San Andres Mountains to the east of the Rio Grande consist of west-dipping Paleozoic sedimentary rocks. Below them are stream-deposited and

ORGAN MOUNTAINS–DESERT PEAKS NATIONAL MONUMENT

The view to the east from Las Cruces includes the Organ Mountains, named for their light-colored, craggy outcrops of vertically jointed Oligocene granite. The Organ Mountains–Desert Peaks National Monument, established in 2014 and managed by the Bureau of Land Management, protects several mountain ranges in discontinuous areas within the Mesilla Valley. Take exit 1 and head east on East University Avenue to reach the popular areas such as Dripping Springs and Soledad Canyon Day Use Area.

The oldest rocks in the Organ Mountains are 1.4-billion-year-old granites exposed on the northwestern flank of Baylor Peak in the northern section of the mountains. Shallow seas that existed during Pennsylvanian and Permian time deposited thick sequences of limestone and dolomite rocks. These carbonates are visible at Bishop Cap in the southern part of the range. These Proterozoic and Paleozoic rocks were uplifted and faulted during the Laramide orogeny. Massive caldera eruptions 36 million years ago deposited three major ash-flow tuffs, which are visible in the hiking destination of Soledad Canyon Day Use Area, at La Cueva, and in the over-2,000-foot-high cliffs at Dripping Springs Natural Area. The Organ batholith is a monzonite, a silicic granitic rock formed from the magma that collected and cooled belowground during and after the caldera eruption. This monzonite forms the vertically jointed face of the central range crest and can also be seen along Pine Creek Trail. The Organ Needle, the high point of the range, is composed of this monzonite and reaches 8,990 feet elevation. The entire range has been faulted and modified by Rio Grande rift extension.

The "organ pipes" of the Organ Mountains are made of 27-million-year-old, vertically jointed Oligocene granite that also composes the steep eastern face. The mountain fronts show the many joints, developed as the granite cooled and shrank, and they were likely enhanced by movement along rift faults.

then stream-dissected fans that originate from these mountains. The Rio Grande rift widens to about 100 miles here, and the floodplain of the Rio Grande widens as well, spreading to support many farms and orchards.

The Robledo Mountains, low, sharp hills west of the river between Leasburg and Las Cruces (from milepost 15 to 0), are an uplifted block of Paleozoic sedimentary rocks that dip 10 to 14 degrees to the south. Most of the sedimentary rock in this range is part of the 280-million-year-old Hueco Group. In 1987, fossils were found in the group, and Prehistoric Trackways National Monument (accessed from the Doña Ana exit) became the one hundredth US National Monument when it was designated in 2009. More than 1,500 feet of Hueco Group sediments were deposited in shoreline, tidal mudflat, and river floodplain environments at the edge of a fluctuating sea. Trace fossils in these red beds provide evidence of land animals, insects, reptiles, and sea creatures. A major mega-trackway is preserved that includes abundant footprints of *Dimetrodon*, *Eryops*, and *Edaphosaurus*, all of which have large sail fins. Fossilized plants and petrified wood indicate an ancient forest existed then.

Despite its name, tracks are not easy to find at Prehistoric Trackways National Monument. In the distance is the Permian-age Robledo Mountains Formation of the Hueco Group.

The climate in southern New Mexico is semiarid. Hot, dry summers and monsoon thunderstorms alternate with cool, dry winters. Total annual rainfall averages less than 10 inches per year. Mesilla Valley Bosque State Park, on the Rio Grande south of I-10 on the western edge of Las Cruces, is an excellent place to view this diverse ecosystem. Desertification has changed the environment in the last hundred years. Desert shrubs like creosote bush, mesquite, and cacti are dominant plants in this hot environment.

LAS CRUCES TO TEXAS ON I-10

From Las Cruces southward to the Texas border, I-10 continues down the east side of the valley of the Rio Grande, with the Organ Mountains to the east and flat, almost level desert punctuated with a few small Quaternary basaltic to andesitic volcanoes of the Potrillo volcanic field to the west. Spring plowing coincides with New Mexico's spring winds, to the detriment of the sparse, sandy soils that line the valley floor.

In places, drifting dunes have developed within the valley, many of them built up around sand-catching vegetation. In the mild, almost-frost-free climate, and with the Rio Grande supplying irrigation water, the floodplain is intensively farmed with alfalfa, cotton, chile, and onion crops, along with pecan, pistachio, and fruit orchards.

Near the Texas border, steeply dipping Paleozoic limestones trace chevrons on ridges of the northern Franklin Mountains. These marine limestones, deposited in Pennsylvanian and Permian seas, extend south and southeast into Texas as well as east and northeast across southern New Mexico. The La Mesa surface, visible to the west of the Rio Grande, represents the maximum sedimentary filling of the Rio Grande Valley before the river began its current and ongoing period of erosion.

I-40
Grants—Albuquerque
72 miles

Grants, north of I-40 between mileposts 82 and 85, lies in a flat-floored valley walled on the south by Quaternary lava flows and on the north by cliffs of Jurassic and Cretaceous sedimentary rock layers capped with basaltic lava flows. A low, almost imperceptible anticline runs north-south through the town and is reflected in the gentle arch of these sedimentary layers. Steep, hummocky slopes along the base of the cliffs are landslides in which lava and harder Cretaceous rock have slid down over soft, slippery Cretaceous shale. Grants lies at the southern edge of the Colorado Plateau, and this section of road transitions from mesas through young volcanism and into the Rio Grande rift.

For many years, Grants was a major center for uranium exploration and mining, with mines mostly in the Morrison Formation, the youngest Jurassic formation in this area. The Grants uranium mining district extends from Gallup eastward to Laguna along the southwest flank of the San Juan Basin. Uranium, primarily used in nuclear reactors, is naturally occurring and has the highest atomic weight of any element. This slightly radioactive element is commonly found as a uranium oxide but is not often present in sufficient concentrations to be economically viable to mine. Uranium mining in New Mexico occurred mainly during the uranium booms of the 1950s and 1970s and ended in 1998. Global economics drive cyclical interest in reopening uranium mines in the Grants mineral belt. New Mexico is the second state, behind Wyoming, for uranium reserves.

Oxidized groundwater picks up soluble uranium minerals as it circulates through volcanic or intrusive rocks. When the groundwater flows into aquifer rocks, such as sandstone, it is reduced upon interaction with organic matter and precipitates as uraninite, a uranium oxide that is the main ore mineral of uranium. Uranium is found in sandstone deposits along the Colorado Plateau, with the Jurassic-age Morrison Formation sandstones particularly productive in the Grants mineral belt. The Morrison Formation, deposited on a river floodplain in richly vegetated surroundings, consists of variously colored mudstone containing clay-rich bentonite derived from volcanic ash. It also contains thin layers of orange sandstone, stream-channel deposits rich in organic matter, within which much of the uraninite is concentrated. Uranium is also found in the Dakota Sandstone, Chinle Group sandstones, and

a broad zone of faults defines the west edge of the Rio Grande rift, separating it from the Colorado Plateau to the northwest

the Albuquerque Volcanoes line up along a fissure that parallels the edge of the Rio Grande rift

Petroglyph National Monument

topped by resistant lava flows, cliffs of Cretaceous sandstone rise above hummocky landslides of weaker shale

Rio Grande Nature Center State Park

CENOZOIC
QUATERNARY
- Qa — alluvial deposits
- Ql — landslide and rockfall deposits
- Qp — piedmont alluvial deposits
- Qw — windblown deposits

QUATERNARY–PALEOGENE
- QTsf — Santa Fe Group, undivided (latest Oligocene to middle Pleistocene)

VOLCANIC AND IGNEOUS ROCKS
- Qb — basalt (Quaternary)
- Tb — basalt (Neogene)
- Tnv — volcanic rocks (Neogene)
- Ti — intrusive rocks; includes dikes (late Eocene to Pliocene)

MESOZOIC
CRETACEOUS
- Kmv — Mesaverde Group
- Ktma — Tres Hermanos Formation, Morena Hill Formation, and Atarque Sandstone
- Kcc — Crevasse Canyon Formation
- Kg — Gallup Sandstone
- Km — Mancos Shale
- Kd — Dakota Sandstone

JURASSIC
- J — sedimentary rocks, undivided
- Jm — Morrison Formation
- Jsr — San Rafael Group

TRIASSIC
- ₮c — Chinle Group

PALEOZOIC
PERMIAN
- P — sedimentary rocks, undivided
- Psag — San Andres Formation and Glorieta Sandstone
- Pya — Yeso and Abo Formations

PENNSYLVANIAN
- ₱m — Madera Group

- ∗ volcanic center
- —···· fault; dotted where concealed
- ⌐ ¬ national monument boundary

Geology along I-40 between Grants and Albuquerque.

Buckled lava on the McCartys lava flow, the youngest flow in the Zuni-Bandera volcanic field. It erupted about 3,800 years ago.

uniquely in the Jurassic-age Todilto Limestone, although these limestone-uranium deposits are rare. Within the sandstone units, uraninite is deposited in curved roll-fronts, which indicate a change from oxidized to reduced aquifer conditions.

A large area of lava known as El Malpais reaches I-40 east of Grants at milepost 89. The flows of El Malpais are part of the Zuni-Bandera volcanic field. Located along the Jemez lineament west of the Rio Grande rift, the Zuni-Bandera is the second-largest volcanic field of the Basin and Range province, covering approximately 950 square miles in a blanket of multiple basaltic lavas that reach close to 500 feet in total thickness in places.

The Zuni-Bandera volcanic field contains the mafic outpourings from at least one hundred vents over the last 700,000 years. Hawai'ian-style basaltic lava flows and cinder cones of the Zuni-Bandera volcanic field show both pahoehoe and aa lava flows, as well as lava tubes. The youngest flow, the McCarty flow, is dated at 3,800 years old and sourced from a shield volcano about 25 miles south of the interstate. Its jumbled, twisted, ropy lava is full of vesicles (gas-bubble holes), and pressure ridges, squeeze-ups, and grooved lava surfaces show where molten lava pushed through its hardening crust. Water-filled sags represent collapsed lava tubes that formed where lava flowed out from under its own hardening crust. Small olivine, plagioclase, and pyroxene crystals are visible in the rock as well as feldspar crystals as much as 2 centimeters in diameter. For more on El Malpais National Monument, see page 69 in the Colorado Plateau section.

East of McCartys, I-40 proceeds down the valley of the Rio San Jose. The flanks of basalt-capped mesas on either side of the highway expose Cretaceous sandstone dipping east off the Zuni uplift. Note the columnar joints of the basalt.

From the rest area at milepost 102, Mt. Taylor (11,301 feet) appears to the north, towering over a lava-capped mesa. Mt. Taylor, along with several other volcanic areas in New Mexico, lies along the Jemez lineament, an ancient suture deep below in the Proterozoic basement rock. The lineament is a zone of faulting and crustal weakness along which the Mt. Taylor magma moved to the surface. Mt. Taylor, a composite stratovolcano built of successive lava and ash-flow tuffs, began to form 3.7 million years ago. The oldest flow, which erupted across Cretaceous sandstone, is composed of basanite, a volcanic lava richer in sodium and potassium than regular basalt. Most of the cone seen today formed between 3 and 2.6 million years ago with initial rhyolite and trachyandesite eruptions followed by increasingly mafic lava flows. Basalts erupted throughout the life of the volcano, but most flowed out from fissures along the flanks of the mountain in later periods of activity. The volcano today is somewhat horseshoe shaped, with an interior valley on its eastern side and Mt. Taylor forming the high point on its western edge. Scientists suggest that due to the many changes in composition of the lavas, Mt. Taylor erupted from a succession of short-lived magma chambers rather than a single, long-filling source. The last eruption occurred 1.5 million years ago.

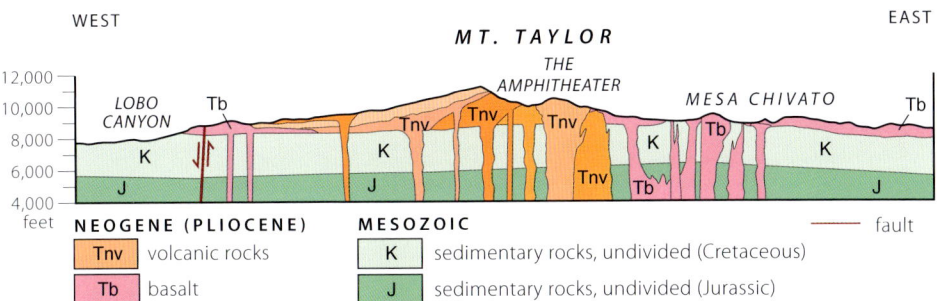

Mt. Taylor, with its alternating lava flows and volcanic debris, is a stratovolcano. The multiple flows of various ages that have together constructed the Mt. Taylor edifice were fed by multiple feeder dikes over time, resulting in a sprawling mountain. A central lava dome, partly destroyed by erosion, is the youngest part of the mountain. —Adapted from Goff and others, 2015

East of Mt. Taylor, I-40 continues through Cretaceous rocks along the Rio San Jose. The boggy area between the highway and the Laguna Pueblo, near milepost 114, is the original lake that led to the Spanish name for the town. East of this, the route crosses another lava flow, older than those of El Malpais and covered with soil and windblown sand.

Between mileposts 130 and 140, the highway leaves the flat-topped tablelands of the Colorado Plateau and enters a broad fault zone that defines the western edge of the Rio Grande rift. Although the faults here cause disruptions in Mesozoic and Cenozoic sedimentary layers, they are hard to see from the highway. East of the Rio Puerco, at milepost 140, Mesozoic rocks are not exposed on the surface. Rio Grande rift faults have dropped the older sedimentary units down, and they are now buried beneath thousands of feet of Santa Fe Group sediments. Although there is significant faulting on the western edge of the Albuquerque Basin, this side is on the hinge of the half graben and faulting is small compared to the eastern margin, which uplifted the Sandia Mountains.

Soft, poorly consolidated sands, gravels, and silts of the Santa Fe Group are exposed in badlands from the Rio Puerco to Albuquerque. These basin-filling sediments and volcanics were deposited over millions of years in closed arid basins from the Oligocene to Pleistocene Epochs as the rift opened. These sediments were brought into the

PETROGLYPH NATIONAL MONUMENT AND THE ALBUQUERQUE VOLCANOES

Petroglyph National Monument lies a few miles north of I-40 on the west side of Albuquerque, and several trailheads give access to the petroglyphs. To reach the visitor center, take exit 154, turn north on Unser Boulevard, and then turn left on Western Trail NW. An estimated 250,000 rock art images are carved into basalt erupted about 200,000 years ago from the Albuquerque volcanic field. Between AD 1300 and AD 1680, ancestral Puebloans used stone tools to cut most of these petroglyphs into the desert varnish that coats the basalt. Spanish settlers later produced younger petroglyphs. Desert varnish is a metallic-looking patina that forms on exposed rock surfaces. The varnish slowly forms as metallic and clay minerals interact with biological processes. The excellent preservation of petroglyphs that are hundreds of years old is evidence for the slow rate of desert varnish formation. When viewing the carvings, keep an eye out for the abundant white, needle-shaped plagioclase crystals that stand out against the black basalt. Also notable are the abundant vesicles, or bubble holes, that form when gas bubbles are trapped within the lava as it cools.

The Ceja Formation, which caps the Llano de Albuquerque, approaches the monument. This Llano, or floodplain, stands between 300 and 700 feet above the modern Rio Grande floodplain and represents the last stage of aggradation (or basin filling) along the Rio Grande rift. The change from basin filling to erosion, or incision, occurred roughly 1.8 million years ago. Since this timeframe, the regional rift setting has been one of incision, driven by downstream river integration, which lowered the Rio Grande's base level. Increased precipitation during the Pleistocene ice ages enhanced this incision.

The five cones of the Albuquerque Volcanoes are aligned en echelon along north-south faults and stand above the Llano. These produced six basalt flows of varying size. They erupted between 190 and 220 thousand years ago and cover about 23 square miles, largely east of the fissure zone, forming a jagged boundary to the westward expansion of the city. During intervals of no volcanic activity, soils developed, which younger flows later covered. The eruptions also produced cones, cinders, lava lakes, and small lava tubes, also preserved within the monument. The largest spatter cone is called Vulcan. Lava tubes 8 to 20 inches in size can be seen on the northeast flanks of the cone.

Petroglyphs etched into basalt at the Rinconada Trailhead in Petroglyph National Monument.

rift basins by side streams and eventually by the through-flowing Rio Grande. Today, the river is cutting a path through its own sediments, revealing older river levels in the terraces that flank the river. A thin veneer of sands, gravels, and windblown deposits marks the modern deposition of sediments.

Broad, gently sloping terraces edge both the Rio Puerco and the Rio Grande Valleys, each terrace representing a time of balance between deposition of the Santa Fe Group and incision. The uppermost terraces are the oldest, with renewed erosion cutting new channels, which in turn fill in with river floodplain deposits that will become lower terraces.

Between mileposts 145 and 150, the highway crosses a particularly broad terrace, the irregular, bumpy surface of which was a sand dune field in Pleistocene time. The Albuquerque Volcanoes, a string of five small cones aligned from north to south along two fissure zones, can be glimpsed to the north.

From the rest area east of milepost 151, Albuquerque rests on several terrace levels. The city straddles the Rio Grande from these western terraces to the eastern flanks of the Sandia Mountains. The rift valley is about 28 miles wide here.

The Rio Grande has been channelized where it passes through Albuquerque, with levees and floodways on either side of the main channel to aid in irrigation. The inner valley the river now occupies is a newcomer in terms of its geologic age. Historically, as the river swung from side to side, silting up its channel and shifting its course, the river formed natural levees that gradually raised the river level higher than its floodplain. Lowlands on either side of the river became swampy and subject to flooding, particularly when tributaries added runoff from sudden severe rainstorms. Today, the flows of the Rio Grande are highly controlled. Twenty-two dams exist along the length of the river; eleven are placed in or north of Albuquerque. Diversion channels, built as part of Middle Rio Grande Conservancy District efforts, have been built to route runoff from the mountains as well as to deliver water for industry and residential use. These diversion channels collect stormwater from across Bernalillo City and Albuquerque, funneling it into the Rio Grande, either at the northern outlet (near Alameda Boulevard) or at the southern outlet (south of Rio Bravo).

The Rio Grande Nature Center State Park in central Albuquerque provides access to the river and its surrounding forested floodplain, or bosque. Offering hiking trails, wetland, bird viewing, and visitor and educational centers, this day-use park highlights the history of the Rio Grande and the ecological importance of a naturally shifting river.

I-40
Albuquerque—Clines Corners
59 miles

I-40 climbs eastward from Albuquerque across large alluvial fans deposited during Pleistocene time by Tijeras and Embudo Creeks. The Sandia Mountains, an uplifted block with a fault on its west side, rise dramatically to 10,678 feet. This jagged, 17-mile-long mountain front rises 5,000 feet vertically above downtown Albuquerque in only about 2.5 miles horizontally. When walking a straight line from foothills to crest, roughly each stride taken horizontally would also gain nearly 1 foot of elevation.

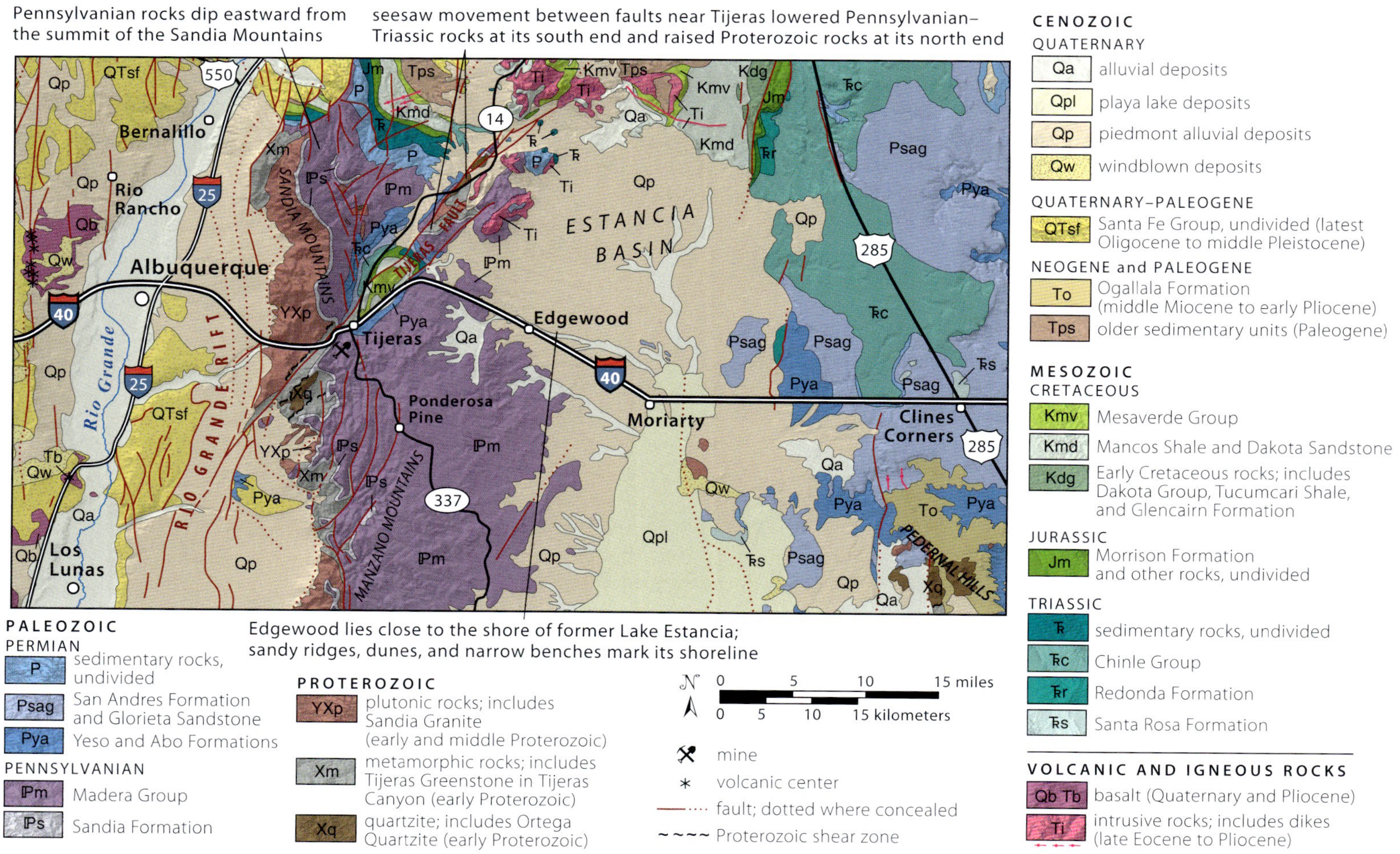

Pennsylvanian rocks dip eastward from the summit of the Sandia Mountains seesaw movement between faults near Tijeras lowered Pennsylvanian–Triassic rocks at its south end and raised Proterozoic rocks at its north end

Edgewood lies close to the shore of former Lake Estancia; sandy ridges, dunes, and narrow benches mark its shoreline

Geology along I-40 between Albuquerque and Clines Corners.

CENOZOIC
QUATERNARY
- Qa — alluvial deposits
- Qpl — playa lake deposits
- Qp — piedmont alluvial deposits
- Qw — windblown deposits

QUATERNARY–PALEOGENE
- QTsf — Santa Fe Group, undivided (latest Oligocene to middle Pleistocene)

NEOGENE and PALEOGENE
- To — Ogallala Formation (middle Miocene to early Pliocene)
- Tps — older sedimentary units (Paleogene)

MESOZOIC
CRETACEOUS
- Kmv — Mesaverde Group
- Kmd — Mancos Shale and Dakota Sandstone
- Kdg — Early Cretaceous rocks; includes Dakota Group, Tucumcari Shale, and Glencairn Formation

JURASSIC
- Jm — Morrison Formation and other rocks, undivided

TRIASSIC
- ℞ — sedimentary rocks, undivided
- ℞c — Chinle Group
- ℞r — Redonda Formation
- ℞s — Santa Rosa Formation

VOLCANIC AND IGNEOUS ROCKS
- Qb Tb — basalt (Quaternary and Pliocene)
- Ti — intrusive rocks; includes dikes (late Eocene to Pliocene)

PALEOZOIC
PERMIAN
- P — sedimentary rocks, undivided
- Psag — San Andres Formation and Glorieta Sandstone
- Pya — Yeso and Abo Formations

PENNSYLVANIAN
- ℙm — Madera Group
- ℙs — Sandia Formation

PROTEROZOIC
- YXp — plutonic rocks; includes Sandia Granite (early and middle Proterozoic)
- Xm — metamorphic rocks; includes Tijeras Greenstone in Tijeras Canyon (early Proterozoic)
- Xq — quartzite; includes Ortega Quartzite (early Proterozoic)

Map symbols:
- N (north arrow)
- mine
- * volcanic center
- fault; dotted where concealed
- Proterozoic shear zone

Scale: 0–15 miles / 0–15 kilometers

SOUTHERN RIO GRANDE RIFT   185

Topographically, the Sandia Mountains are a relatively young mountain range, but the steep, rugged western face exposes the 1.45-billion-year-old Sandia granite and older metamorphic rock. The mountains are capped with thin layers of Pennsylvanian and Permian limestone, shale, and sandstone of the Madera Group and Sandia Formation.

The Four Hills neighborhood south of the interstate at milepost 167 is built at the junction of alluvial fan deposits and granite bedrock. The Kirtland Air Force Base stretches to the south of Four Hills here, notably bereft of housing.

East of milepost 167, I-40 quickly leaves the spread of the city and climbs into Tijeras Canyon, which divides the Sandia Mountains to the north from the Manzano Mountains to the south. Immediately east of exit 167, Sandia Granite appears in roadcuts on the north side of the road. The Sandia Granite is famously pink, perhaps the source of the name Sandia, meaning "watermelon" in Spanish. The pink hue of the granite comes from its abundant potassium feldspar minerals. Aplite and pegmatite dikes and smaller, sparkling-white quartz veins crosscut the Sandia Granite, which cooled 1.45 billion years ago during the Picuris orogeny.

Between mileposts 168 and 169, a giant aluminum sculpture of a yucca cactus towers above the foothills of Sandia Granite bedrock. In 2003, Artist Gordon Huether built this sculpture using recycled fuel tanks from retired military aircraft. The goal of this artwork was to honor the history of the Old Route 66, Kirtland Air Force Base, and the native vegetation of New Mexico, including its state flower, the yucca.

The Sandia Granite weathers into rounded boulders separated by masses of coarse sand, called grus, composed of quartz and feldspar crystals that have broken apart as the granite weathers. This photo was taken along a trail in the Sandia Foothill area, just north of the highway between mileposts 168 and 169.

Near the entrance to Tijeras Canyon, the Sandia Granite weathers into steep, angular crags and pinnacles. In some highway cuts, the normally blocky, jointed granite has clearly weathered, in places many feet below the modern land surface. Physical and chemical weathering along joints combine to attack the granite, breaking it down over time. Dikes stand out as dark masses in the roadcuts while quartz veins weather out in white relief.

Tijeras Canyon developed in weakened rock along the northeast-trending Tijeras fault. Movement in the Tijeras fault zone has been recurrent, likely since its formation during the Precambrian. Erosion removed broken and crushed rock along the fault zone and transported it west into the rift and east onto the plains. The interstate winds through the canyon, which is filled with a series of alluvial and Quaternary stream deposits. South of Tijeras Canyon, Paleoproterozoic metamorphic rocks are exposed in roadcuts and canyon walls. These include mafic metavolcanics that originally were basalts and andesites. These rocks and the interlayered schists were metamorphosed 1.65 billion years ago in the Mazatzal orogeny.

On the north side of the interstate between mileposts 171 and 172, the interstate crosses the Seven Springs shear zone, east of which is the Cibola Gneiss, marked by steep reddish slopes. The Cibola Gneiss (also called the Cibola Granite) is a coarse-grained, two-mica granite with a strong metamorphic fabric of aligned micas. The pluton that became the Cibola Gneiss is thought to have been emplaced during the Mazatzal orogeny more than 1.65 billion years ago and was associated with massive caldera magmatism. Many younger dikes and veins cut the granite, testament to its long history. The mountains are laced with faults from different episodes of mountain building and deformation. Veins within the Sandia Granite and Cibola Gneiss hold precious minerals that precipitated out of magmatic fluids, including small quantities of quartz-pyrite-gold veins. These quantities were not sufficient to spur any notable mineral production.

Near milepost 173, the road passes in quick succession from the Proterozoic metavolcanics to the overlying claystone, limestone, and sandstones of the Pennsylvanian-age Sandia Formation and then the Madera Group, and all are cut by faults. In places, granitic clasts are incorporated within overlying sandstone, limestone, and claystone of the Sandia. About 170 feet thick, the Sandia Formation has a gradational contact with the overlying Madera Group. The Sandia limestone is distinguished from limestone in the overlying Madera Group by being thinner-bedded and greenish, with abundant clastic material eroded from the continent and transported into the sea. The first thick bed of limestone marks the transition into the Madera Group.

The Madera Group is broken into a lower cherty, cliff-forming gray limestone and an upper arkosic limestone, a carbonate unit with an abundance of grains or crystals of feldspar. Discontinuous layers of sandstone and mudstones appear gray to reddish and often contain pebbles. Gray, greenish-gray, or tan, the Madera is fossil-rich and has excellent cliff-forming exposure throughout the Sandia and Manzano Mountains. The Madera Group fossils include brachiopods, corals, bryozoans, clams, snails, trilobites, and button-like sections of crinoid stems. Mudstone layers in the formation locally contain fossil ferns and leaf impressions of other land plants. The Madera Group sediments were deposited in a shallow marine platform and swampy coastal environment.

At milepost 174, roadcuts and canyon slopes take on a green hue derived from the Tijeras Greenstone, a rock unit composed of metasedimentary and metavolcanic rocks. These Proterozoic rocks have dropped to roadside level between two faults east of the Tijeras fault zone. The narrow block in which they occur moved in seesaw fashion—one end down, the other up. Rocks near the faults are bent in little folds and broken by many smaller faults.

From milepost 174 at Tijeras, NM 337 heads south to the small community of Ponderosa Pine, where ponderosa pines take over from the scrubby bushes and grasses that dominate the plains below. Ponderosa Pine is situated on one of many north-south-trending faults. NM 14, Turquoise Trail National Scenic Byway, heads north from Tijeras (See NM 14: Tijeras—Santa Fe on page 155).

A large, open-pit mine and cement processing plant is visible on the south side of I-40 at milepost 175. The Tijeras Mine opened in 1982 and produces limestone and shale for cement and other uses. These commodities are sourced from the Pennsylvanian Madera Group.

East of Tijeras there are views of the "back" side of the Sandias. The long, tree-covered slope is composed of east-dipping Paleozoic sedimentary rocks. The majority of this sloping mountainside is surfaced with Madera Group limestones.

Visible from near Tijeras, tilted Pennsylvanian sedimentary rocks, primarily Madera Group limestones, dip gently eastward off the eastern slope of the Sandia Mountains. From this vantage point, they are all tree covered.

Near milepost 177, the road swings to the southeast, following a valley along the fault that edges the southern side of a syncline. To the north lie Cretaceous rocks. In the valley adjacent to the highway is gray Mancos Shale, here covered with reddish alluvium, and beyond that are the rough, tree-covered hills of the Mesa Verde Group. To the south are the dark-reddish-brown sandstones and shales of the Permian Abo and Yeso Formations. These thinly bedded units contain short, straight cross-beds, indicating moving water deposited them. In places, the red rocks contain petrified wood, leaf impressions, and footprints of Permian reptiles, all of which demonstrate that the units were deposited above sea level, probably on a floodplain or delta where streams spread out on a relatively flat surface. The color of the rocks is due to tiny particles of iron oxide that coat the sand grains or compose part of the clay of the mudstone.

At the small community of Edgewood (milepost 187), the highway enters the Estancia Basin, with the Sangre de Cristo Mountains in the distance to the north. Edgewood obtains its water from Pennsylvanian rocks of the Madera Group, replenished from the east slope of the Sandias.

The Estancia Basin is a north-south-trending structural basin that developed during the formation of the Ancestral Rocky Mountains in the Pennsylvanian Period. The basin covers more than 1,500 square miles. Proterozoic igneous and metamorphic basement rocks lie up to 8,500 feet below the surface at its deepest extent in the eastern parts of the basin. The western part of this asymmetrical basin is much shallower, with basement rocks found less than 1,000 feet below the surface.

Pennsylvanian and Permian sedimentary rocks fill this basin beneath its veneer of Quaternary sediments. During and after the Laramide uplift, erosion removed Jurassic and Cretaceous rocks. At the surface, the Madera Group is between 1,300 and 1,900 feet thick, and records deposition in the sea and near the seashore. The character of the rock changes throughout this thickness: the lower, gray sections (older) are marine limestone and shales, while the upper section of the Madera contains more clastic beds, sediments sourced from rocks above sea level. Groundwater travels through fractures, joints, and solution cavities in the limestone, and in some places through porous sandstone layers. The groundwater contains high amounts of calcium carbonate, as well as high levels of magnesium, sodium, potassium, and fluoride.

In the Estancia Basin, the Madera Group is overlain by the Permian Abo Formation or by Quaternary stream and lake deposits, which partly fill the basin. Fine, dark-red shales in the Abo Formation are relatively impermeable and help to confine water within underlying layers. Hydrologically, the Estancia Basin is an internally drained basin, meaning it has no stream outlets. Surface water collects in more than sixty playa lakes that exist seasonally. Clay and gypsum, known as the Dog Lake Formation, have been deposited by these temporal lakes, which first began to form 45,000 to 40,000 years ago and eventually joined to become a single large lake during the peak of the final Pleistocene glaciation.

The Pleistocene lake fluctuated in size and depth. At its largest, 20,000 to 16,000 years ago, the lake reached nearly 300 feet deep and was freshwater. At later, lower levels, the lake was a mere 20 feet deep and distinctly saline. The final major lake here existed approximately 10,000 years ago. Beach ridges, narrow benches cut by lapping waves, appear clearly on aerial photographs; they formed during this last high stand. A drier climate in the region since has prevented any major lake formation. The sands, gravels, silts, and clays of Quaternary deposition are generally less than 250 feet thick, though in places they reach 450 feet in thickness. Gypsum dunes, formed from mineral precipitation as the lake dried, stretch discontinuously between Clines Corners south toward Willard, on US 60.

East of the Estancia Basin, I-40 climbs gradually through Pennsylvanian and Permian sedimentary rocks. Watch for red shales and sandstone of the Abo Formation and, east of them gray, San Andres Formation. To the south at milepost 217 are the Pedernal Hills, blocks of Proterozoic granite and quartzite that were first uplifted during the rise of the Ancestral Rockies in Pennsylvanian time.

US 54
Texas—Carrizozo
124 miles

US 54 north of El Paso, Texas, follows the down-dropped Tularosa Valley, the easternmost basin of the Rio Grande rift in southern New Mexico. The broad valley is a graben, a block of crust dropped down along faults between mountain ranges, in this case, the Sacramento Mountains to the east and San Andres, Organ, and Franklin Mountains to the west. Prior to the down-dropping of the Tularosa Valley, Paleozoic sedimentary rocks in the two ranges formed a single large anticline, or arch. As the crust pulled apart with the onset of rifting, the center of the arch collapsed, forming the valley and leaving faulted edges of tilted sedimentary rock exposed in the mountains. The Tularosa Valley was not carved by a river, so it has no outlet. Water entering it from surrounding mountains infiltrates into the porous surface gravel or temporarily pools in low areas southwest of White Sands National Park.

Between mileposts 30 and 35, US 54 passes the Jarilla Mountains, a small range to the west just north of Orogrande. An Eocene granodiorite body surrounded by younger monzonite intrusions forms the southern part of the Jarilla Mountains. Emplacement of these intrusions warped up the surrounding Pennsylvanian carbonate rocks, which were also metamorphosed by contact with the intrusion.

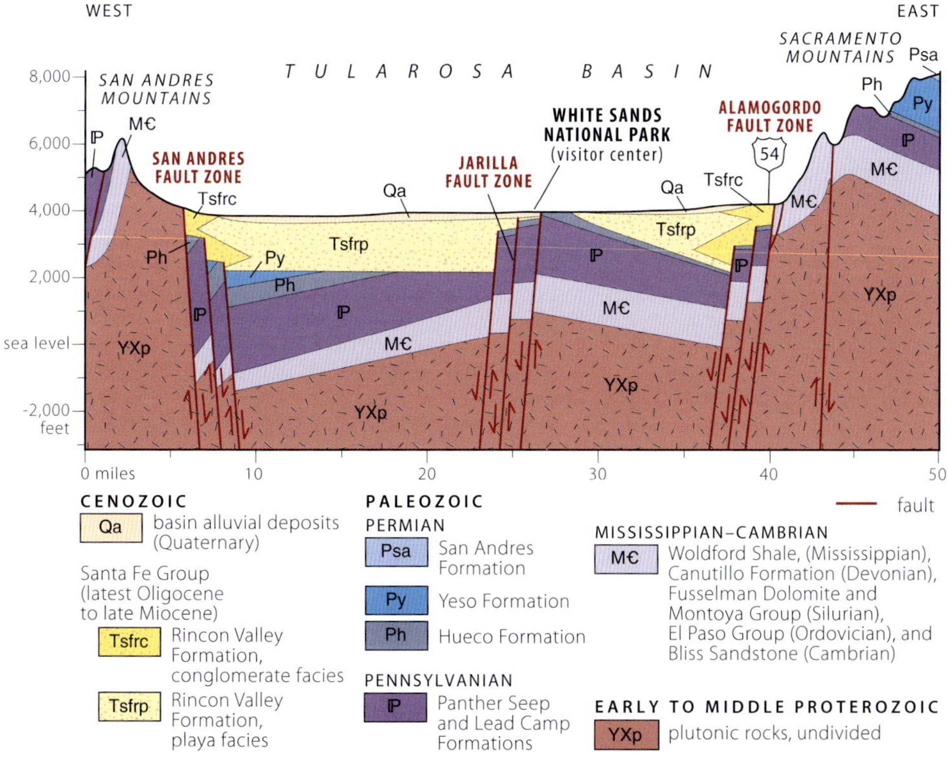

Cross section of the Tularosa Basin between the San Andres Mountains to the west and the Sacramento Mountains to the east. —Modified from Huff, 2004; Newton and Land, 2016

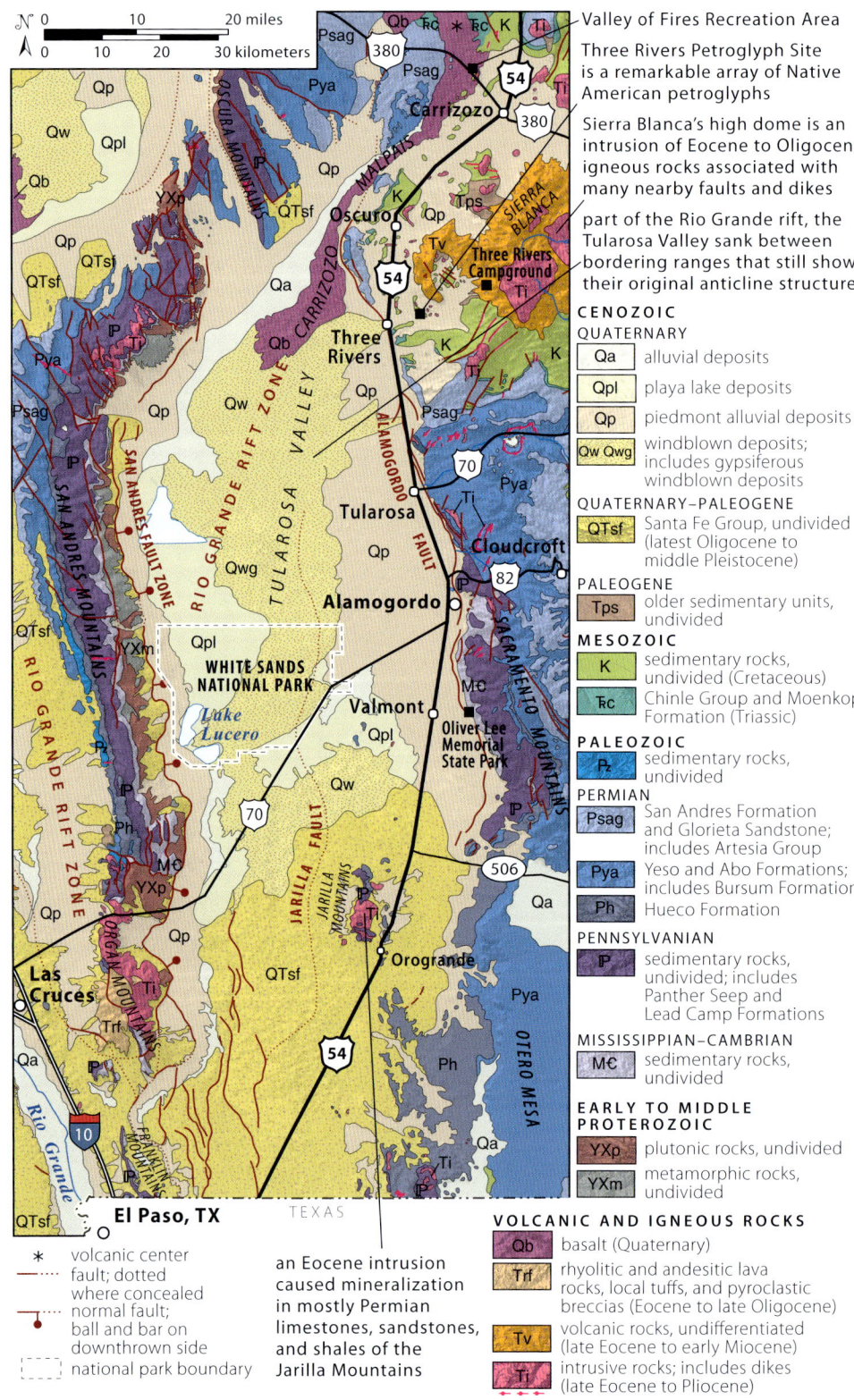

Geology along US 54 between the Texas border and Carrizozo.

Limestones outcrop in particular on the northern, southern, and eastern parts of this range. When gold was discovered in the Jarilla Mountains in the late 1800s, the town of Jarilla changed its name to Orogrande, or "big gold." Mining of gold, iron, copper, and lead in the Orogrande mining district ended by 1930.

The down-dropping of the Tularosa Valley exposed Permian rocks to erosion and, in the case of the evaporite minerals, to solution. Some of the gypsum that was dissolved out of these Permian rocks was redeposited in lake sediments of the closed basin, while the rest remained in the groundwater, contributing to the very hard, mineral-rich local groundwater. Sand-size particles of gypsum are entrained by the wind and transported to the White Sands dune field. Driving north along US 54, the real expanse of the White Sands can be seen. Less than half of the dune area is included within White Sands National Park, directly west at milepost 55. See US 70 on page 206 for more information about the dunes.

Beyond the dunes to the west are the San Andres Mountains, marked with stripes of Paleozoic sedimentary rock. The layered rock stands out on the mountain slopes, emphasized by differential erosion of soft shale and hard limestone layers. A complete Paleozoic sequence—Cambrian to Permian—is preserved in the northern Sacramento Mountains to the east, but alluvial fans cover the Cambrian, Ordovician, and Silurian rocks here. The oldest visible rock layers are the Devonian Oñate and Sly Gap Formations, which are dolomitic sandstone and black shale. Above the Devonian are massive limestones and shales of the Mississippian Caballero, Lake Valley, and Rancheria Formations. At Oliver Lee Memorial State Park near Valmont (milepost 55) there is a hike into Dog Canyon on the Dog Canyon National Recreation Trail in the Sacramento Mountains. Interpretive signs along trails and in the visitor center explain the geologic setting there.

The formations in Dog Canyon rise above the visitor center (bottom of the photo) at Oliver Lee Memorial State Park.

Most of the Ordovician through Pennsylvanian rocks of this face of the Sacramento Mountains are fossiliferous limestones and dolomites deposited in the sea.

Alamogordo (milepost 65) provides access to both White Sands National Park to the west (see page 206 in US 70: Las Cruces—Alamogordo) and the high country to the east near Cloudcroft (See page 255 in US 82: Alamogordo—Artesia).

At Tularosa (milepost 79), Sierra Blanca is visible to the northeast. Volcanic rocks surround this complex of Eocene to Oligocene intrusions. This volcano was active 38 to 28 million years ago, emitting lava flows, volcaniclastic sedimentary deposits, and minor welded ash-flow tuffs. Two main episodes of volcanism resulted in voluminous outpourings. The 38- to 34-million-year-old Walker Group contains numerous debris flows, suggesting it erupted from stratovolcanoes. Parts of this unit are red, indicating a humid environment at the time of deposition. Minor volcanism continued, and the emplacement of the Three River stock at about 27 million years ago concluded volcanism in the Sierra Blanca region. The stock can be seen up close at the Three Rivers Campground, 13 miles east of Three Rivers (north of milepost 96) on County Road B030.

Five miles east of Three Rivers (milepost 96) on county road B030 is a remarkable mile-long array of prehistoric rock art. The pictures and geometric designs at the Three Rivers Petroglyph Site were pecked into the desert varnish on volcanic rock on a lava ridge. The designs resemble those found in pottery of the Mimbres culture of southwestern New Mexico, dated between AD 900 and AD 1300.

From Three Rivers north to Carrizozo, a rough, barren, black lava flow occupies the center of the valley to the west. Known as the Carrizozo Malpais, or Valley of Fires, lava flow, it erupted 5,200 years ago from Little Black Peak northwest of Carrizozo. These olivine basalt flows are best seen from Valley of Fires Recreation Area on US 380 west of Carrizozo. (See US 380: Carrizozo—San Antonio on page 209 for more information about the flows.)

View east toward Sierra Blanca over petroglyph chiseled into basalt at the Three Rivers Petroglyph Site. Note the dark weathering rind where it has broken off at the top of the boulder.

Small mine dumps near Oscuro, remnants of its mining days, are visible from the highway at milepost 108. There's not as much gold in the hills here as there is farther north near White Oaks, Nogal, and Lincoln, but enough to be worked. The earliest strikes were placer deposits, stream gravels that contained flecks and nuggets of gold that could be retrieved by washing the gravels over riffle boards designed to catch the heavy bits of gold. The desert setting hindered this water-intensive process because water had to be carried in barrels from nearby mountains. Later mining involved more digging but still used water to concentrate the gold.

Hidden beneath the lava and other valley fill are faults that edge the east side of the rift. The fault zone curves eastward here, passing through the saddle between the San Andres and Oscura Mountains (east at milepost 120). Desert washes coming from the Sierra Blanca highlands convey storm runoff from mountains to the valley. Little of their water ever reaches the low part of the valley, however; most sinks into the broad alluvial aprons that surround the mountains.

Carrizozo, originally a railroad stopping point, became the supply and shipping center for mines near Oscuro, White Oaks, and the Nogal-Lincoln area to the east. While Carrizozo's transportation importance has dwindled, the town is now nationally famous for its cherry cider and associated festival.

US 60
Bernardo—Mountainair—Encino
100 miles

From I-25, US 60 heads east across the fertile, well-watered Rio Grande floodplain. As basins of the Rio Grande rift subsided in late Oligocene to early Miocene time, they were filled with the Santa Fe Group—easily erodible sediments and concurrently erupted volcanic units. In this location in the Albuquerque Basin, sediments are fine- to-coarse sandstones, siltstones, conglomerates, and volcanic materials. While much of the history of the Rio Grande has been one of deposition, drainage reorganizations within the last 500,000 years have shifted the riparian areas to an erosional setting in which the river is cutting down into its own valley fill sediments. Abandoned terraces on either side of the modern river document this downcutting.

East of the river, US 60 rises over several broad alluvial terraces or benches between mileposts 168 and 170. Each step represents a former (higher) level of the valley floor. The surface of each terrace formed during a period of relative stability when downcutting ceased for a time and the river deposited sand and gravel to form a floodplain. The lowest terrace is the youngest, and higher terraces are successively older. Terrace surfaces are almost level, though the highest shows some of the hummocky topography of former sand dunes. The gravel that forms the terraces is derived from the Rio Grande and side streams flowing from the valley-bounding mountain ranges. The alluvial fans extending from the Manzano Mountains interfinger with the river deposits. East of milepost 182, the uppermost terrace blends indiscernibly with a surface cut into the granite bedrock (a mountain pediment) of the Manzano Mountains.

Turning sharply south at milepost 183, US 60 soon crosses the rift-bounding Manzano fault zone, the boundary between the Albuquerque rift basin to the west and the upthrown fault block mountains to the east. The Manzano fault has at least 5,000 feet of throw and has been active as recently as the Pliocene Epoch. The faults are largely hidden, buried beneath the extensive alluvial fans on the western slope of the Manzano mountain front, but some fault activity can be seen in the disruption of young river terraces.

The broad, gently sloping bajada west of the Manzano Mountains formed from a series of coalescing alluvial fans. Each canyon along the mountain front deposited its own fan. As fans grew overtime, they converged to form the enormous slope seen today.

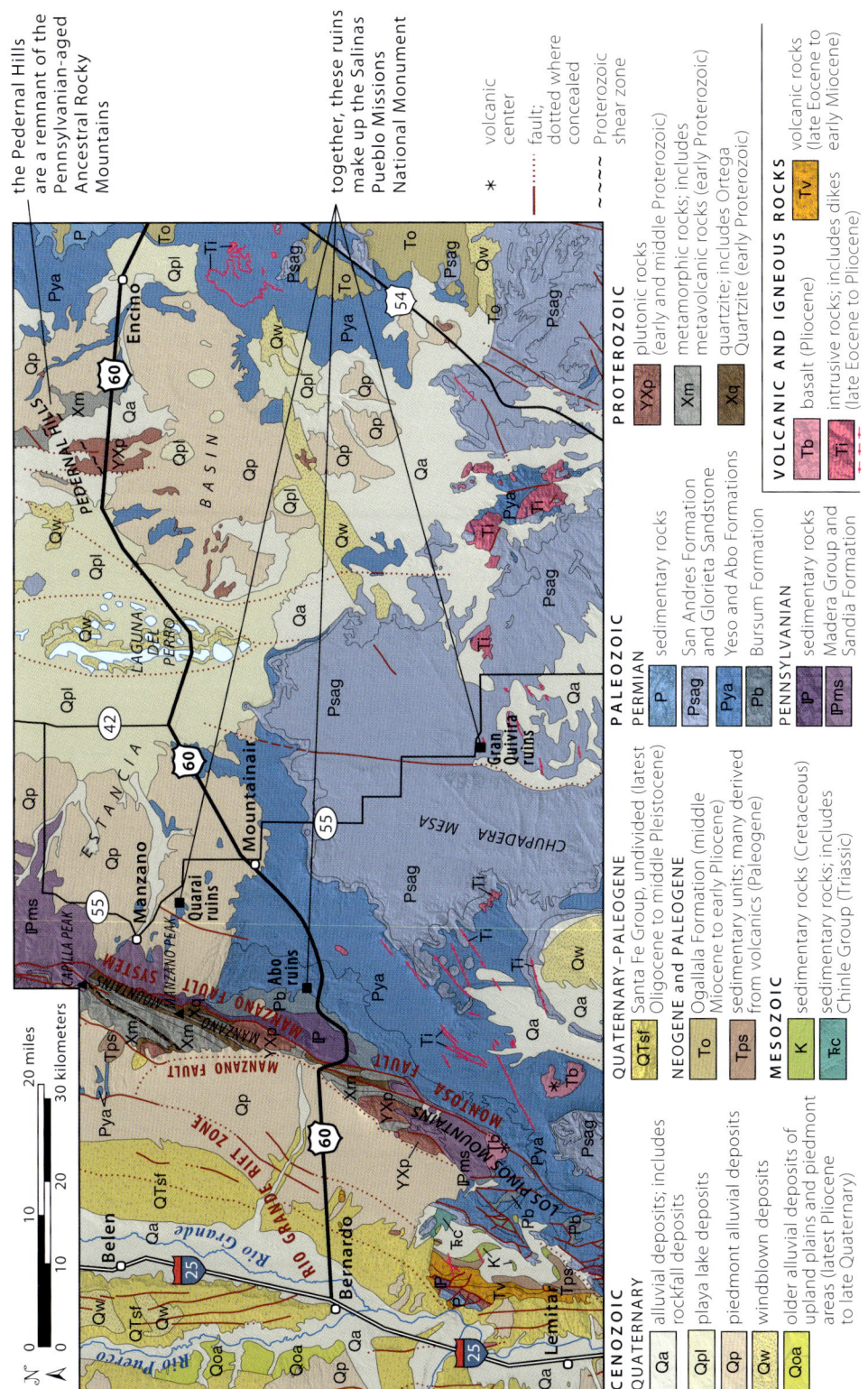

Geology along US 60 between Bernardo and Encino.

On the east side of the Manzano fault, US 60 climbs to the divide between the Manzano Mountains to the north and the Los Pinos Mountains to the south. At milepost 184, the western craggy face of this fault-bound range is composed of Proterozoic quartzites, schists, and metavolcanics. Several plutons intruded these basement rocks between 1.7 and 1.5 billion years ago during the Mazatzal orogeny.

Proterozoic bedrock between mileposts 184 and 185 takes many forms along this highway, with visible faults, folds, and unconformities. Some of what is seen are brightly iridescent schists formed by metamorphism of fine-grained shale and mudstone, dark greenstones formed by the metamorphism of volcanic rocks, and chunky white quartzites formed by metamorphism of sandstone. The complexity of the Proterozoic bedrock in the Los Pinos Mountains has been the focus of several studies because the rocks are particularly well exposed and accessible. The rocks contain evidence of at least two episodes of mountain building that occurred during the Proterozoic, associated with the collisions of microcontinents with the early North American continent.

East of the pass, US 60 descends the slope of a large alluvial fan from Abo Canyon. This older Pleistocene-age fan, originally deposited by east-flowing streams, is deeply channeled near the mountains by an ephemeral stream that flows only seasonally, and now, to the west, toward the Rio Grande. The fans along the western fronts of the Sandia and Manzano Mountains coalesce with neighboring fans to form an alluvial apron, or bajada. A similar bajada feature, though smaller in scale, can be seen on this southeastern slope of the Manzanos.

Near milepost 185, the Montosa fault system bounds the eastern flank of the Manzano Mountains. This Laramide-age fault system dips 50 degrees to the west, and it is a west-side-up reverse fault. Rocks have moved nearly 5,000 feet along the fault, bringing Proterozoic rocks above the Pennsylvanian Sandia Formation and Madera Group and the Permian Bursum and Abo Formations to the east.

Near milepost 186, the Abo Formation appears in roadcuts. This dark-red and brown mudstone and sandstone was deposited on a river floodplain or a delta near the

The Abo Formation near milepost 186. —Photo by Kent Budge

SALINAS PUEBLO MISSIONS NATIONAL MONUMENT

The Abo, Gran Quivira, and Quarai ruins are protected in Salinas National Monument. The ruins are located at three disparate sites accessed from Mountainair (milepost 204): Quarai and Abo are north of US 60 and Gran Quivira is to the south. Built by Native Americans in the twelfth to fourteenth centuries, the ruins lie in the Estancia Basin east of the Manzano Mountains. They and the churches added by sixteenth-century Spanish padres are built of local rock—Permian San Andres Formation at Gran Quivira and red sandstone of the Permian Abo Formation at Abo and Quarai. The Abo Formation dips distinctly eastward, with more-resistant beds forming cuestas west of Abo and extending north and south along the base of the mountains, while the softer shale and mudstone form slopes.

The ruins of Mission of San Gregorio de Abo, a half mile north of the highway at milepost 194.5, show that pre-Spanish as well as Spanish builders appreciated the way that sandstone layers of the Abo Formation break naturally along joints, providing rectangular blocks suitable for masonry.

The structures of Gran Quivira are built with the familiar gray limestone, sandstone, and gypsum of the San Andres Formation. This mostly limestone unit was deposited in shallow, restricted seas from about 272 to 260 million years ago and forms the resistant caprock surface of Chupadera Mesa.

Gran Quivira lies on a small mesa capped by the San Andres Formation. The San Andres Formation contains gypsum and salt, both of which flavor local water. Spanish historical documents tell that water for the village was obtained from rooftop runoff and from thirty-two walk-in wells. It was stored in rock and clay cisterns. There are no nearby springs. The pueblo of Quarai was more fortunate, obtaining freshwater from several seeps at the edge of the Manzano Mountains.

Oligocene volcanism affected this area 34 to 23 million years ago, when intrusions warped the rock layers beneath Gran Quivira upward. Mafic, fine-grained, basaltic andesite dikes are poorly exposed within the area. Young alluvium and gravel are eroded from the Manzano Mountains and other highlands.

sea from 300 to 272 million years ago. Sandstone layers show through cross-bedded conglomeratic sands, typical characteristics of the channels and bars of a meandering river. Ripple marks are found in silty layers, while mud cracks, formed by shrinkage as the sediments dried, can be seen in mudstone sections. Permian-age trace fossils, bivalves, gastropods, and conifer fossils can be found in the Abo Formation. Copper oxides were mined from 1915–1961 in the Scholle mining district near milepost 192 from some of the channel sandstone. Mining is currently not economical.

Between mileposts 195 and 210, the widespread Yeso Formation is frequently at the surface with pinkish, cross-bedded sandstone, thin-bedded brown mudstone, and abundant gypsum. It was deposited on a coastal plain about 282 to 272 million years ago.

Mountainair lies at the western edge of the Estancia Basin, a topographically low area that held Lake Estancia in Pleistocene time. Beaches, bars, and spits can still be distinguished around the margins of the valley, particularly on aerial photographs. Ancient campsites on the old lakeshores show that hunter-gatherer peoples were in this area as early as 12,000 years ago. As the climate became drier after the close of the ice age, residents of pueblos cultivated their staple foods of corn, beans, and squash on the fine, loamy lake deposits.

After the Pleistocene Epoch, the lake dried up, and only an area of salt ponds called Laguna del Perro remain. These small, ephemeral lakes fill depressions, or blowouts, that were scoured out by high winds. Native Americans, Spanish missionaries, and settlers obtained salt from these lakes for trade and domestic use. Some of the salt made its way to Mexico for use in smelting silver. Sand hills around the margins of the valley are the remains of old dunes. The high spring winds common to this location continue to erode and remove remaining lake sediments because vegetation is absent.

The Pedernal Hills, small upfaulted hills east of Laguna del Perro and approximately north of milepost 40, are the southernmost expression of the Sangre de Cristo

Aerial view looking west at the ephemeral lakes that form in the Estancia Basin after significant precipitation. The Manzano Mountains rise in the background.

Mountains uplift. They include Proterozoic metamorphosed sedimentary, volcanic, and plutonic rock. They were also an uplift of the Ancestral Rocky Mountains, which were eroded before the overlying, flat-lying Permian sediments were deposited. Windmills along this stretch west of Encino take advantage of winds blowing east off the mountains. A roadcut at milepost 250 shows reddish Permian Yeso Formation that was deposited directly on the Proterozoic rocks.

MANZANO AND QUARAI RUINS

To access Manzano and the Quarai ruins, take NM 55 north from Mountainair. The highway skirts the smooth, east-dipping ramp of the Manzano Mountains, crossing coalesced alluvial fans. Thick crusts of caliche on the alluvial fans suggest they are of Pleistocene age. The Mountainair municipal airport is built upon one of these Pleistocene to Holocene alluvial deposits.

A sharp, westward turn brings the mountains into view, including Manzano Peak (10,098 feet) directly ahead, which is part of Manzano Mountain State Park and accessed from the village of Manzano (milepost 75). While the flank of the mountains is sloping Pennsylvanian and Permian rock layers, Manzano Peak is Paleoproterozoic metasedimentary rocks. The high points here are of the White Ridge Quartzite and Sais Quartzite.

North of the Arroyo Manzano crossing (the stream in Cañon Colorado), the road roughly follows the scalloped contact of the Madera Group and the piedmont alluvium. Manzano is at milepost 75. The Quarai ruins of Salinas Pueblo Missions National Monument are located nearby, built where the Abo Formation dips eastward off the Manzano Mountains. North of Manzano, Capilla Peak is the high point to the west at milepost 80, standing 9,368 feet.

Snow enhances the multiple triangular facets visible on the eastern slope of the Manzano Mountains. A triangular facet forms as a fault plane is eroded by multiple streams, the canyons of which leave triangular fault plane remnants standing in high relief.

US 70
LAS CRUCES—ALAMOGORDO
66 miles

North of its intersection with I-25 at the north edge of Las Cruces, US 70 climbs three poorly defined terraces built at prior elevations of the Rio Grande. The younger river has incised into its own ancestral floodplain here. Normally hidden by a thin veneer of gravel, sediments that compose the terraces can be seen in roadcuts about 1 mile east of I-25. They are part of the Santa Fe Group, an extensive sequence of sediments found throughout the Rio Grande rift corridor. Here, the Santa Fe Group consists of old alluvial fans intercalated with river sediments.

The porous sands and gravels of the Santa Fe Group serve as important aquifers for the city's water wells. Due to the sediment's complex interfingering, waters stored within the unit are discontinuous and challenging to map. In places, the Santa Fe sediments contain petrified bones of fossil mammoths, mastodons, horses, camels, and other vertebrates that wandered across this area between 10 million and 500,000 years ago.

Between I-25 and Organ (milepost 162), US 70 crosses the southern end of the Jornada del Muerto, a broad, dry valley that was the preferred route for Spanish travelers going north to Santa Fe. Even though small faulted and intruded ranges separate it from the main channel of the Rio Grande, the Jornada is considered part of the Rio Grande rift. The ancestral Rio Grande flowed through the Jornada del Muerto before shifting to its present channel farther west.

As the road crosses the south end of the Jornada del Muerto, there are good views of the Organ Mountains to the southeast. The range gets its name from craggy, organ-shaped outcrops of 27-million-year-old granite exposed in its steep eastern face. On the mountain, the shape of the intrusion is almost perfectly outlined. Cooling joints,

Organ Mountains viewed from US 70. The light-colored spires at the left are 27-million-year-old granite. The darker mountains at the far right are 36-million-year-old ash flows.

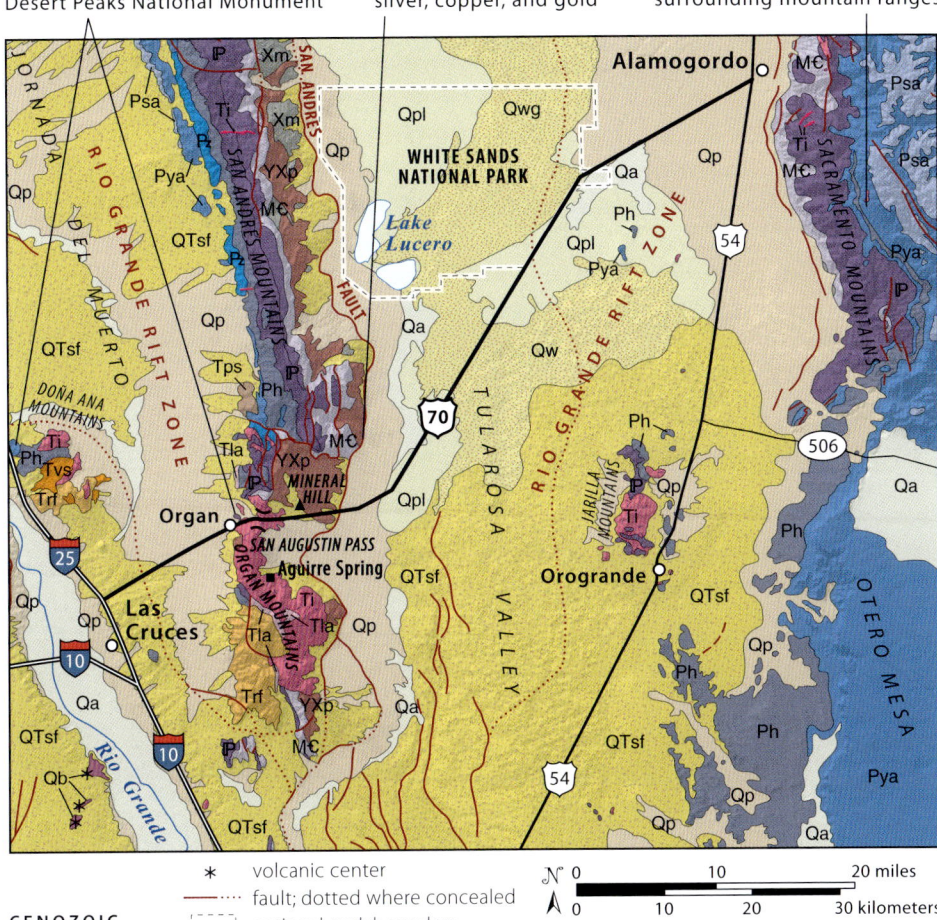

Geology along US 70 between Las Cruces and Alamogordo.

developed as the granite cooled and shrank, were enhanced by movement along rift faults. See more on the Organ Mountains in the guide for I-25: Truth or Consequences—Las Cruces on page 171.

A few slivers of Paleozoic sedimentary rock are exposed on the western side of the mountains far south of the highway, and a fluorspar mine exists near their line of contact with the granitic intrusion. Fluorspar is the principal source of fluorine and is used as a flux in smelters, in the preparation of glass and enamel, and in manufacturing hydrofluoric acid and fluorocarbons. Fluorspar is known to occur in association with calderas. Two such collapsed volcanoes have been identified here, ringed with faults and associated with Oligocene intrusive and volcanic rocks of the Organ and Doña Ana Mountains.

Old mines near Organ (milepost 162) are situated along a fault where quartz veins cut through Proterozoic rocks. The mines produced copper, lead, silver, gold, and zinc.

As the highway climbs more steeply, look north into the San Andres Mountains to see their west-dipping bands of Pennsylvanian and Permian rock layers. In the foreground, the west-dipping pattern is somewhat jumbled, but farther north it becomes quite systematic. There are 4,700 feet of Paleozoic rocks here; the top 3,000 feet of them are Pennsylvanian and Permian. Almost all the formations are marine,

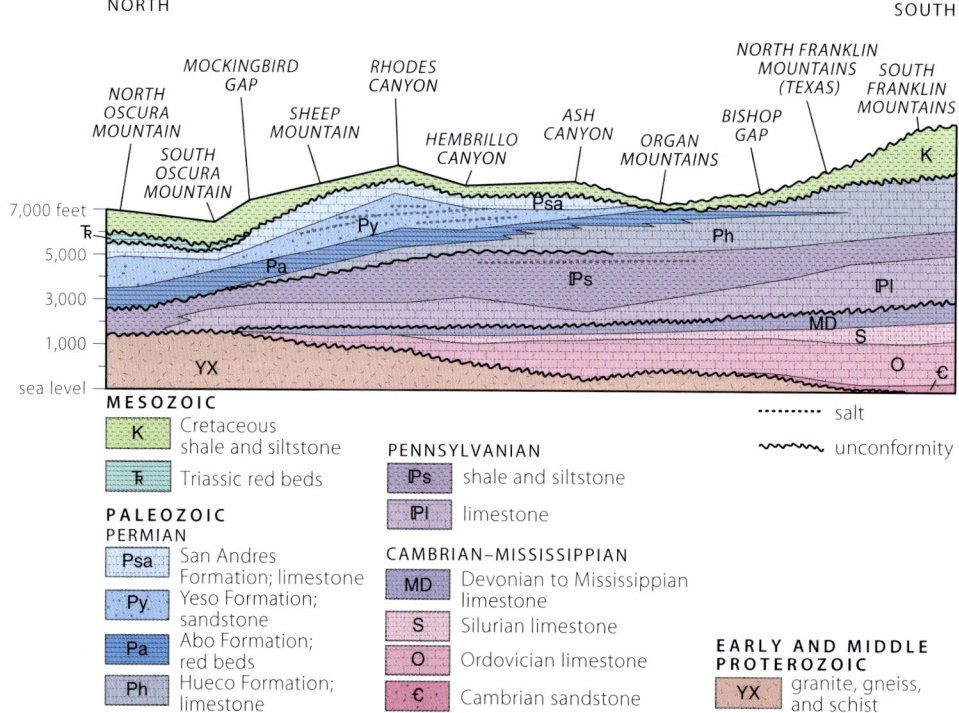

By diagramming rocks identified in successive canyons of the Oscura, San Andres, Organ, and Franklin Mountains, geologists have clarified relationships between Paleozoic and Mesozoic formations. Wavy lines show unconformities. Limestones older than Permian age thicken southward in the direction of the open seas at that time. Permian and Triassic rock layers thicken and coarsen northward, in the direction of the newly risen Ancestral Rocky Mountains.

deposited in shallow seas that advanced over a nearly level land surface. Some of the Paleozoic rocks were quarried near milepost 160 for crushed stone used in the flood-control revetments in Las Cruces.

At San Augustin Pass (5,725 feet; milepost 164), the highway cuts through the Oligocene intrusion of the Organ Mountains. Roadcuts offer a good chance for a close look at the unweathered granite. Elsewhere, the densely jointed rock is iron stained from weathering of pyrite grains within it. The mineral pyrite, a brassy yellow iron sulfide, is sometimes called fool's gold.

A roadcut through Oligocene granite at San Augustin Pass.

The San Augustin Pass Viewpoint at the pass looks into the Tularosa Valley, an enclosed desert basin with no external drainage. White Sands National Park is also visible from here. Any water that falls within the boundaries of the Tularosa Valley stays within the basin. Rain and snow that fall onto surrounding limestone-cored mountain ranges carry dissolved gypsum into the Tularosa Valley, where it pools in ephemeral playa lakes. As the lake water evaporates, gypsum precipitates in white crusts. The white sand dunes are composed of the gypsum that comes from the floor of this playa lake, which existed as a much larger lake during the last ice age. At the southern end of the playa, Lake Lucero holds moisture after heavy rains even today.

North from milepost 167, Proterozoic-age basement rock is exposed in Mineral Hill. Mineral Hill is considered to be the location of the mythical Lost Padre Gold Mine, worked in the late 1790s by a French priest. At the time, its location was described as "two days north of El Paso del Norte, east of El Camino Real, in a lone mountain surrounded by a large basin, with a spring nearby."

At the base of the mountains proper, the highway crosses the alluvial apron that surrounds them. Stream gravel and sand and the finer sediments of a succession of lakes, as well as layers of salt and gypsum, fill the fault-edged valley to a depth of

AGUIRRE SPRING RECREATION AREA

At milepost 165 on US 70 is the turnoff to Aguirre Spring Recreation Area on the northeastern slope of the Organ Mountains. Aguirre Springs Road climbs an extensive alluvial fan toward an abrupt topographic change at the mountain front, where vertical, gray-white cliffs expose monzonite, a silicic intrusion similar to granite. Aguirre Spring emerges from and cascades over the monzonite of the Organ Mountains. A few ephemeral streams are present within the Aguirre Spring area. The high points of the range, known as the Needles, tower more than 1,500 feet above the Aguirre Springs trailhead. The Pine Creek Loop Trail, which starts at the popular Aguirre Spring Campground, affords excellent viewing of the monzonites.

Looking northeast from Aguirre Spring toward the southern end of the San Andres Mountains. Note the evenly sloping alluvial fans that drop down toward the Tularosa Valley at far right.

At Aguirre Spring, water flows along joint fractures in the Oligocene granite of the Organ Mountains. Close-up shows the square feldspar crystals in the granite.

about 4,000 feet. Soils on the floor of the valley are high in salt and gypsum, which discourages the growth of all but salt-tolerant plants, such as the creosote bushes that dominate local vegetation.

As the highway descends into the Tularosa Valley, large blocks of Paleozoic rock are visible to the north, their eastward dip contrasting with the predominant westward dip of rocks in the San Andres Mountains. At milepost 168, the highway crosses east onto Proterozoic bedrock. The 1.45- to 1.35-billion-year-old plutonic granite exposed largely to the north of the highway underlies the Paleozoic rock layers and is exposed here along the fault that edges the eastern side of the San Andres Mountains. The ancient granite is much darker than the younger granite of the Organ Mountains.

Across the valley, the Sacramento Mountains also contain Paleozoic sedimentary rocks underlain by Proterozoic granite. There, the rock layers dip east. Along with the San Andres Mountains, the Sacramento Mountains form an immense anticline, with the Tularosa Valley lying where the fold's crest should be. The crest dropped more than 7,000 feet between bordering and central valley faults. In the process, it produced the mountains on either side: two parallel, asymmetrical half grabens. The down-dropping occurred between 17 and 8 million years ago due to extension in the Earth's crust related to the formation of the Rio Grande rift.

On the west wall of the Sacramento Mountains, sliced-off edges of Paleozoic sedimentary rocks lie above Proterozoic granite. The sedimentary rocks range in age from Cambrian to Permian; the highest exposure is Mississippian limestone. Geologists studying these rocks along the cut faces of the San Andres Mountains have found that Cambrian, Ordovician, Silurian, Devonian, and Mississippian rock layers thicken southward, while Pennsylvanian and Permian layers thicken northward, reflecting the geometries of the ancient basins in which they were deposited.

Variations in rock types are also well exposed here. The Paleozoic layers were deposited in a succession of seas that swept across a nearly flat coastal plain, a region much like the Gulf Coast today. As along modern shores, sandy beaches gave way to muddy bays, and muddy bays to gravelly river estuaries. Conglomerate and sandstone were deposited near the shore, whereas shells of marine animals accumulated offshore to become limestone. Unconformities developed when the marine sediments

Layers of Paleozoic sedimentary rocks form the Sacramento Mountains that rise above Alamogordo.

were raised above sea level for a time and eroded. Thus, geologists can reconstruct the Paleozoic history of the area, telling when seas came and went, which direction they came from, where bays indented the shoreline or river deltas extended it, even the direction of far-off mountains.

Small buttes east of milepost 192 are lone outcrops of Paleozoic limestone, lifted along faults and jutting up through the valley fill. Gypsum dunes appear along the roadside between mileposts 195 and 199. Some of these dunes are relatively stable and have developed growths of yucca, saltbush, and other plants. Others move too actively for vegetation to become established. Between mileposts 199 and 200 is the turnoff to White Sands National Park to the west of the highway.

WHITE SANDS NATIONAL PARK

In the Tularosa Valley, between the San Andres and Sacramento Mountains, three factors necessary for development of dunes come together: a source of sand, plenty of wind, and a place where the wind is forced to release whatever sand it carries. Here, a unique fourth factor serves to create world-class dunes: gypsum. Most of the world's dunes are composed of silica (quartz) sand, but those at White Sands form the largest gypsum dune field in the world. Gypsum is a soft mineral (softer than a fingernail) and easily breaks down when wind, rain, and freeze/thaw events attack it.

To forestall its own burial, a yucca growing on a dune constantly lengthens its roots.

Small playas edge some of the dunes. When the water table is high enough, they fill with water, providing a unique environment for the animal species that inhabit them and the areas nearby.

To create sand dunes, wind must blow 15 miles per hour or more—as it does here in February, March, and April. These spring winds blow from the southwest, so the sand source must be to the southwest of the dunes. A broad white playa, the bed of ancient Lake Otero, lies southwest of the present dunes in one of the deepest parts of the Tularosa Valley.

The Tularosa is an internally drained basin, meaning that any water that falls within its boundaries stays within the basin. Rain and snow that fall onto surrounding limestone-cored mountain ranges carry dissolved gypsum into the Tularosa Valley, where it pools in ephemeral playa lakes. As the lake water evaporates, gypsum precipitates in white crusts. Most of the gypsum that ends up in dunes comes from the floor of this playa lake, which existed as a much larger lake during the last ice age. At the southern end of the playa, Lake Lucero holds moisture after heavy rains even today.

In silty lake deposits that fringe Lake Lucero, large, clear, daggerlike selenite crystals seem to grow right out of the soil. These crystals, cracked and shattered by desert temperature changes and blasted by wind, break down into sand-size particles that the wind can pick up and bounce across the lake flats to the dunes.

The ultimate source of the gypsum is in Permian rocks of the surrounding mountains, whose striped scarp marks the sides of the Tularosa Valley. The San Andres Formation was deposited about 282 to 272 million years ago in a constricted arm of a Permian sea. At times, the seawater evaporated, leaving behind thick deposits of salt, gypsum, and other soluble minerals. Up to 1,300 feet thick, the San Andres Formation consists of limestone, dolostone, and gypsum (or anhydrite, the nonhydrated form of the calcium sulfate mineral).

The collapse of the rift basin and the corresponding uplift of the mountains exposed Permian rocks to erosion and, in the case of the evaporite minerals, to solution. Some of the gypsum dissolved from Permian rocks was redeposited in lake sediments of the closed basin, and some remained in the groundwater, making the local groundwater very "hard" and mineral rich. From both lake deposits and selenite crystals, particles of sand ride the wind to become part of the White Sands dune field.

Dunes have been classified by shape into several easily recognized types, four of which occur at White Sands. Where winds are strongest, on the southwest side of the dune field, low dome dunes occur. These are the closest dunes to Lake Lucero and move up to 30 feet per year. In the center of the dune field, crescent-shaped barchan dunes develop. Crescent barchan dunes have an arced, gentle face into the wind and appear convex with the two horns or crescent ends downwind and leading the way as the dune migrates. Long ridges of transverse dunes occur in parallel lines in areas of ample sand and are characterized by a long, wavy top ridge that can form by the

The dune sand is composed of gypsum. Feet for scale.

Cross-bedding within a sand dune.

Tracks of beetles and other animals and insects are soon covered by new windblown sand.

alignment of several barchan dunes in a row. At the margins of the dune field and on the sand sheet, parabolic dunes form, their long, trailing arms caught and anchored by desert vegetation. These dunes migrate slowly toward the center of the dune field due to vegetation.

White Sands National Park is home to the oldest known human footprints in North America. A study published in 2021 documents the footprints and provides a preliminary date of 23,000 to 21,000 years ago. The human habitation occurred along the shores of Lake Otero, a body of water that occupied the Tularosa Valley during the cool climate of the Pleistocene. The footprints were likely made during an abrupt and short-lived warming event that lowered lake levels and exposed muddy shorelines, which were preserved for the geologic record. The footprints were buried in gypsum for centuries before erosion recently uncovered them. These human footprints add to the already established tracks of saber-toothed cats, mammoth, dire wolf, and other known ice age animals. They show a longer coexistence of ancient humans and ice age megafauna than previously hypothesized.

US 380
Carrizozo—San Antonio (I-25)
64 miles

Carrizozo's name is derived from an abundant reed grass that grows in the surrounding desert grassland. The town was established in 1899 as a stop on the expanding El Paso & Northeastern Railroad and grew as mining and ranching became established in the Tularosa Valley. Today, the town is a tourist attraction for those looking to find a taste of the Old West. West of Carrizozo, US 380 almost immediately comes to rugged, dark basalt lava flows known as the Valley of Fires, or Little Black Peak, lava flows, best seen from Valley of Fires Recreation Area on US 380 west of Carrizozo. Ridges of older sedimentary rock jut through the lava flow and are visible rising above the

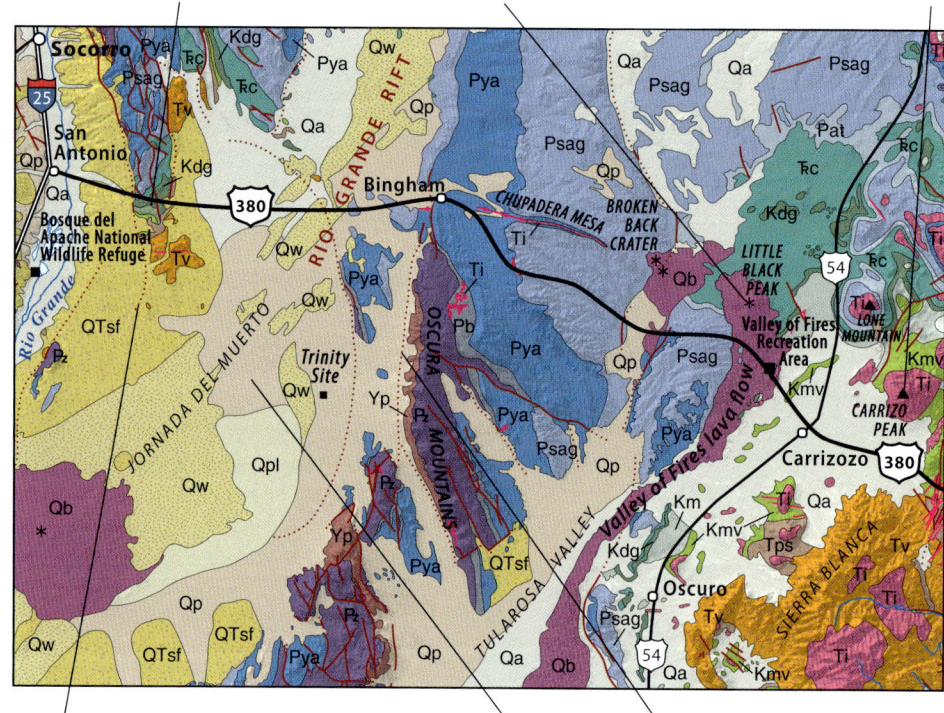

coal from the Carthage coal field supplied local needs and served as smelter fuel in Socorro, El Paso, and northern New Mexico

The Valley of Fires lava flow came from a small volcano called Litttle Black Peak

ringed with upturned sedimentary rocks, Carrizo Peak is a laccolith

some of the volcanic rocks in this area are derived from the Pleistocene eruption that created the Valles Caldera, over 150 miles to the north, west of Santa Fe

the Rio Grande formerly flowed through the Jornada del Muerto, which is part of the Rio Grande rift

a sharp ridge of Proterozoic and Pennsylvanian rocks marks a fault along the west side of the Oscura Mountains

CENOZOIC
QUATERNARY
- Qa — alluvial deposits
- Qpl — playa lake deposits
- Qp — piedmont alluvial deposits
- Qw — windblown deposits

QUATERNARY–PALEOGENE
- QTsf — Santa Fe Group, undivided (latest Oligocene to middle Pleistocene)

PALEOGENE
- Tps — older sedimentary units

MESOZOIC
CRETACEOUS
- Kmv — Mesaverde Group
- Km — Mancos Shale
- Kdg — Dakota Sandstone and Early Cretaceous rocks

TRIASSIC
- ₮c — Chinle Group and Moenkopi Formation

PALEOZOIC
PERMIAN
- Pat — Artesia Group
- Psag — San Andres Formation and Glorieta Sandstone
- Pya — Yeso and Abo Formations
- Pb — Bursum Formation

PENNSYLVANIAN–CAMBRIAN
- Pz — older Paleozoic sedimentary rocks, undivided (Pennsylvanian and Cambrian to Mississippian)

MIDDLE PROTEROZOIC
- Yp — plutonic rocks, undifferentiated

VOLCANIC AND IGNEOUS ROCKS
- Qb — basalt (Quaternary)
- Tv — volcanic rocks, undifferentiated (late Eocene to early Miocene)
- Ti — intrusive rocks, undifferentiated; includes dikes (late Eocene to Pliocene)

* volcanic center
— fault; dotted where concealed

Geology along US 380 between Carrizozo and San Antonio on I-25.

valley alluvium. One of them, a hogback of Dakota Sandstone, forms the Valley of Fires National Recreation Area campsite halfway between mileposts 62 and 61.

About 44 miles long, only a few miles wide, and 5,000 years old, the Valley of Fires lava is composed of several long, thin, overlapping flows. The lava is between 30 and 45 feet thick and covers an area of approximately 127 square miles. The surface of the young lava exhibits many of its original features. A rumpled, ropy, shiny surface formed from lava that flowed with the consistency of taffy. Caves and tubes were created as molten lava flowed out from beneath the solidified crust. Some lava tunnels extend for hundreds of feet. Pressure ridges formed where thickening lava pushed up through breaks in hardened crust. Domes, blisters, and tiny bubblelike vesicles formed where gases, inherent in the fluid basalt, existed. Many deep fissures resulted from the shrinkage of the basalt as it cooled. Little Black Peak, the small volcano from which the lava erupted, can be seen to the northwest.

These olivine basalt flows lie along the Capitan lineament, a line of igneous features that exists due to a zone of crustal weakness along which magmas have risen. The line also includes the Capitan pluton and Broken Back Crater (west of Little Black Peak).

Though the lava has not yet decomposed into soil, desert plants grow in pockets of sand and silt brought in by the wind. Small desert animals, many of them wearing darker-than-normal camouflage in adaptation to their dark surroundings, live among the fissured rocks and pockets of vegetation. In White Sands National Park, only 70 miles farther south, the same animals are whiter than normal.

At the western edge of the lava flow (milepost 58) the highway climbs onto light-gray limestone of the Permian San Andres Formation, on which it remains until about milepost 38, except for dry stream valleys filled with recent gravels. The San Andres Formation was deposited about 272 to 260 million years ago in a constricted, Permian-sea arm. During dry times, the seawater evaporated and left behind thick deposits of salt, gypsum, and other soluble minerals.

A small, dark cave leads to a lava tunnel where fluid lava flowed out from under a cooling crust. Note the ropy pahoehoe lava rock along its right side.

Following fractures in the Abo Formation, mineral-rich fluids deposited gypsum veins in a roadcut at about milepost 32.

In the desert climate, groundwater saturated with calcium carbonate from the limestone is drawn toward the surface, where it evaporates, depositing its calcium carbonate in the form of caliche, the whitish, chalklike material visible in shallow roadcuts. Between mileposts 33 and 32, US 380 cuts through several red roadcuts of the Permian Abo Formation as the highway curves northward around the Oscura Mountains. Unlike most other Paleozoic formations in New Mexico, the Abo is continental, having been deposited on land rather than in the sea. The climate on Pangaea during the deposition of the Abo was tropical, with seasonal monsoons. Extensive muddy floodplains were subjected to periods of dryness, during which mud cracks formed. Trace fossils—tracks of numerous species including freshwater amphibians and lungfish—can be found in the deposits, along with fossils of conifers and freshwater shells.

The Oscura Mountains separate the Tularosa Valley from the Jornada del Muerto, another extensional basin. Not very high as mountains go, the range is an east-tilted fault block of Proterozoic, Pennsylvanian, and Permian sedimentary rocks that are lifted by faults along its western flank. The fault passes under the town of Bingham, where alluvial fans and sediment bury it. The abrupt edge of the mountains, the fault scarp, can be seen to the south from just west of that community near milepost 29. A smaller fault block, also with east-tilted rock layers, lies farther southwest of Bingham.

Between mileposts 20 and 11, US 380 crosses the northern end of the Jornada del Muerto, the dry, desolate rift basin between Las Cruces to the south and Socorro to the north. The basin, formed during rifting, is a graben. The bedrock buried beneath its surface tilts to the west, though the valley floor today is largely flat. A gravelly veneer of stream and alluvial deposits covers Santa Fe Group sediments, which fill the basin to a depth of more than 3,000 feet. The Rio Grande flowed through the valley a few million years ago, but its course shifted to the modern, more western route in

TRINITY SITE NATIONAL HISTORIC LANDMARK

The world's first atomic weapon, known as Gadget, was detonated in a test at 5:30 on the morning of July 16, 1945, at the Trinity Site on the White Sands Missile Range, about 10 miles directly south of the highway at milepost 22. This was a final test of the work of the Manhattan Project during World War II. The light flash of the test was visible at least 160 miles away, and the cloud surge reached 38,000 feet into the atmosphere in less than 7 minutes. The drop tower, a 100-foot-high steel tower, was instantly replaced by a crater 8 feet deep and a half mile in diameter. Trinitite, a deep-green glass unknown to science before being discovered in the crater, is composed of sand fused by the energy of the nuclear explosion. Today, a monument exists at the location of the test, and Trinity Site was designated a National Historic Landmark in 1965. Public tours are held biannually. In Pleistocene time, the Trinity Site was covered by a shallow lake, and it is now a seasonally wet playa.

---

the early Pleistocene. With no exit route, the basin flooded, creating a shallow playa lake known as Lake Jornada approximately 1 to 0.3 million years ago. It was 5 to 45 feet deep and stretched about 25 miles north to south. It filled with runoff from the bordering limestone-rich mountains; when it periodically evaporated, it left behind gypsum-rich evaporites. Selenite crystals can be found in these deposits. The Santa Fe Group sediments within the valley are a discontinuous, highly faulted series of units. Groundwater levels associated with the ancestral Rio Grande that once supported the lake dropped after the Rio Grande migrated west and entrenched down into its own valley. Today, groundwater is found hundreds of feet below the valley floor.

Pinkish-brown sand dunes along US 380, partly anchored by vegetation, derive their sand and color from the Abo Formation. Stop and pick up a handful of this sand and notice the uniform size and rounded shape of the quartz grains, characteristic of dune sand everywhere. Wind picks up and transports sand, and where the topography changes and the wind slows down, it drops its load of sand, forming dunes. Here, winds from the west slow when they encounter the flanks of the Oscura Mountains, and dunes form on the western side of the range.

Distant mountains on the western side of the Rio Grande Valley are part of the Mogollon-Datil volcanic highlands that extend west into Arizona and south almost to I-10. The highway descends toward the Rio Grande on Cenozoic sedimentary rocks composed of river deposits interbedded with volcanic ash from eruptions near and southwest of Socorro. At milepost 8, Cretaceous-age Dakota Sandstone and Mancos Shale are exposed in a narrow swath of hills. The highway then returns to the sands and gravels of the Santa Fe Group at milepost 7. The highway steps down river terraces and crosses the Rio Grande at milepost 2. The Rio Grande floodplain is dotted with trees on the east and divided into agricultural fields on the west. To the south at milepost 1, the Bosque del Apache is a nationally renowned wildlife refuge famous for its critical wildlife migration habitat. On the west side of the river, here entrenched in Nogal Canyon, US 380 ascends a massive alluvial fan that forms the eastern toes of Chupadera Mountain to the west. This part of the Rio Grande rift is still geologically active; occasional tremors can be felt or seen on seismometers.

An abandoned church sits on New Mexico's plains in Taiban, east of Fort Sumner.

THE PEACEFUL PLAINS

The plains of New Mexico extend from the Oklahoma and Texas borders westward to the ranges that edge the Rio Grande rift. Low bluffs and shallow river valleys interrupt the broad plains and rolling hills here. This area marks the transition onto the North American Craton, the stable interior of the continent. In this part of the southern Great Plains, the geology is far simpler than that of the Rio Grande rift and the highlands of western New Mexico.

Here, most sedimentary rocks lie flat, with very little in the way of geologic structures: a few faults, a few folds, and some very gentle warping of the sedimentary layers. Most of the relief is a product of erosion by the Pecos and Canadian Rivers and their tributaries and of collapse over caverns dissolved in limestone, salt, and gypsum. Mountains in this region, where they occur, are nearly all volcanic. Well to the south, an unusual 250-million-year-old limestone reef lends a change of scene as it angles southward into Texas.

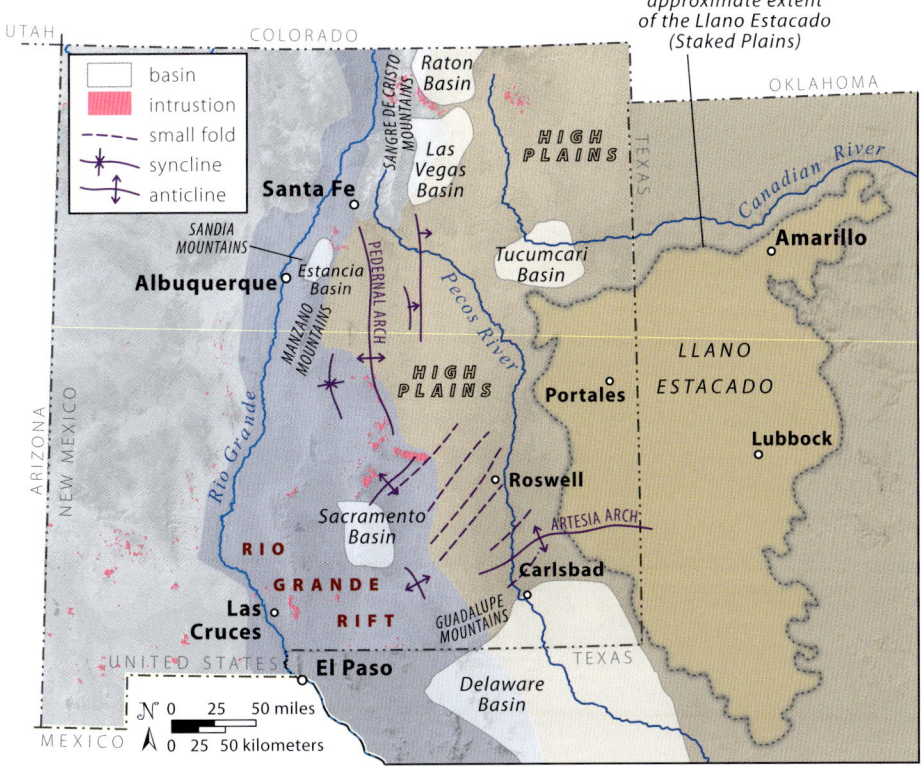

The plains of eastern New Mexico slope toward Texas and Oklahoma, rising along the well-defined escarpment that demarcates the Llano Estacado, or Staked Plains.

Because it resists erosion, the limestone of the Permian San Andres Formation provides a surface for much of the region, with more surface exposure than any other rock unit in New Mexico. The San Andres was deposited in the sea on a stable carbonate shelf. During the Permian, several pulses of tectonic movement occurred. The associated uplifting and downdropping are recorded by sedimentation sequences within the San Andres Formation that show rising and falling sea level. Evaporites were more common during regressive cycles, when sea levels fell, leaving lagoons and embayments isolated from full ocean flooding. As the mineral-rich ocean water dried up, it left behind thick deposits of evaporites. Limestone filled with marine fossils such as bryozoans, corals, and crinoids is evidence of a normal marine setting, with good ocean circulation when sea levels rose during transgressive cycles.

The San Andres Formation is riddled with solution passages and caverns. Rock units beneath it, notably the Yeso Formation, contain lots of gypsum and once included thick beds of salt. Because both salt and gypsum are even more soluble than limestone, sinkholes result from solution of these minerals. Topography that developed because of solution of underlying rocks—whether limestone, gypsum, or salt—is called karst, named for a cavernous limestone region in Slovenia. Karst topography dominates the landscape of eastern New Mexico, even when younger alluvium and sedimentary rocks cover the Permian rocks in which it developed.

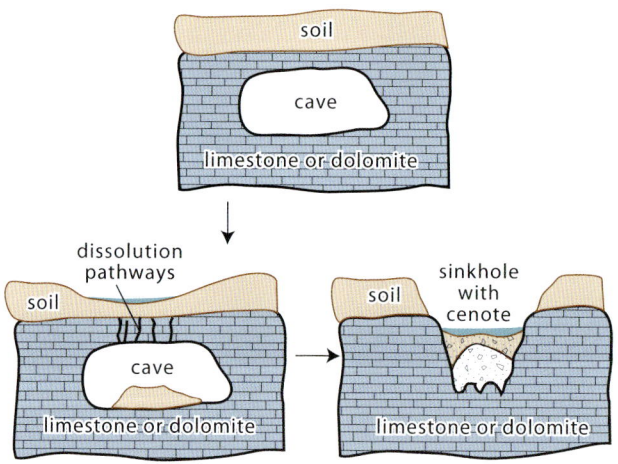

Sinkholes form when salt, gypsum, or limestone dissolve in one layer and the overlying layers cave in. A cenote is a lake that developes above a sinkhole.

Other prominent units are red Triassic sandstones and siltstones, mostly of the Chinle Group, and the Cretaceous Dakota Sandstone and Mancos Shale, all of which are well exposed in the northeastern part of the state. In places, the Great Plains are capped with patches of the Miocene to Pliocene Ogallala Formation, a widespread gravel unit that continues east and north far beyond New Mexico's borders.

The Ogallala Formation, a unit of overlapping alluvial fans and stream gravels, was deposited when sediment was eroded and washed down the eastern slopes of the southern Rocky Mountains and the eastern sides of rift-bounding ranges to the south. The Ogallala deposits span from the edge of the Rocky Mountains in Colorado

NORTHEASTERN PLAINS STRATIGRAPHIC COLUMN

ERA	PERIOD (mya)*	Rock Unit	Rock Color/Composition	Depositional Environment
CENOZOIC	QUATERNARY — HOLOCENE — 0.01	recent sediment		soil-covered sediments deposited in basins
	PLEISTOCENE — 2.6	Raton-Clayton-Capulin volcanics		gray to black basalts, cinder cones, and other volcanics
	TERTIARY — NEOGENE — 23	Ogallala Formation		soil-covered plains with stream deposits
	PALEOGENE — 66	Raton Formation		
MESOZOIC	CRETACEOUS	Vermejo Formation / Trinidad Sandstone		alluvial stream deposits
		Pierre Shale / Niobrara Formation		shallow marine with millions of coccoliths (small phytoplankton)
		Carlile Shale / Greenhorn Limestone / Graneros Shale		near shore marine shale / marine limestone / near shore marine shale
	— 145	Dakota Group / Glencairn Formation / Lytle Sandstone		rivers, beaches, lagoons
	JURASSIC	Morrison Formation		broad floodplain and tidal flats with rivers, streams, lakes, and sand dunes
		Bell Ranch Formation		
	— 201	Exeter Sandstone (eastern extension of Entrada Sandstone)		sand dunes; streams washing dune sand
	TRIASSIC	Dockum Group		continental streams, floodplains, with occasional volcanic ash
	— 252	Artesia Group Grayburg Formation / San Andres Formation		
	PERMIAN	Glorieta Sandstone		coastal sand, mostly dunes
		Sangre de Cristo Formation		alluvial fans sloping down to basins from Ancestral Rocky Mountains uplifts
PALEOZOIC	— 299	Alamitos Formation		
	PENNSYLVANIAN	Madera Group Porvenir Formation		shallow water bottom mud interfingering with nearshore arkosic
		Sandia Formation		complex setting with nearshore sand and mud and deeper marine bottom mud; pulses of mountain building result in pulses of arkos sand and pebbles
	— 323 MISSISSIPPIAN — 359 — 541	Arroyo Peñasco Group		marine limey mud
PROTEROZOIC		basement		metasediments, metavolcanics

*mya = millions of years ago

Legend: river sandstone, dune sandstone, mudstone or shale, siltstone, limestone, coal, salt/evaporite, conglomerate, unconformity

Important units in the eastern plains.

and New Mexico eastward as far as Nebraska and South Dakota, and their ages vary due to differences in erosion. The Ogallala used to form a single intact surface sloping eastward from the mountains to the plains, but now streams have dissected it closer to the mountains. The cemented remnants of the exposed unit along New Mexico's eastern border form a highland known as the Llano Estacado, or Staked Plains.

The Ogallala plays a critical role in agriculture because it is the main water-bearing unit in the southern Great Plains. The porosity, or amount of holes, that exists in the spaces between the coarse gravels of the unit allows for the storage of water. Permeability, the interconnectivity of these holes, allows the water to flow within the unit. The Ogallala is both porous and permeable and has collected water for millions of years.

Streams flowing east off the rising Rocky Mountains deposited the Ogallala Formation in Miocene and Pliocene time.

The prehistoric groundwater contained within the Ogallala is pumped for agricultural, residential, and industrial use. Removal of water outpaces recharge, however, and the aquifer is gradually running dry. Scientists have dated the age of the water in the Ogallala aquifer using radiocarbon and tritium age dating. Many water samples show that while there is a small amount of modern recharge, the vast majority of these waters filled the aquifer between 15,600 and 1,800 years ago. As human usage pulls out more and more water from the upper parts of the Ogallala unit, wells must be deepened—an expensive operation—to continue producing water. All across this eastern edge of New Mexico, dead and dying trees that nearby windmills once watered shade small, deserted farms.

Several of the valleys that lie close to the rift-bordering ranges held lakes in Pleistocene time, notably Lake Estancia east of the Sandia Mountains. The lakes left behind fine, silty lake deposits, wave-cut shorelines, and, in many cases, dune ridges along their eastern edges. When full, Lake Estancia may have drained eastward to the Pecos River. When not at maximum depth, the waters may have been saline, as were the playa lakes of many nondraining, enclosed basins of southern New Mexico.

Dotted over the northern part of this region are at least 125 small volcanoes of the Raton-Clayton volcanic field. The eruptive features include cinder cones, silicic volcanoes, and small lava domes and flows. The field covers 7,500 square miles of northeastern New Mexico and adjacent areas in Colorado and Oklahoma. It is the easternmost Cenozoic volcanic field in the United States. The volcanoes erupted in three phases: the olivine basalts of the Raton phase, from 9 million to 3.6 million years ago; the Clayton phase, from 3.6 to 2.6 million years ago; and olivine basalts and andesites of the youngest Capulin phase, from 2 million to 50,000 years ago. The older volcanics are deeply eroded, and these basalt flows cap many of the visible mesas. Northeast of the Canadian River, broad lava flows cover and protect part of the old Ogallala surface. North of Raton, remnants of older flows form high ramparts along the Colorado border. These flows are an excellent example of inverted topography: the basalts here once flowed into and along the lowest parts of the landscape. As the region rose and canyons eroded through the deposits, the resistant basalt formed what are now the high points.

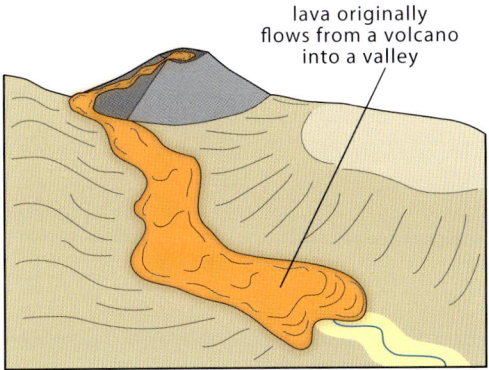

lava originally flows from a volcano into a valley

Lava is more durable than most types of sedimentary rock. When it flows down a valley, hardening there, bordering ridges may wear away, leaving a new ridge to mark the position of the former valley. Topography is "inverted," so that the former low point in the landscape (the streambed) is now the high point (the mesa).

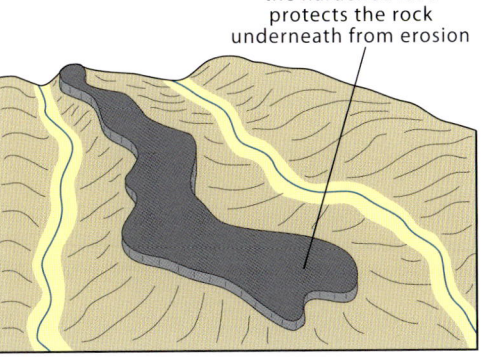

the hardened lava protects the rock underneath from erosion

The Raton-Clayton volcanic field lies at the northeast end of the Jemez lineament, a deep subduction zone-scar separating the basement rock of the 1.8-billion-year-old Yavapai and 1.7-billion-year-old Mazatzal Provinces. Magma exploited the weakness in the Earth's crust, so the Jemez lineament is manifested at the surface by a chain of Miocene to recent volcanic fields extending from the White Mountains of Arizona northeast to Raton, New Mexico.

The Canadian and Pecos Rivers are the two main rivers in the plains of eastern New Mexico. The Canadian River drains more than 17,000 square miles, about 14 percent of the land surface in New Mexico. From its headwaters in the southern Rockies near the Colorado–New Mexico border at elevations of over 12,000 feet to near the eastern New Mexico–Texas border, the river drops to about 3,600 feet. The Pecos River begins in the Sangre de Cristo Mountains near Santa Fe and flows generally south, paralleling the Rio Grande, which is several mountains ranges to the west.

One of the interesting aspects of the Canadian River, and in places of the Pecos River as well, is that its course has been, to some extent, governed by underground solution of salt and gypsum in the Permian rocks. Flowing almost due south from Raton to Conchas Lake, the Canadian turns abruptly east, following the former margin of a Permian sea in which, as it evaporated, gypsum and salt were deposited. Near Santa Rosa, the Pecos River twists through a curving line of collapsed caverns, some of them now drowned by Santa Rosa Lake. The effectiveness of solution in creating collapse structures is well demonstrated by the many sinkholes that dot the region around Santa Rosa and other areas where Permian rocks are at or near the surface.

Several archeological sites exist in this part of New Mexico. Ancient campsites and spear points and other artifacts found in association with bones of now extinct Pleistocene bison and mammoths in late Pleistocene deposits and in caves indicate hunter-gatherers roamed these plains at least as early as 13,200 years ago, right at the tail end of the last ice age. More recent sites include villages built in the twelfth to fourteenth centuries by ancestors of today's Puebloan peoples, groups with complex cultures based on agricultural crops of corn, beans, and squash. Many of these villages were thriving when the Spaniards arrived, and some Spanish churches and convents were located near these occupied settlements.

Several of the highway descriptions in this section begin or end within the Rio Grande rift, not because the eastern plains extend that far, but because population centers in New Mexico are concentrated along the gentle waters of the Rio Grande and make convenient starting and stopping points.

I-25

Colorado—Raton—Las Vegas
117 miles

At the summit of Raton Pass, I-25 enters New Mexico from Colorado. On one of the main historic routes from the north and east, Native Americans, Spanish explorers and settlers, Union soldiers, and emigrants on the Bents Fort branch of the Santa Fe Trail used the pass. In 1879, a railroad was completed across the pass, and in 1922 the first highway—not much more than a narrow dirt track—was built.

On the south side of the pass, look to the east to see Bartlett Mesa, where a thick, resistant cap of olivine basalt protects layers of Cretaceous and Cenozoic sedimentary rocks. Most of these lavas came from volcanic centers near La Junta, Colorado, between 9 and 3.6 million years ago and spread in horizontal sheets well into New Mexico. Erosion has gradually whittled them down, and the cap on this high mesa is the largest remnant. Columnar cooling joints are visible in some of the exposed lava cliffs.

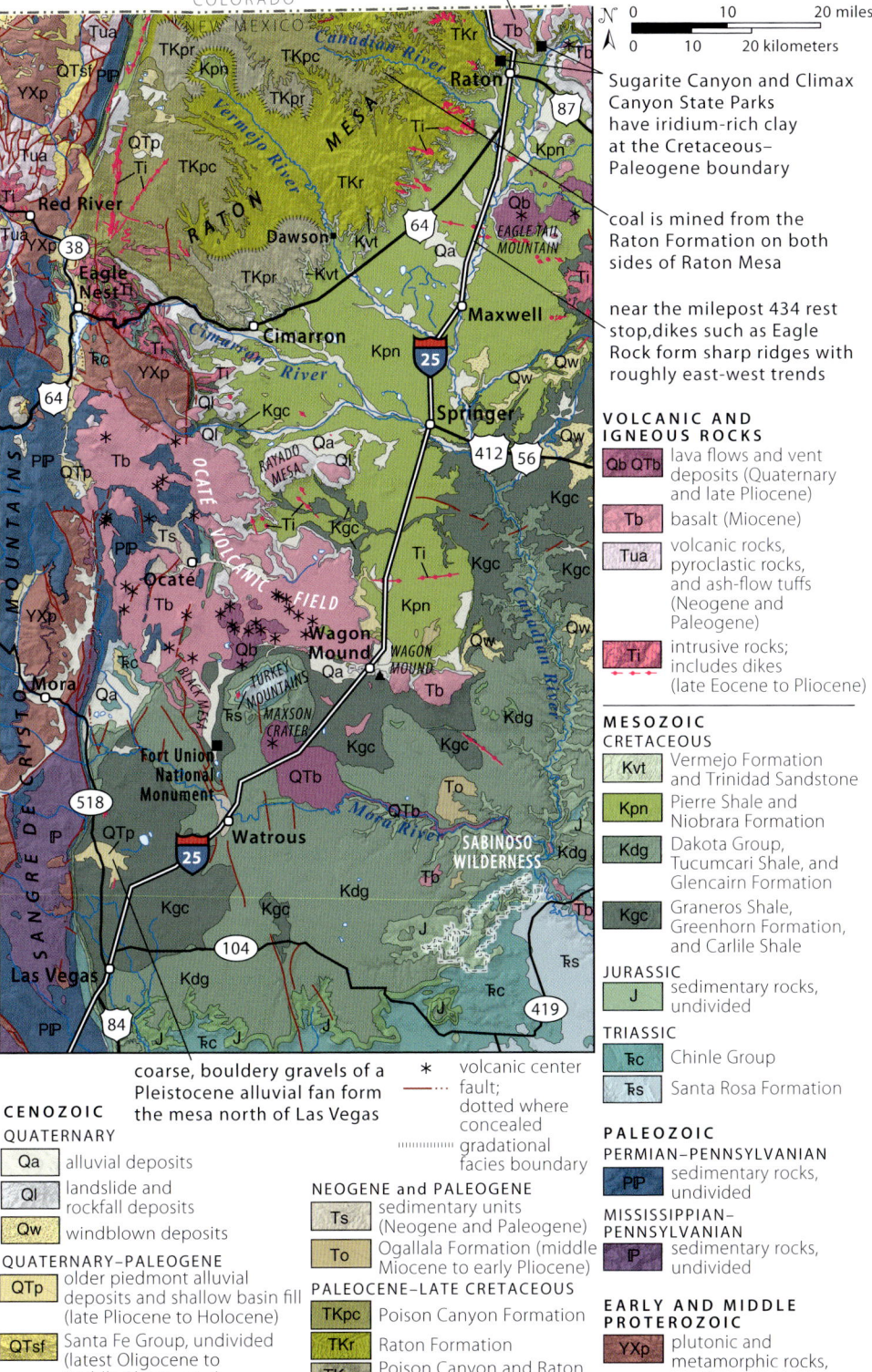

Geology along I-25 between the Colorado border and Las Vegas.

View south from Raton Pass at milepost 460. Bartlett Mesa lies at the left. On the skyline in the distance is Tinaja Mountain (left) and Eagle Tail Mountain (right). —Photo by Alex Nereson

Sedimentary rocks that underlie the basalt appear in highway cuts along I-25. At milepost 457, look on the east side of the highway for light-brown sandstone, gray shale, and seams of coal, part of the Raton Formation. Coal from the Raton and the underlying Vermejo Formation were historically mined, and the last coal mine in the area, the York Canyon Mine along the Vermejo River about 30 miles west of Raton, closed in 2003.

The coal was deposited within the Raton Basin, which formed during the Laramide orogeny. While the Rocky Mountains are the spectacular product of this continental-scale mountain building, the basins that formed as crust downwarped and buckled are equally important, especially in terms of understanding sedimentary histories. When geologists speak of "basins," they are referring not to surface features but to the shape of the rock layers beneath the surface. In the Raton Basin, sedimentary rocks bow downward between the Sangre de Cristo Mountains to the west and the Sierra Grande arch, an anticline, to the east. Cretaceous sedimentary rocks that have been worn off both mountains and the arch dominate the units within the basin.

The Raton Basin is floored deep below the surface with Proterozoic basement rock overlain by the Jurassic-age Morrison Formation and the Cretaceous-age Dakota Sandstone, Benton Shale, Niobrara Formation, and Pierre Shale. The Pierre Shale, deposited as mud on the floor of a receding Western Interior Seaway, was one of the last marine deposits to cover the western interior of the continent because the landscape rose above sea level due to the Laramide orogeny.

This area preserves the most complete Late Cretaceous–Paleogene rock sequence on the eastern side of the Rocky Mountains, making this an excellent location to follow the post-Cretaceous seaway depositional history of the region. The marine Pierre Shale is overlain by the Trinidad Sandstone and Vermejo Formation, remnants of the Cretaceous coastline, where mudflats and sandbars alternated with marshes and swamps. They record the final gradual recession of the Cretaceous seaway. The layers vary in thickness, in places filling shallow channels. Elsewhere they thin and disappear, as expected along a land-sea boundary, where beaches and lagoons are discontinuous.

As the coastline migrated, the environment became swampy and full of stagnant water, ideal conditions for the deposition of more than 1,000 feet of peat and alluvial plain deposits such as conglomerates, gravels, sands, and muds. These sediments, the coal-bearing Raton Formation, were deposited spanning the Cretaceous–Paleogene boundary. The layers of the Raton Formation—rich with fossilized ferns, leaves, and other plants—record the transition from swamps to river channels to alluvial plains. As the landscape was elevated during the Laramide orogeny, sediments eroded from the highlands were deposited across the plains. The unit was later metamorphosed due to deep burial, converting the plant material to coal.

SUGARITE CANYON STATE PARK

Sugarite Canyon State Park (exit 452) was once the site of a large coal mining camp. The medium-grade bituminous coal came out of the Raton Formation and was shipped by rail to communities throughout the Southwest. The coal mines were closed in 1941. An excellent visitor center has many photos of Sugarite during its heyday, and interpretive signage along trails introduces the town and mines. The bituminous coal is present in nearly 6-foot-thick layers. Mining occurred underground with tunnels (adits) bored into hillsides. Mules transported the coal to the Rio Grande railway, which stopped in Raton.

The volcanic rock visible from Sugarite Canyon State Park is basalt of the Raton-Clayton volcanic field. The fluid basalt once flowed down river bottoms, but the basalt stands high in the landscape because it is more resistant than the underlying sedimentary units. When the much softer sedimentary rocks erode, they cause collapse of the overlying resistant caprocks, resulting in hillslopes covered in landslide deposits.

Ruins of a coal mining camp at Sugarite Canyon State Park. House and building foundations are still in place and document a large community that climbed the sides of the canyon to the coal mines.

The cliffs and slopes to the north of Raton are composed of Cretaceous to Paleogene sandstone, shale, and coal of the Raton Formation. Photo taken near milepost 455.

The Cretaceous–Paleogene (K–Pg) boundary, also called the Cretaceous–Tertiary or K–T boundary, lies within the Raton Formation. This boundary is dated at 66 million years ago and marks the end of the Mesozoic Era, where the Cretaceous Period ended and the Paleogene Period (the first of the Cenozoic Era) began. The mass extinction event that occurred at this time coincides with the meteorite impact that formed the Chicxulub Crater in modern Mexico. This impact is thought to have caused the extinction of all dinosaurs except birds. The boundary is marked here, and in many parts of the world, by a thin layer of clay rich in iridium, an element rare on the Earth's surface but more abundant in asteroids. Tektite, glass associated with meteor impacts, is also found at the boundary. The K–Pg boundary can be seen at Climax Canyon Park west of Raton. See *New Mexico Rocks! A Guide to Geologic Sites in the Land of Enchantment* for more on this site.

Dust from the impact and sulfuric acid ejected upon impact would have enveloped the Earth, blocking sunlight and preventing photosynthesis. The Earth's surface would have become cold and dark for up to ten years as these aerosols and particulates settled out. Under these conditions, most plants would perish, and many animals would freeze or starve. At nearly the same time, the eruption of the extensive Deccan Traps flood basalts in India may have contributed to the global extinction events by also releasing dust and sulfuric aerosols into the atmosphere. A sufficiently large impact could have triggered the eruptions of these massive flood basalts. Scientists note that other factors also may have contributed to the global extinction: many other meteorite craters seem to have formed at nearly the same time as the Chicxulub Crater, and the notable drop in sea level at the end of the Cretaceous may have reduced the area of life-rich continental shelf enough to cause a marine extinction.

The town of Raton lies in the basin of the Canadian River. To the west, foothills hide the Sangre de Cristo Mountains, the southernmost range of the Rocky

THE PEACEFUL PLAINS   225

Mountains. To the east, the Great Plains are hidden by the Sierra Grande arch, a broad anticline, and by many lava-capped mesas, buttes, and small volcanoes of the Raton-Clayton volcanic field.

The dark-gray Late Cretaceous Pierre Shale floors the Canadian River's valley. An easily recognized unit, the Pierre Shale derives its name from its key occurrence along the Missouri River near Fort Pierre, South Dakota. The unit is widespread and can be found along the eastern edge of the Rocky Mountains from Canada to New Mexico. In places, this shale contains beautifully preserved fossil shells of cephalopods (shell-bearing relatives of octopuses and squids) and other invertebrates, shark teeth, and skeletons of marine reptiles. The Pierre Shale and the mustard-yellow soils derived from it can be seen in the steep banks of the Canadian River. The soils are yellow due to the presence of oxidized iron minerals.

Ranches and towns within the Raton Basin harvest water from porous and permeable units within the Cretaceous Dakota Sandstone. Water enters the Dakota Sandstone in highlands along the mountain front, then flows downgradient eastward through the porous sandstone, which is sandwiched between impermeable shale layers above and below. In the Raton Basin, the Dakota Sandstone also contains small amounts of oil, natural gas, helium, and carbon dioxide.

The highway slices through two prominent dikes between mileposts 436 and 435. The hard, igneous rock of these dikes resists erosion, and they jut up in sharp ridges. On either side, well exposed in the highway cuts, the Pierre Shale is baked and hardened by the heat of the molten rock that formed the dikes. These are feeder dikes, part of the extensive Raton-Clayton volcanic field.

The flat-topped basalt mesa to the east between milepost 439 and 435 rises up to Eagle Tail Mountain, a shield volcano that erupted around 1 million years ago and is part of the Raton-Clayton volcanic field.

The vertical dike on the east side of the interstate at exit 435 is known as Eagle Rock. It baked the Pierre Shale on either side of it.

The flat-topped mesa south and west of Springer (exit 414) is capped with Miocene basalt that erupted from small volcanic centers not visible from the highway. The dikes, flows, and cinder cones visible along this route are part of the large Raton-Clayton volcanic field. To the east from high points along the highway between Springer and Wagon Mound, a low mesa capped by sandstone and conglomerate of the Ogallala Formation is visible. The Ogallala Formation, deposited in Miocene to early Pliocene time, blanketed much of the eroded plains east of the Laramide ranges. The Canadian River has since carved its valley down through these gravels and into the underlying Cretaceous rocks. Here, the vastness of the plains, with their gently rolling topography and shallow stream courses, can be observed.

CANADIAN RIVER CANYON IN SABINOSO WILDERNESS

In the Sabinoso Wilderness Area in northeastern New Mexico, the canyons of the Mora and Canadian Rivers converge. The Canadian River cuts a gorge, more than 1,200 feet deep in places, that extends nearly 60 miles, from Springer to Sabinoso. The canyons are cut largely in sedimentary rocks that vary in age from the Triassic Santa Rosa Formation and Chinle Group to Jurassic Entrada and Morrison Formations to the Cretaceous Dakota Group. Along the length of the Mora River, basaltic to andesitic lava flows associated with Jemez lineament volcanics have been deposited upon mesa tops and cling to canyon walls. These basalts originally flowed down low areas of the topography. The basalt now forms a resistant cap that protects the underlying softer sedimentary units from erosion.

View looking west-northwest toward the black remnants of a mesa-capping basalt flow. Tumbled basalt boulders have fallen down the wall of the Canadian River Canyon, about half a mile downstream from the Mora River confluence in the Sabinoso Wilderness. The large, tan boulders in the river are Chinle Group sandstones. The reddish cliff walls are Jurassic sedimentary units, with the Dakota Group capping the distant high point. —Photo by Alexandra Priewisch

The lava caps of the Wagon Mound, the butte east of the town of the same name (milepost 388), were a famous landmark on the Santa Fe Trail. When travelers saw them, they knew it was "six days by wagon to Santa Fe." The two thick lava flows capping the Wagon Mound are 6 million years old. They are part of the Ocaté volcanic field, which has at least sixteen flows and fifty small vents and was active from about 8 million to 800,000 years ago. It is one of many volcanic fields that lie along the Jemez lineament. The heat of molten rock baked soil under the flows to a warm brick red. The flows that form the Wagon Mound also appear in patches of lava to the west and east and form excellent examples of inverted topography (the lava originally flowed down a valley).

Visible to the northwest from milepost 380 are the Turkey Mountains, dark with trees. The small, symmetrical dome is a 29-million-year-old Oligocene laccolith, a small intrusion that squeezed between sedimentary layers, pushing and doming up the Triassic and Jurassic rocks above. Some of the domed-up rocks have eroded from

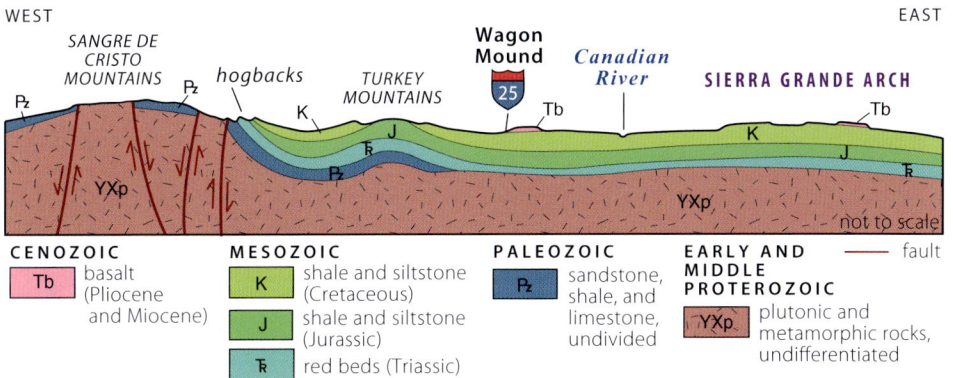

Section across I-25 at Wagon Mound.

The Wagon Mound is part of two lava flows that, for a time, filled the valley. Ridges that bordered the original valley have now eroded away. This photo is taken from the east along NM 271.

the summit, leaving a hogback ring of Dakota Sandstone surrounding a valley of soft Jurassic rock.

Fort Union, which is now a national monument located 8 miles north of exit 366, was built in 1851 and operated for forty years. It lies at the junction of the Mountain Branch of the Santa Fe Trail and the Cimarron Cutoff, near the transition from the Great Plains to the east and the Rocky Mountains to the west. The fort is built on the Graneros Shale, part of the Mancos here, deposited as a muddy, migrating shoreline to the Western Interior Seaway in Cretaceous time. The volcanic Black Mesa to the northwest of the Fort Union National Monument exposes more than 100 feet of basalt, a flow of the larger Ocaté volcanic field. Much of the geology seen from the national monument is also visible from the Fort Union rest area (milepost 376).

The adobe ruins at Fort Union National Monument are set in a broad valley of Quaternary fill. The rock pathway adjacent to the monument has ripple marks. Adjacent hills are capped with the lavas of the Ocaté volcanic field.

Maxson Crater, due east of Fort Union National Monument, is the vent and source of a young basalt flow that erupted about 1.6 million years ago. The vent followed the Mora River drainage to its confluence with the Canadian River.

At Watrous (milepost 365), the interstate is traveling on soil-covered Dakota Sandstone, a water-bearing unit of the Raton Basin. The Dakota Sandstone is visible in roadcuts and stream escarpments as a yellowish, blocky, and fairly resistant sandstone. Look for it in low roadcuts near exit 366. Incredibly widespread, the Dakota Sandstone extends north into Wyoming and the Dakotas.

South of Watrous, the highway draws closer to the mountains while remaining on an even floor of Cretaceous shales. Close to Las Vegas, the faults that edge this side of the southern end of the Rockies peter out, although those on the west side continue southward as the margin of the Rio Grande rift.

A low, flat-topped mesa just north of Las Vegas (west at milepost 356) is a remnant of a Pleistocene alluvial fan. Erosion of the mountains reached all-time highs during Pleistocene time, when continued uplift of these mountains was coupled with heavy precipitation of the ice ages. Torrential streams flooding from the mountains likely built many such fans, but few still exist. The fan sits above the valley, which is floored in Cretaceous shales and limestones. Jurassic and Triassic units form the mountain foothills to the west, while basement rocks rise above.

THE PEACEFUL PLAINS

I-40
CLINES CORNERS—SANTA ROSA
57 miles

The pale-gray rock around Clines Corners (the junction of US 285 at milepost 218) is part of the Permian San Andres Formation, a marine rock that covers more of the surface of New Mexico than any other single formation. The San Andres is thin here but thickens southward, which was seaward in Permian time. The road stays on its surface—a grass-covered tableland spotted with juniper trees—for several miles, passing or crossing a number of undrained silt-floored depressions. These are sinks formed by collapse in and below the limestone.

Mesas visible to the north are capped with light-tan, cross-bedded Glorieta Sandstone. This fine-grained quartz sand was swooshed around on a Permian beach just before the San Andres Formation was deposited. The Glorieta Sandstone is quite near the highway at milepost 225.

In the distance to the south from milepost 219, the Proterozoic rocks of the Pedernal Hills stand above the plain. The Pedernal Hills were once a mountain range of the Ancestral Rocky Mountains. Eroded over time, the Pedernal Hills were a source of the siliciclastic sediments that collected in surrounding sedimentary units.

How and why the Ancestral Rockies formed in the middle of the North American plate is an ongoing debate for geologists, but at least they agree that the mountains' formation relates to the continental collision that formed the supercontinent Pangaea. The ancient Gondwana plate (South America–Africa) collided with the Laurentian plate (North America) during Pennsylvanian and early Permian time in what is known as the Ouachita-Marathon orogeny. This collision was similar to the deformation and mountain building seen today in the Himalayas, resulting from the collision of the Indian and Asian plates.

Near mileposts 228 and 230, low arroyo washes expose red and yellowish-green shale and whitish limestone of the Yeso Formation, with the Glorieta Sandstone on top. Veins of gypsum are visible in this rock. The Glorieta Sandstone appears again east of the Yeso exposures and on a mesa to the south at milepost 225.

Near milepost 237, the highway crosses a high, level surface, which is a small outcropping of the Ogallala surface that extended all the way west to the mountains. This Miocene to early Pliocene gravel washed eastward from the various ranges of the Rocky Mountains and the Basin and Range areas south of them. The Ogallala Formation is thin here, and along this highway it is mostly covered with silty soil or with hummocky sand dunes left over from Pleistocene time, now stabilized by vegetation.

One of the few places along this highway where some of these rocks can be seen is near the rest area at milepost 251, where part of the Santa Rosa Formation is exposed on the walls of a collapsed valley, best seen from the top of a small hill within the eastbound rest area. The same rocks can be seen in a nearby highway cut and edging the mesa to the southeast. Near the rest area, the Santa Rosa Formation is mostly light-brown or gray sandstone with brightly sparkling feldspar grains. Between the sandstone layers, the formation also includes bands of dark-reddish-brown mudstone and some conglomerate. Weathered surfaces are generally dark. Watch for more exposures of these rocks in roadcuts between the rest stop and the junction with

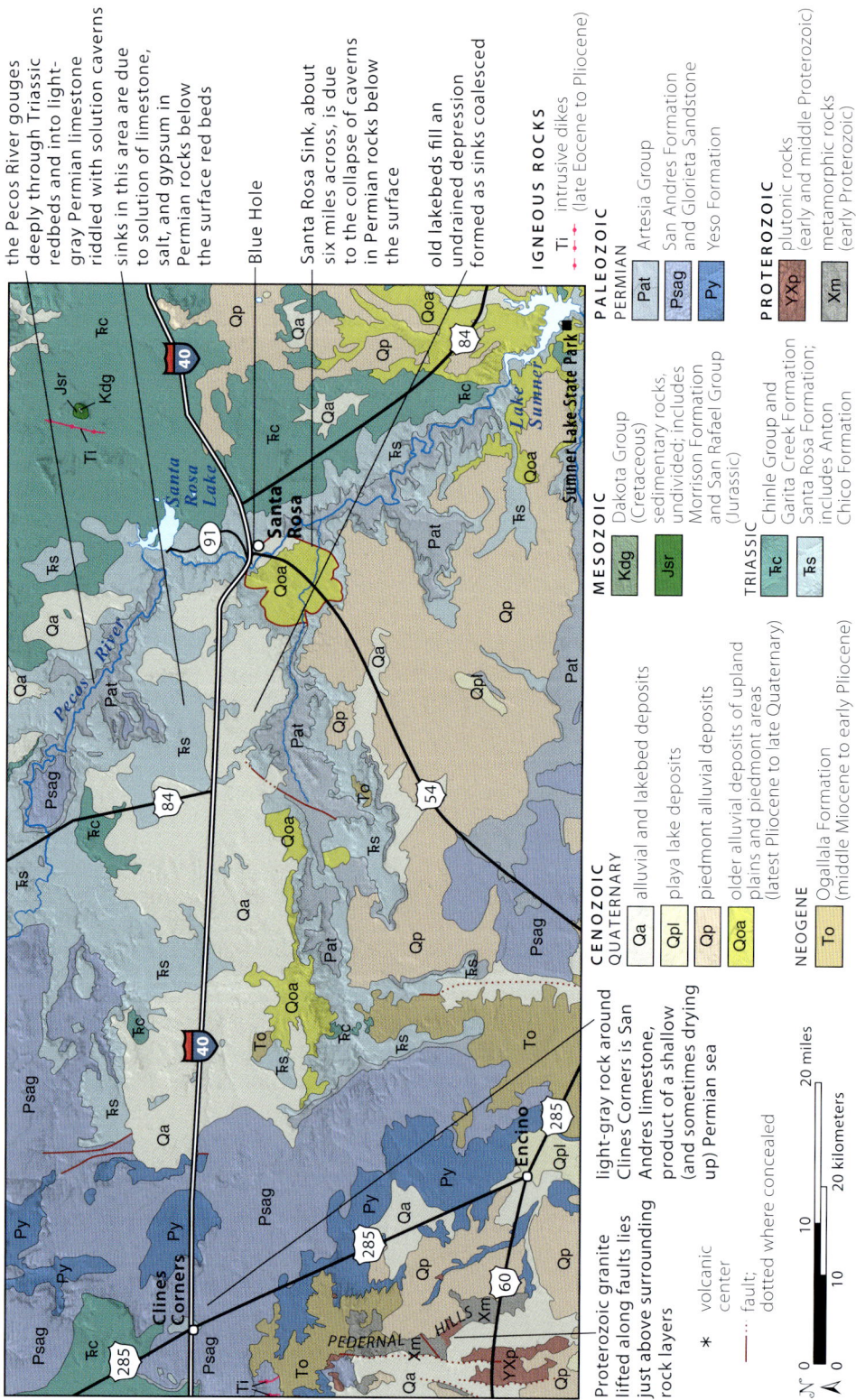

Geology along I-40 between Clines Corners and Santa Rosa.

US 84 (exit 256). Both the Santa Rosa Formation sandstone and overlying Chinle Formation contain petrified wood and other fossil plants, as well as invertebrate and vertebrate fossils. Notable among the latter are phytosaurs, crocodile-like reptiles that inhabited the floodplains of Triassic rivers.

East of the US 84 junction, the highway crosses more Triassic rocks, partly hidden by silty Pleistocene soils and sand. This whole area is pocked with sinks caused by solution of limestone and evaporites in the Permian Artesia Group and San Andres Formation, both beneath the surface here.

Just west of Santa Rosa between milepost 270 and 271, the highway drops rather suddenly into the Santa Rosa Sink, a large, flat-floored collapse area, circular in outline and some 6 miles across. The sink is edged by faults formed by the subsidence of rocks within the sink. The sink initially filled with alluvium as it formed, but much of the alluvium has since been eroded. The reddish rocks in roadcuts at the top of the hill are the Santa Rose Formation.

A sinkhole in the San Andres Formation north of the road just east of milepost 268.

Many of the lakes near Santa Rosa are natural, although Santa Rosa Lake north of town is a manmade reservoir on the Pecos River. The natural lakes form as the result of the karst topography. Groundwater seeps into the limestone, gypsum, and evaporites of the Permian-age units and dissolves pockets within the soluble units. Over time, the overlying rock collapses, and the depressions, called sinks (short for sinkholes), fill with water, forming small lakes. There are several smaller sinks in the town of Santa Rosa, among them the Blue Hole, an extremely clear, 61-degree pool that is a favorite of scuba divers. The Blue Hole, with a diameter of 60 feet and a depth of 81 feet, is fed by a spring discharging at a rate of 3,000 gallons per minute.

At Santa Rosa, I-40 crosses the Pecos River. In places, the Pecos River and its tributaries follow collapsed valleys, which are abundant in this karst topography. In 1541, Francisco Coronado and his followers built a log bridge to cross the river (about 10 miles south of town) while searching for the legendary seven cities of gold. Santa Rosa

Dam, on the Pecos River about 6 miles upstream from town, is 1,950 feet long and just over 200 feet high and serves for conservation of irrigation water, flood control, and sediment control.

Many good exposures of the 220-million-year-old Santa Rosa Formation occur near NM 91 (the road to the dam), particularly in the walls of Pecos Canyon below the dam and along the spillway. The Santa Rosa Formation is composed of

The Blue Hole at Santa Rosa is a sink formed by the partial collapse of an underground cavern in the Santa Rosa Formation. Scuba divers explore its crystal-clear depths.

A scenic walking loop on the east side of Santa Rose Dam exposes horizontally bedded Santa Rosa Formation that has eroded into distinctive ledges.

maroon-to-brown sandstone, mudstone, and siltstone deposited in a continental semiarid environment in stream channels and as overbank deposits on floodplains. Local cross-bedding reflects deposition in channels. Underlying the Santa Rosa Formation is the Triassic Anton Chico Formation, reddish-brown and gray sandstone and conglomerates that were deposited in stream channels. This formation, visible in the canyon bottom, contains fossils from several vertebrates. The red color is from the hematite that forms when iron minerals oxidize. The Santa Rosa Formation has been a source of heavy oils, and in places tar sands were mined from it.

Sumner Lake State Park, 43 miles southeast of Santa Rosa off US 84, lies along another reservoir on the Pecos River. The lake is surrounded by Permian and Triassic rocks that have a veneer of Quaternary sediment on top of them. The Triassic rocks, part of the river-deposited Anton Chico and Chinle Group, form broken red cliffs in the vicinity of the reservoir.

The youngest geologic deposits in the area are river gravels, silts, and clays deposited over the last 2 million years by the ancestral Pecos River. Before Santa Rosa Dam was built, archeological studies identified many archeological sites in the area, including those of Native Americans, Spanish, and Americans, that range in age from 5000 BC to AD 1870.

I-40
Santa Rosa—Tucumcari—Texas
100 miles

East of Santa Rosa, Sunshine Mesa comes into view to the south at milepost 280. The mesa's mudstone and sandstone slopes of the Late Triassic Garita Creek Formation of the Chinle Group are capped with a resistant layer of caliche, an impure mixture of calcium carbonate and soil developed since Pleistocene time. On this mesa, caliche deposits are thick enough to be mined for road material. Look for small mine dumps near the edge of the mesa.

Near Newkirk are small buttes and mesas of the red Chinle Group protected by a cap of caliche. The Late Triassic rocks and their fossils formed where streams entering a large freshwater lake deposited sand and mud. Shallow stream estuaries and marshy lakeshores supported clusters of rushes and other marsh plants. Large, sluggish, flat-headed amphibians like *Apachesaurus gregorii* and reptiles like the archosauriform *Vancleavea campi* swam in the seas. The phytosaur *Redondasaurus gregorii* and other agile, crocodile-like reptiles inhabited the mudflats and splashed through the water to capture fish. Small clams and snails burrowed in the mud or clung to the rushes. Ferns and a few trees grew where hills rose above the general floodplain surface. Far to the west and northwest were the remains of the Ancestral Rocky Mountains, eroded to low, rolling hills by Triassic time.

North of Newkirk (milepost 300) and 400 to 800 feet below the surface, the Triassic Santa Rosa Formation contains an estimated 62 million barrels of heavy oil. Because of difficulties in extracting this thick oil, it is not being removed. However, it is considered part of the nation's oil reserve, obtainable when the price of oil exceeds the cost of extracting. Much deeper Pennsylvanian rocks are currently producing some gas, including small amounts of helium, northeast of Santa Rosa.

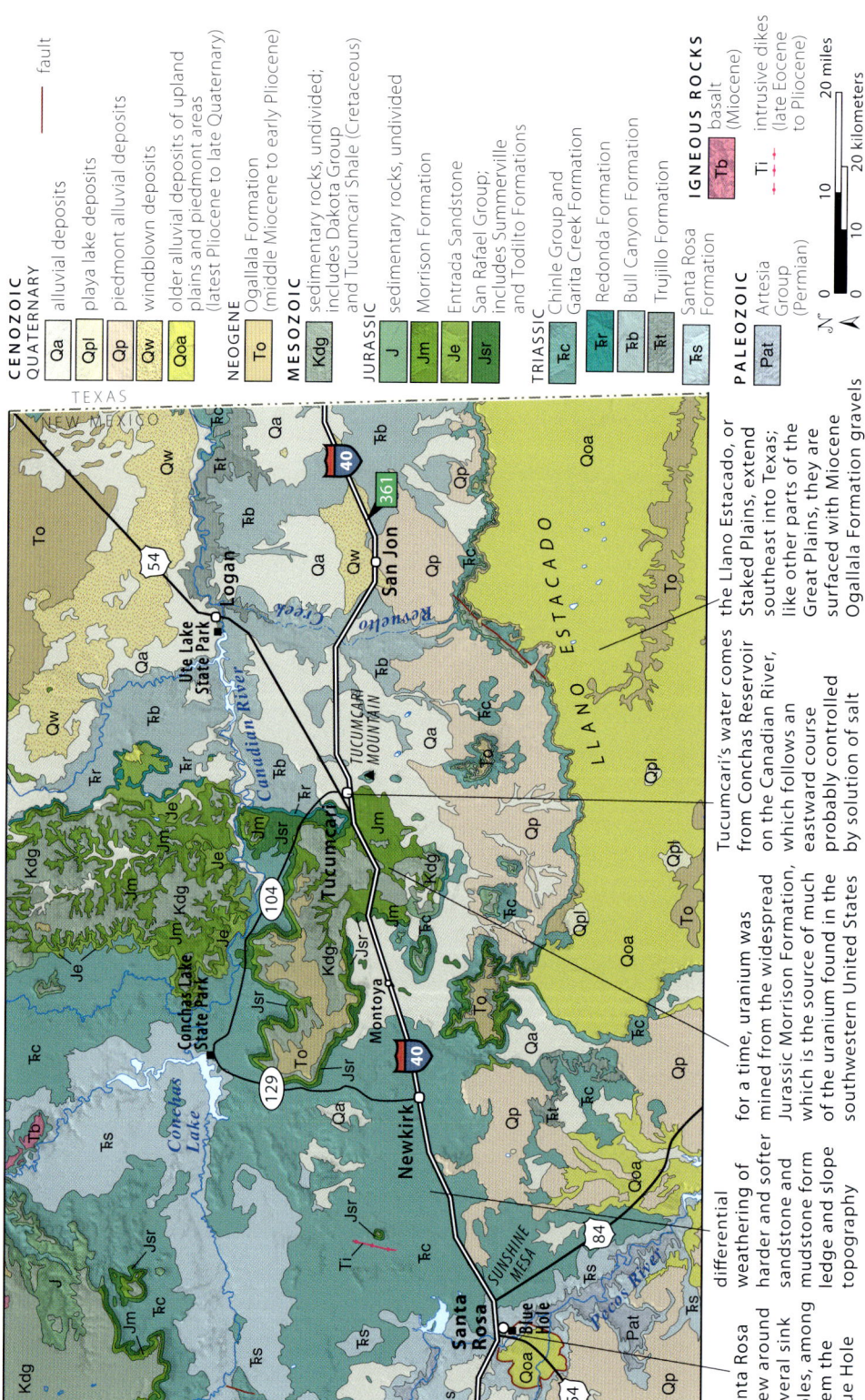

Geology along I-40 between Santa Rosa and the Texas border.

Caliche develops in arid climates where soil moisture containing dissolved calcium carbonate is drawn to the surface and evaporated, leaving crusty deposits of calcium carbonate in the soil.

In a butte west of Newkirk, near milepost 298, are red sandstones and mudstones of the late Triassic Chinle Group. A hard cap of caliche on the mesa top protects the softer red rocks.

Conchas Lake State Park lies 23 miles north of Newkirk on a reservoir of the Canadian River. The dam was built in 1939 in the 220-million-year-old Chinle Group sedimentary rocks. The red coloring that tints many of the layers within the Chinle is due to the oxidation of iron within the minerals of the sandstones. This rusting indicates that the environment during sand deposition was seasonally arid, allowing for the wetting and drying of the particles. The Chinle, a widespread group, was deposited by a river system comparable in size to the modern Mississippi River.

Both northeast and southeast of Newkirk, and visible from the rest area east of town, are higher mesas capped with the Ogallala Formation, a widespread but

The Llano Estacado is capped with the caliche-hardened Ogallala Formation. Before erosion of the Pecos Valley, this remarkably flat surface was regionally extensive. This photo was taken south of Montoya.

relatively thin (up to 60 feet) gravel unit. It originated as eroded material from the newly rising Rocky Mountains in Miocene to Pliocene time. Today, these mesas are isolated or nearly isolated remnants of a once vast surface that formerly extended without interruption all the way to the Rocky Mountains. Eastward travelers will see that the southern of these two remnants connects with the main surface, here called the Llano Estacado, or Staked Plains. The caliche-hardened Ogallala Formation surface overlies Cretaceous and Jurassic rocks, which in turn rest on the Triassic rocks described above. These form the embankments seen below the much thinner Ogallala.

East of Montoya (exit 311), the interstate dives into Jurassic rock units, the lowest of which is the Entrada Sandstone, a unit composed of thick layers of cross-bedded, fine-grained sandstone formed from sand dunes. The Entrada is beveled near the top by limestone and gypsum of the Todilto Formation, deposited in a large, shallow lake separated from the main sea. Influxes of freshwater, with occasional flooding or seepage of seawater, maintained the lake's water level. The Todilto is part of the San Rafael Group.

An interesting aspect of some of the Todilto limestones is that they accumulated in thin varves, each varve consisting of three paper-thin layers: limestone, dark organic material, and fine clay. Varves are produced by yearly variations in runoff, water temperature, and abundance of lake organisms. Varves can be counted like tree rings to determine lengths of time. Accurate counting of the paper-thin varves in the lower part of the Todilto Formation tells us that the Todilto accumulated over 14,000 years. The upper part of the Todilto Formation contains gypsum and is less regularly varved. Fossils are rare, though some locations of the formation contain fossil fishes and insects.

Near milepost 319, the highway climbs onto a small hill where roadcuts to the south expose soft, rounded, uniformly beige rocks, part of the Summerville Formation, with blocks of sandstone fallen from above. The Summerville Formation, above the Todilto Formation, is composed of sandstones, siltstones, and mudstones, all cyclically deposited in a river floodplain environment.

Just west of milepost 320 and on the north side of the highway is a roadcut in which the variegated red, yellow, and green shales of the Morrison Formation are easily recognized, even at highway speeds. The Morrison Formation, which lies above the Summerville Formation, is composed of soft, easily eroded shale and siltstone, with a few sandstone and conglomerate beds, also deposited in a river floodplain environment. The shales include several clay minerals known to be derived from volcanic ash, as well as uranium. The Morrison Formation is famous for its large dinosaur bone fossils, particularly those located in Colorado and Utah. Here, in northeastern New Mexico, the unit contains a few dinosaur bones and rare fragments of petrified wood.

At milepost 323, large, lengthy roadcuts expose the contact between the rocks of Jurassic and Cretaceous time: a channel cut in the surface of the light-colored Morrison Formation is filled with the younger, orange Dakota Sandstone. The sandstone was originally deposited as a thin but widespread beach and nearshore sandstone, indicating the return of the sea in Cretaceous time.

Tucumcari Mountain, rising above the plains to the south at milepost 333, has been a distinctive landmark for millennia. The mountain's stratigraphy spans from the Triassic Period to the Miocene Epoch. From oldest to youngest: Triassic Chinle Group and Redonda Formation, Jurassic Exeter Sandstone and Morrison Formation, Cretaceous Tucumcari Shale and Mesa Rica Sandstone, and the Miocene to Pliocene Ogallala Formation.

Approaching Tucumcari from the west, the highway drops down through these Cretaceous, Jurassic, and Triassic rocks again, into the wide valley of the Canadian River, which lies about 6 miles to the north. In Pliocene time, when erosion was spurred by uplift of this entire region, this river broke through the Ogallala Formation caprock of the Great Plains to form what is known as the Canadian Breaks, a broad erosional westward curvature in the west margin of the Llano Estacado. Within

Tucumcari Mountain, south of the town of Tucumcari the town, is capped with the Ogallala Formation.

this indentation, terraces developed in response to cyclic climate changes of the Pleistocene ice age. All around Tucumcari are salmon-colored soils derived from Triassic rocks. These soils, with the addition of irrigation water from the Canadian River, support local agriculture.

The highway continues eastward over these salmon-colored soils, with portions of the Llano Estacado escarpment visible in the distance to the south and north. White caliche marks low spots and shows where water has collected and evaporated.

Because the irregular edge of the Llano Estacado forces winds to rise, and therefore to lose some of their carrying power, wind-deposited sand dunes are common along this stretch of the highway and elsewhere along the edge of the Llano Estacado. The dune fields, developed during Pleistocene time, are now occasionally stabilized by vegetation. Where streams have cut through the pink soil, Triassic units are exposed. Between mileposts 346 and 347, for instance, the ephemeral Revuelto Creek has incised a particularly deep wash in the Bull Canyon Formation of the Chinle Group.

South of the highway, just a few miles along Revuelto Creek, are fossil vertebrate sites in the Bull Canyon Formation from which skulls and bones of Middle Triassic phytosaurs, fish, amphibians, and reptiles have been recovered. These include fragments of skeletons of the phytosaur *Rutiodon*, the archosaur *Desmatosuchus*, and the crocodilian raptor *Poposaurus*. Unfortunately, no complete skeletons have been found there.

Ute Lake near Logan to the northeast of Tucumcari on US 54, impounds the waters of the Canadian River. The river and lake deposits of the Bull Canyon Formation of the Chinle Group are the dominant bedrock in the vicinity of this park.

I-40 continues to converge with the edge of the Llano Estacado. East of San Jon (milepost 356), the highway rises onto a surface covered with redeposited gravel derived from the older Ogallala Formation that forms much of the Great Plains surface. As the road approaches the Texas border, travelers get a good view of this remarkably flat surface of the Great Plains, which has remained virtually unchanged, except for erosion nibbling at its edges, for several million years. Reinforced with caliche, the caprock absorbs water readily, preventing development of streams strong enough to erode channels into it. Ranches and farms pull water from the underlying Ogallala aquifer.

US 54
Carrizozo—Santa Rosa
82 miles

From Carrizozo, the black lava flows of the Valley of Fires can be seen to the west, with Little Black Peak, the source of the lava, near its northern end. The lava erupted about 5,000 years ago and flowed 47 miles from its source. The ropey lava is and called pahoehoe for similar flow-styles in Hawai'i. These flows lie along the Capitan lineament, a line of igneous features that exists due to a zone of crustal weakness along which magmas have risen. For more on the Valley of Fires lava, see US 380: Carrizozo—San Antonio on page 209.

A short distance northeast of Carrizozo, hidden between Carrizo Mountain and the Jicarilla Mountains, is White Oaks, a gold-mining boomtown of the 1880s. Placer

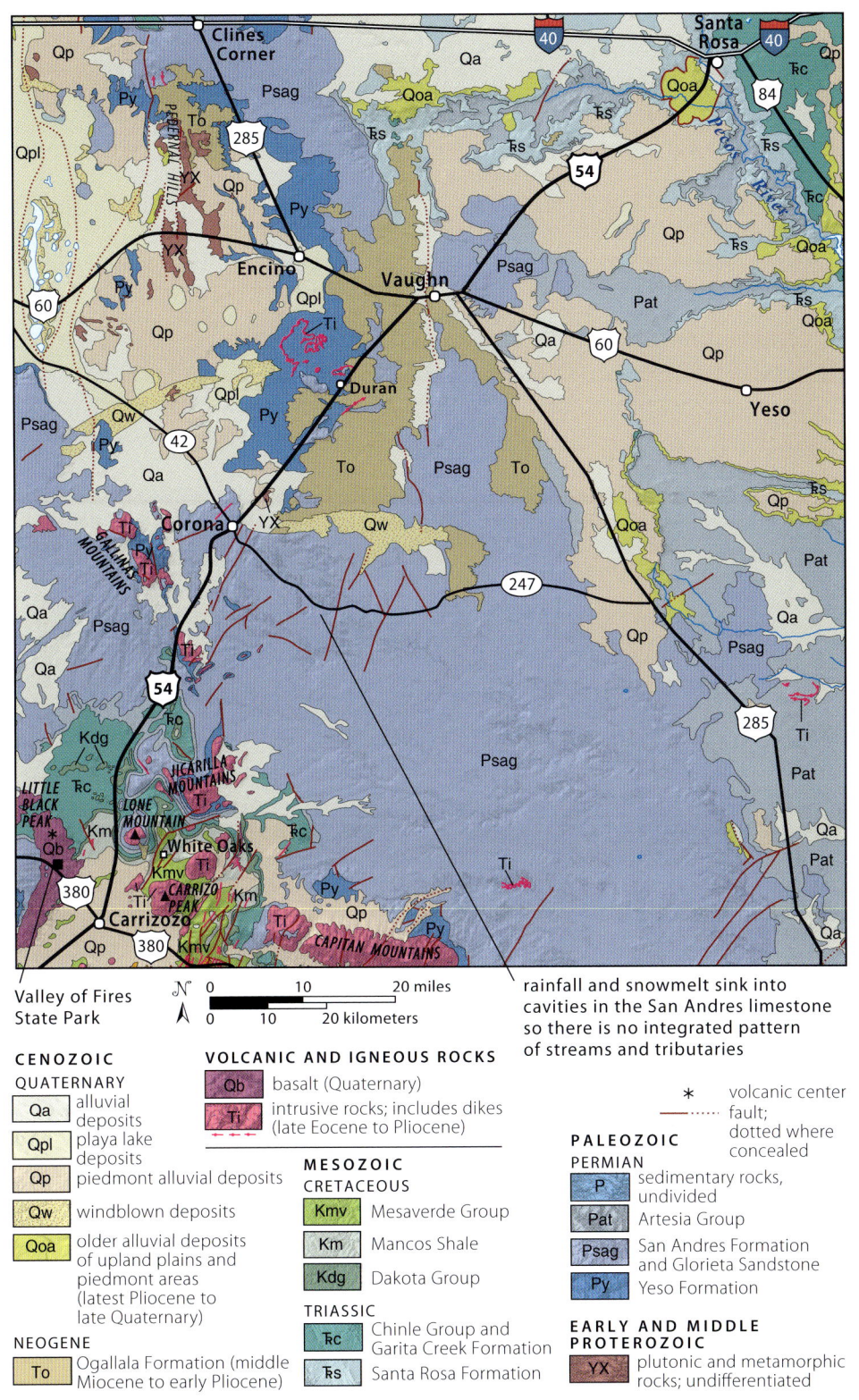

Geology along US 54 between Carrizozo and Santa Rosa.

gold was discovered there in 1850, but mining at the time was sporadic due to its isolated location, Native American populations, and the imminent Civil War. The town didn't really develop until thirty years later, when around 150,000 ounces of gold was mined from intrusive rocks west of town. But by the turn of the twentieth century, White Oaks was a ghost town, its riches forgotten. Some of its more elegant homes were moved to Carrizozo.

The highway moves from Quaternary alluvium onto Mancos Shale at about milepost 131. Lone Mountain, another Cenozoic stock that intruded through and domed up Permian to Cretaceous units, is to the east. The highway proceeds north across a large alluvial fan containing gravel from the Jicarilla Mountains. Lines of desert shrubbery mark its lowest watercourses. The alluvial fan probably began forming in Pleistocene time. Low roadcuts between mileposts 135 and 136 show that part of the fan was deposited over the top of dark basalt and whitened with caliche near the surface.

The Jicarilla Mountains are a cluster of irregularly shaped, 45- to 39-million-year-old intrusions. Like Lone Mountain, these igneous masses domed up the overlying sedimentary rocks that range from Permian sandstone and limestone to Cretaceous sandstone and shale. These sedimentary rocks eroded off the central core of the mountains but are present in concentric rings around the edges of the intrusive rock.

US 54 comes upon some of the Cretaceous rock—notably the ridge-forming Dakota Sandstone—just north of milepost 136. Farther north, the highway passes onto deep-red Triassic siltstone and sandstone of the Triassic Chinle Group, in which it remains for some distance. Hills on either side of the road are bright with these colorful rocks.

Still farther north, the highway crosses a roller-coaster surface of pale-gray limestone of the Permian San Andres Formation. The terrain is dotted with sinkholes where the limestone has collapsed into underground caverns. North of milepost 152, US 54 converges with light-colored hills of Yeso Formation, a Permian rock unit slightly older than the San Andres Formation. The contact between these two units is marked by a thick yellowish or reddish soil layer developed before the sea that deposited the San Andres advanced across the older unit. The Yeso Formation contains both gypsum and bentonite, the latter a group of clay minerals commonly derived from volcanic ash. Gypsum is relatively soluble, and bentonite tends to expand when it gets wet and shrink again when it dries. As a result, portions of the Yeso Formation slough off easily and, in places, work slowly downslope as gray, surface-covering mats. The formation originally contained thick layers of salt, much more easily dissolved than gypsum. Solution of the salt is responsible for many sinks in this area.

West of milepost 16, another fairly large group of Cenozoic igneous intrusions juts up as the Gallinas Mountains. These mountains are being studied for their potential to provide several rare-earth elements that are increasingly important in a technological world. The elements occur in veins and amongst breccias and were deposited by hydrothermal fluids during phases of the Tertiary intrusions. A few examples of such elements include lanthanum, which is used in catalytic converters and rechargeable batteries, and erbium, which is used in fiberoptic cables and lasers.

From the railroad overpass between mileposts 172 and 173, several rounded rocky outcrops in the distance to the north look like the Loch Ness Monster surfacing through waves of juniper trees. These hills are isolated outcrops of coarse-grained

Reddish soil surrounds a hill of Proterozoic granite near Corona. —Timlewisnm, CC SA 2.0

Proterozoic granite, exposed here where Permian sedimentary rocks that normally cover them have worn away. They display a characteristic weathering pattern of granite (and some other rocks) in which the gradual rounding of sharp-edged blocks forms large boulders. Look for a small mound of Proterozoic rock just east of the highway north of milepost 174.

Permian rocks of the Yeso Formation rest directly on the Proterozoic granite. No Cambrian, Ordovician, Silurian, Devonian, Mississippian, or Pennsylvanian rocks exist here today. If older Paleozoic rocks ever covered this area, they were worn away in late Pennsylvanian and early Permian time, when a north-south ridge stretched from here to Colorado, the southernmost part of the Ancestral Rocky Mountains.

Between Corona and Vaughn, much of the surface is patched with gravels of the Ogallala Formation. The surface strongly resembles that of the Great Plains—fertile, rolling hills with summits all at about the same elevation. The true Great Plains, capped with Miocene gravel, are farther east.

Hills near Duran (milepost 189) are held up by sills, sheetlike horizontal intrusions that forced their way between layers of Permian rocks. Sinks between Duran and Vaughn formed when gypsum dissolved from the Yeso Formation. The large collapse valley just north of milepost 197 that the highway dips into and then climbs back out of formed the same way.

Collapse occurs, too, in limestone of the Permian San Andres Formation, which lies above the Yeso Formation. Limestone dissolves much less easily than gypsum or salt and only in slightly acid water. Rain picks up enough carbon dioxide from the atmosphere and soil to become weakly acidic and, over long periods of time, can carve out extensive cavern systems. Northeast of Vaughn are many sinks formed by limestone solution, most of them shallow. Some expose the blocky tan limestone.

Circular features in this satellite image of the area north of Vaughn are sinks and solution pits.
—Google Earth base image

A few of these sinks hold water. There is no regular drainage pattern on the limestone surface—no well-defined streams or stream valleys. When snow melts or storms occur, the runoff flows toward individual sinks and then disappears underground to etch more caverns in the limestone or to enlarge those already that elready exist.

US 64
OKLAHOMA—CLAYTON—RATON
90 miles

US 64 enters New Mexico in the Kiowa National Grassland, an area the government bought from bankrupt landowners when drought and poor land management practices in the Dust Bowl era caused soil to dry up and blow away. Now, this unique area is managed by the US Forest Service, and some of the grasslands they once were have been restored. The grasslands are on the Great Plains physiographic province, a gently rolling upland capped in gravel and sand that has existed with very little change since Pliocene time.

Near Clayton, Rabbit Ear Mountain punctuates the landscape north at milepost 426. The two-pronged mountain, which rises several hundred feet above the Great Plains, reaches an elevation of 6,036 feet and was created by basaltic lava flows of the Raton-Clayton volcanic field. A National Historic Landmark today, it was a critical landmark along the Cimarron Cutoff of the Santa Fe Trail, which approximately paralleled the present US 64. Hundred-year-old wagon ruts are still apparent in places, a measure of the slow erosion rate in this semiarid land where the gravelly

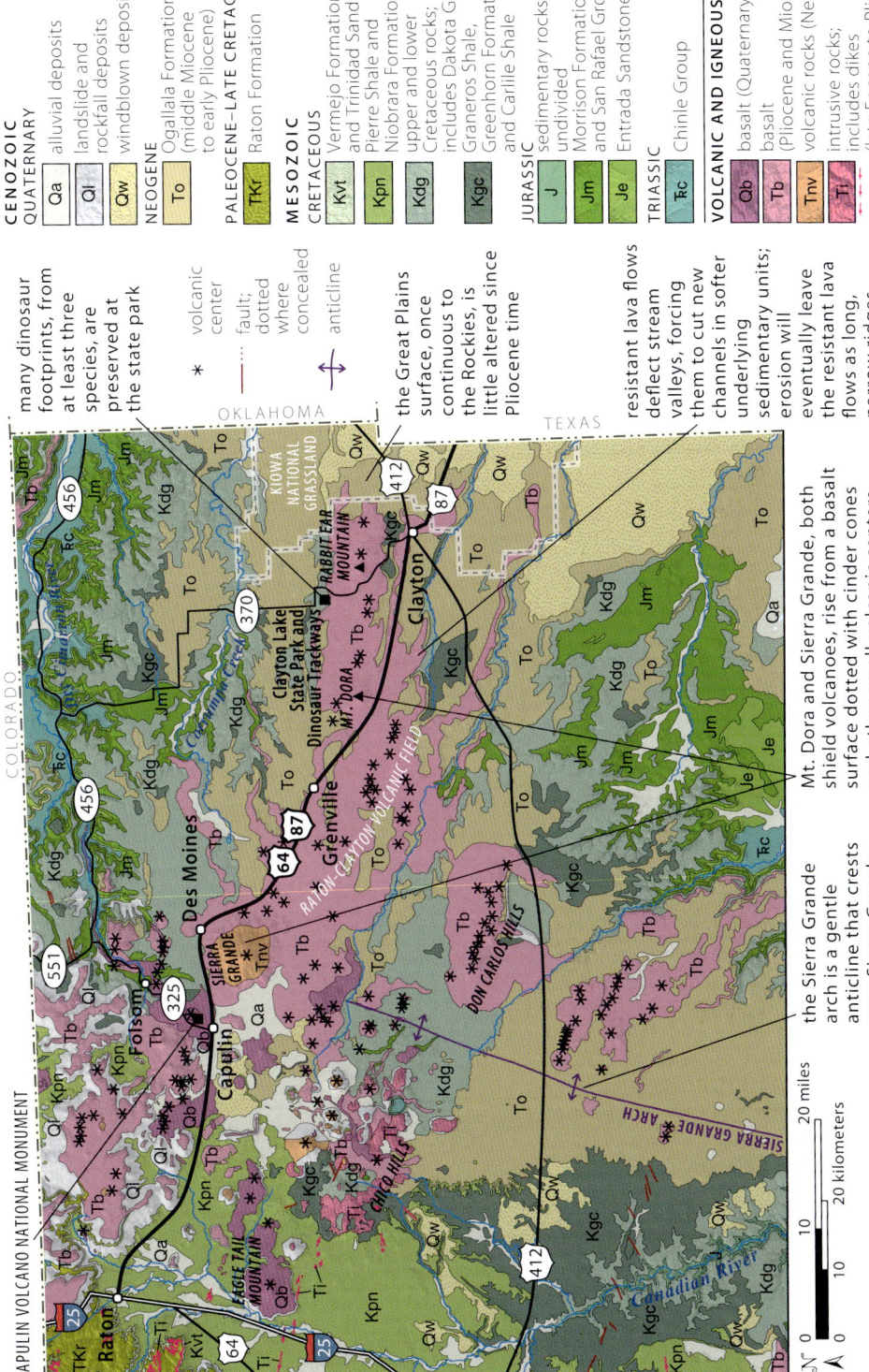

Geology along US 64 between the Oklahoma border and Raton.

CLAYTON LAKE STATE PARK AND DINOSAUR TRACKWAYS

At Clayton Lake State Park and Dinosaur Trackways, about 12 miles northwest of Clayton on NM 370, more than 560 dinosaur footprints are concentrated in a 2-acre expanse of Cretaceous Dakota Sandstone. Exposed in 1982 during excavation for the dam spillway, the rock surface has prints of predators and other dinosaurs, as well as sedimentary features like mud cracks.

The dinosaur trackway preserved at Clayton Lake State Park is about 2 acres in size and lies within this fenced-in enclosure.

The different styles and sizes of the tracks indicate that at least three species and both adults and babies used this location. Kite-shaped tracks, like the big ones in this photo, are from Iguanodons.

Mud cracks indicate a nearshore plain that was occasionally flooded.

At Clayton Lake State Park, young basalt flows overlie the 12- to 4-million-year-old Ogallala, which in turn sits on the gray, slope-forming, 97- to 91-million-year-old Graneros Shale and the underlying coastline deposits of the Dakota Sandstone. The dinosaur footprints, worm burrows, ripple marks, and mud cracks are preserved in the shale, siltstone, and sandstone of the Dakota. Footprints of at least three species of dinosaurs include ornithopods and theropods. Herbivores made these bipedal tracks as they walked on muddy areas, with feet facing forward rather than sideways, as today's reptiles' feet do. They used their tails for balancing.

---

surface of the Plains absorbs rain and snowmelt. Windmills pump water from the Ogallala Formation gravels below.

Clayton sits upon a lava flow surface. West of town, US 64 travels for many miles on the crest of a higher tongue of lava, now covered in grass and shrubs. The fluid, basaltic lava flowed down the valley of an east-flowing stream, displacing it about 2.5 million years ago. Later, as ridges that edged the valley eroded away, the lava flow remained as a resistant ridge, an example of inverted topography. A glance at the geologic map reveals many other flows, now ridges, parallel this one and have a similar history.

The road follows the basalt tongue. Some of the dark-gray basalt blocks along the railroad and fence line are coated with white caliche, a mixture of calcium carbonate and other soluble minerals. Moisture, drawn upward through joints in the rock, deposited the minerals as it evaporated. Occasional small depressions in the lava surface mark collapses of lava tunnels formed where molten lava flowed out from under its own hardened crust.

Rabbit Ear Mountain at right. The smaller hill at left is Bible Top Butte. —Photo by Alex Nereson

The dome-shaped mountain northwest of milepost 412, with radio relay towers, is Mt. Dora, a shield volcano formed by lava slightly less fluid than that in the flows previously seen on the route. These flows were erupted about 3 to 0.15 million years ago. West and northwest of Mt. Dora are other volcanoes, some steep-sided and partly eroded, others with the classic low profiles of shield volcanoes. Most have craters, though erosion breaches some of them. Seen from above or on a topographic map, streams draining such shield volcanoes show a characteristic radial pattern.

At milepost 395, Sierra Grande (8,720 feet) is straight ahead. One of New Mexico's largest volcanoes, and the largest in the Raton-Clayton volcanic field, it rises above a long anticline known as the Sierra Grande arch. The broad, north-south-trending fold separates the area from the Raton Basin, a geologic sag that flanks the Sangre de Cristo Mountains.

The highway curves north around Sierra Grande, through the town of Des Moines (milepost 386). Near here, archeologists have found many skillfully made stone spear points similar to those originally discovered near Folsom, about 6 miles to the north.

Sierra Grande, a stratovolcano that erupted from 3.8 to 2.8 million years ago, viewed from the summit of Capulin Mountain. The lava flow at the bottom of this picture is only 60,000 years old and comes from Capulin.

CAPULIN VOLCANO NATIONAL MONUMENT

Capulin Mountain, the centerpiece of Capulin Volcano National Monument, is a cinder cone that erupted 60,000 years ago. Fragments of foamy lava, hurled up to 1,500 feet into the air from a volcanic vent, fell back to the ground and piled into a cone-shaped hill. Lava flows around its base came from low on the west side of the cone. One flow extends right to the highway, presenting a rough, irregular surface. When such lava moves, it rolls along like a military tank, laying down a "tread" of cooled and broken pieces that fall from the front of the flow, then slowly pushing forward onto these pieces as it carries more cooled and broken blocks forward on its upper surface.

At the top of Capulin is a view of the surrounding volcanic field and down into Capulin's crater. The crater has partly filled with sliding cinders from its walls, and coarse blocks surround the original vent at the bottom.

Layers of ash and cinders deposited or collapsed along the side of the cone due to gravity and water acting on the steep slopes.

Capulin viewed from the southwest along the entrance road. The basaltic lava that was blown into the air broke apart and solidified into cinders (inset) before landing. The prevailing wind blew from the west, blowing the erupting cinders to the east; thus, the mountain is higher on the east.

Hunter-gatherers that camped in this region 9,500 to 8,000 years ago produced the Folsom points. A leaf shape, a shallow groove that runs the length, and neatly chipped edges characterize the points.

Capulin Mountain, a cinder cone with a road spiraling up it, comes into view to the northwest of Des Moines. Head north on NM 325 to reach Capulin Volcano National Monument to learn more about the Raton-Clayton volcanic field.

From milepost 355 west, the sawtooth profile of the Sangre de Cristo Mountains, the southernmost part of the Rocky Mountains, marks the western skyline. Below their snowy crest are foothills of soft Cretaceous marine shales capped with the Cretaceous to Paleogene Raton Formation, a sandstone-shale sequence that represents withdrawal of the last sea to cover this part of the continent.

High cliffs of Miocene to Pliocene volcanic rocks jut skyward north of milepost 351. They are part of a once-broad lava plateau that covered the Colorado–New Mexico border. Now on the west side of the Sierra Grande arch, the highway descends into the Canadian River valley, where the Ogallala surface, well preserved farther east, has been cut through and eroded away as streams were strengthened by uplift and the added rain, snow, and meltwater of Pleistocene time. The Canadian River and its tributaries, issuing from the mountains to the west, carried the rock debris southward and then eastward across Texas, Oklahoma, and Arkansas to the Mississippi River, helping to build the Mississippi floodplain and delta.

US 70
Clovis—Roswell
109 miles

Clovis sits upon the Llano Estacado ("Staked Plains"), a surface formed by the deposition of resistant caliche in the top layers of the Ogallala Formation. This plateau within the Great Plains covers an area of more than 32,000 square miles and reaches from the Pecos River eastward to Palo Duro Canyon in Texas, all the way south to Hobbs, New Mexico. Here in eastern New Mexico, the ruffled edge of the Llano Estacado can be traced along a cliffy rim more than 150 feet above the plains. In the Clovis-Portales area, the hardened surface of the Llano Estacado ranges in thickness between 10 and 40 feet, sheltering underlying Chinle Group units. This impermeable layer of calcium carbonate formed in arid conditions during the Miocene and Pliocene Epochs as surface water percolated upward through the soil and evaporated, depositing the minerals over time. This caliche, also called calcrete, is found mostly in the Ogallala, but in places it extends upward into the Pleistocene-age Blackwater Draw Formation and even some overlying younger dunes of windblown sand.

In a sequence of unconsolidated stream and lake deposits in Blackwater Draw, north of Portales on NM 467, an amazing assemblage of tools and animal bones has become one of the premier archeological sites in North America. Discovered and first explored in the 1920s and 1930s, this 640-acre locality is situated in what used to be a spring-fed lake environment in Pleistocene time. This location is the namesake of the Clovis culture, even though this hunting-and-gathering culture was widespread across North America and northern parts of South America. The Clovis people were the first widely spread prehistoric culture in the Americas.

70 THE PEACEFUL PLAINS 249

Geology along US 70 between Clovis and Roswell.

CENOZOIC
QUATERNARY
- Qa — alluvial deposits
- Qpl — playa lake deposits
- Qp — piedmont alluvial deposits
- Qw — windblown deposits
- Qwp — windblown and piedmont deposits
- Qoa — older alluvial deposits of upland plains and piedmont areas (latest Pliocene to late Quaternary)

NEOGENE
- To — Ogallala Formation (middle Miocene to early Pliocene)

MESOZOIC
CRETACEOUS
- K — sedimentary rocks, undivided

TRIASSIC
- ₸c — Chinle Group and Garita Creek Formation
- ₸s — Santa Rosa Formation

PALEOZOIC
PERMIAN
- Pat — Artesia Group
- Psa — San Andres Formation

VOLCANIC ROCKS
- Ti — mafic intrusive dikes (Late Eocene to Pliocene)

--- fault; dotted where concealed

Radiocarbon dating suggests the artifacts at Blackwater Draw are approximately 13,200 to 12,900 years old. These chipped-stone implements exemplify some of the oldest toolmaking technologies in North America. The Clovis people hunted mammoths, bison, mastodons, and smaller animals near the end of the last glacial period. Blackwater Draw also contains a variety of stone and bone tools and weapons, including cutting tools and stone blades, preserved in many layers. This indicates that this location was used for many years. These tools are found in close proximity to and within the same stratigraphic levels that contain fossils of Pleistocene animals such as camels, bison, horses, Columbian mammoths, saber-toothed cats, sloths, and dire wolves. It also contains, in the layers above the Clovis-age artifacts and bones, artifacts of at least four other cultures.

Geologists and archeologists have worked closely at Blackwater Draw to delineate the stratigraphy, or layers of rock and soil, that contain the artifacts and bones. Sediments change depending on climate, precipitation, and where they eroded from. Fire pits and other man-made features provide additional information regarding ages of use and habits. Fire pits also offer the opportunity to use radiocarbon dating on any charcoal left in the site. Radiocarbon dating uses the known half-life and decay pattern of the carbon-14 isotope.

At Blackwater Draw National Historic Landmark, owned and operated by Eastern New Mexico University, travelers can visit both the Archeological Interpretive Site, where excavations are going on, and a nearby museum. The Blackwater Draw site has a self-guided tour through modern and historic excavations.

South of Clovis, vegetated Quaternary dunes have a pink tone to them. For most of the long, straight stretch southwest of Portales, the road sits on the Ogallala Formation. This widespread gravel unit, eroded in Miocene to Pliocene time from the newly risen Rocky Mountains, was deposited on Triassic and Permian sedimentary rock, which occasionally has a surface presentation of reddish soil.

At Oasis State Park, within the Portales Valley, the Pleistocene Blackwater Draw Formation overlies the Ogallala Formation. The Blackwater Draw presents as thin,

In Oasis State Park, a clear pool of water among vegetated dunes has been stabilized by sandstone blocks and stocked with fish.

silty, clayey soils. This unit was deposited cyclically over 1 million years. Windblown deposition was followed by subhumid conditions that helped to cement soils. Three ashes erupted from the Jemez caldera help date the unit: the 1.61-million-year-old Guaje Pumice, several separately dated layers in the 1.59- to 1.22-million-year-old Cerro Toledo tephra, and the 1.22-million-year-old Tsankawi pumice.

Between mileposts 401 and 402, look south to see windmills of the Roosevelt Wind Project, powered by 125 wind turbines, each generating 2 megawatts. This wind farm became operational in 2015 and is one of many throughout the state.

Northeast of Roswell near milepost 365, US 70 lies just north of the Railroad Mountain dike, which looks like an elevated railroad. The dike is about 30 miles long and 150 feet wide. US 70 crosses its subdued western end halfway between milepost 365 and 364, but it is difficult to see from the highway. Railroad Mountain Road (Chaves County Road 38) follows it for part of the road's length. This fine-grained gabbro dike intruded 28 million years ago.

Haystack Mountain off-road vehicle area at milepost 357 accesses bare rock, gullies, and sand washes along a subtle escarpment of the Santa Rosa Formation.

At the Pecos River crossing (milepost 348), Triassic Chinle Group layers form upfolded ridges, the first sign of deformation related to uplift of the Sacramento Mountains to the west. In places, the cross-bedded and red-gray-buff variegated sandstone remnants of younger Triassic-age Santa Rosa Formation are visible in canyon walls to the north.

Bitter Lake National Wildlife Refuge, east of Roswell and part of the Salt Creek Wilderness, is home to one of the most diverse assemblages of dragonflies and damselflies in North America. It contains both saline and fresh artesian waters in natural lakes, sinkholes, springs, and wetlands. The collection of mostly circular water bodies is known as the Bitter Lake Group sinkholes. These waters are stored in cavities eroded in the gypsum bedrock and limestone of the San Andres Formation. The water that slowly replenishes the aquifer comes from the Sacramento Mountains, an up-thrown mountain range to the west. Streams that flow across the Pecos Slope west of Roswell lose water into the ground. Precipitation falls on the up-faulted San Andres Formation at the crest of the range and the underlying Grayburg Formation and percolates down and to the east, staying within the same porous formations.

US 82
Alamogordo—Artesia
111 miles

Alamogordo lies at the eastern edge of the Tularosa Valley, an internally drained low area that was not carved by a river. As the easternmost expression of the Rio Grande rift, it is a broad graben, a block of crust dropped down on faults between the Sacramento Mountains to the east and San Andres Mountains to the west. The structural basin contains more than 6,000 feet of sedimentary fill, including stream-deposited sand and gravel, rockslides, alluvial fans from mountains on either side, and playa lake deposits rich in salt and gypsum derived from the sedimentary rocks of both ranges. Streams draining from the surrounding mountains enter the valley and sink into porous alluvial deposits or pool in the deepest part of the valley, forming shallow,

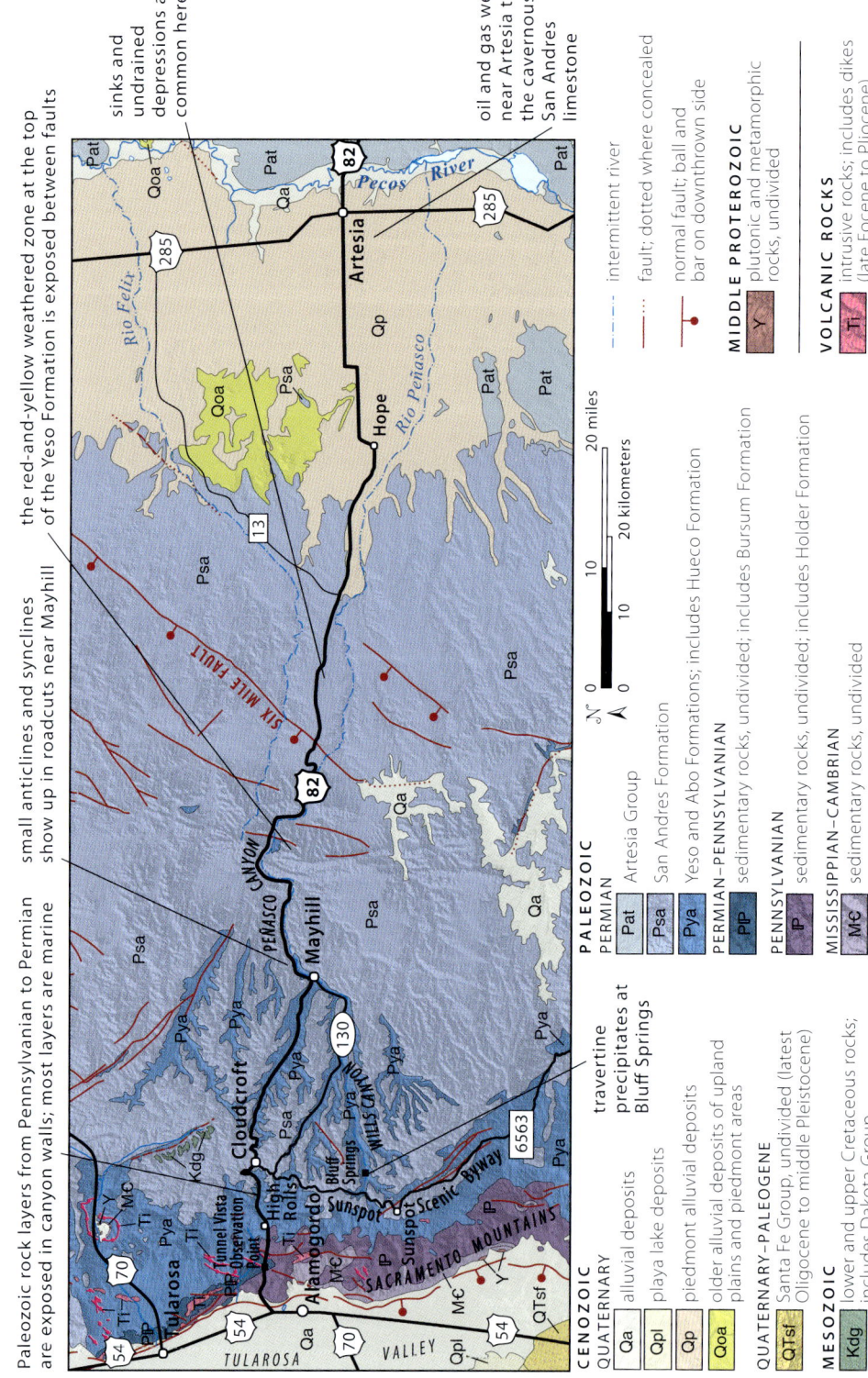

Geology along US 82 between Alamogordo and Artesia.

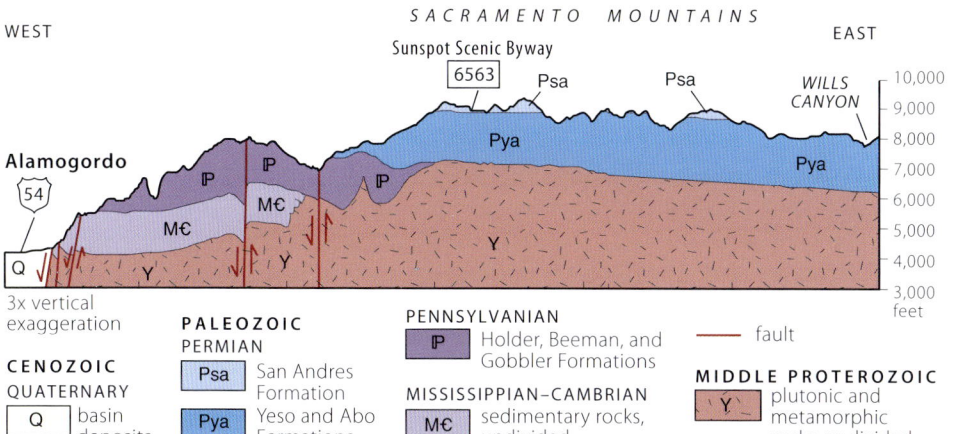

Cross section of the Sacramento Mountains east of Alamogordo. —Modified from Rawling, 2012

Layers of Pennsylvanian Gobbler Formation limestone show the line of a fault in the Sacramento Mountains. The beds on the south (right) have a higher dip angle than the ones on the north side of the fault. Viewed east from milepost 6.

ephemeral playa lakes. As these lakes dry, they leave behind their mineral burden of gypsum, salt, and other evaporite minerals in broad, white crusts.

The San Andres Formation, one of the sources of the salt and gypsum, shows up well in the Sacramento Mountains. As US 82 heads east from US 54/US 70, the stratified nature of the rocks becomes increasingly apparent in the view eastward toward the mountains. The layered rocks stand out on the mountain slopes, emphasized by differential erosion of soft shale and hard limestone layers. Shaly sections are much more visible in roadcuts than they are in the hillslopes. On the north side of the road between mileposts 5 and 6, a 44- to 36-million-year-old volcanic dike related to the earliest Rio Grande rift activity cuts through some of the limestone, baking adjacent rocks like reddened pottery. The highway crosses many north-south-trending faults as it progresses up this mountain slope.

A complete Paleozoic sequence—Cambrian to Permian—is preserved in the northern Sacramento Mountains, but along US 82, the Cambrian, Ordovician, and Silurian rocks are covered by alluvial fans. The oldest visible rocks are Devonian Oñate and Sly Gap Formation dolomitic sandstone and black shale, and massive Mississippian limestones and shales of the Caballero, Lake Valley, and Rancheria Formations.

At milepost 7, vertical cliffs of layered limestone units rise above the highway as it follows Fresnal Canyon upstream. Caves of many sizes show up in these limestones. Solution of these caves began in Late Mississippian time as the marine limestone was lifted above sea level. The surface is rough and irregular with patches of reddish silt and sand, evidence of ancient sinks and development of rugged karst topography similar to those that form today in limestone exposed to warm, humid climates.

The Fresnal Canyon tunnel is the only highway tunnel in New Mexico, and the Tunnel Vista Observation Point just west of the tunnel (milepost 7) affords good views of Pennsylvanian rock layers. The Magdalena Group showcases the limestones and shales of the Gobbler, Beeman, and Holder Formations, all of which were

The tunnel on US 82, viewed from the Tunnel Vista Observation Point west of the tunnel, passes through tilted layers of the Pennsylvanian Holder Formation. The Holder Formation, up to about 800 feet thick, forms cliffy outcrops of cliff-forming limestone, shale, and sandstone.

deposited on a stable marine shelf. These units display remarkable regional continuity: the Pennsylvanian and Permian layers here match up with those on the San Andres Mountains on the other side of the Tularosa Valley. Here, layers dip east; on the west side of the valley they dip west. At one point, these rocks arched completely across what is now the Tularosa Valley. As the Rio Grande rift developed, however, the central part of this arch dropped 5,000 feet or more along bounding faults, creating the present valley.

The Permian rocks of the great arch formerly included the thin Laborcita Formation and the more notable Abo, Yeso, and San Andres Formations. These units show changing depositional environments. Red shales merge into the arkosic red beds of the Abo Formation and into the shales and gypsums of the Yeso Formation, deposited in an isolated arm of a restricted seaway. When these confined, mineral-rich waters evaporated, both salt and gypsum were deposited, as well as smaller quantities of other evaporite minerals such as anhydrite (nonhydrated calcium sulfate) and borax (sodium borate). The deposited salt is so soluble that it was easily eroded away by percolating groundwater soon after the rocks were lifted above sea level. Erosion continues to remove less-soluble minerals more gradually. Much of the salt may have returned to the sea during the Mesozoic Era. The less-soluble gypsum is still being carried down into the Tularosa Valley, where it has concentrated in and around ephemeral Lake Lucero, becoming the source of the snow-white dunes of White Sands National Park.

Just west of the tunnel, the road crosses the buried north-south Fresnal fault. This fault is partly responsible for the formation of the canyons that extend to the north. Look for more caves and several small springs in the vertical roadcuts, first in the Pennsylvanian Beeman Formation (north), then the Gobbler Formation (both sides of road) and the capping Permian Bursum Formation (north side of road). East of the tunnel, US 82 continues up the canyon, first through more Pennsylvanian sedimentary rocks, then Permian rocks beginning at milepost 8. The rock is extremely porous, partly because the salt and gypsum it once contained have leached out and partly because the limestone itself is slightly soluble and is also highly fractured. The town of High Rolls rests on the Permian Abo Formation.

Thinly bedded limestone and reddish shale of the Yeso Formation occur between about milepost 10 and nearly to Cloudcroft (milepost 15), where the overlying San Andres Formation takes over. This latter rock, a resistant limestone, surfaces the long, gentle east slope of the Sacramento Mountains and eventually plunges eastward under the valley of the Pecos River.

The east side of the Sacramento Mountains is much less abrupt than the faulted western scarp. The sloping layer of San Andres Formation is cut by stream canyons, in which the underlying Yeso Formation is exposed. Near milepost 34, just west of Mayhill, beds of east-dipping Yeso Formation are particularly well exposed in a long roadcut. Yellow and orange-brown sandstone and siltstone separate the limestone layers. Small anticlines and synclines in the Yeso Formation show up in roadcuts east of Mayhill.

The gentle dip of the capping limestone and underlying Yeso layers can be seen in the walls of Peñasco Canyon, which US 82 follows east of Mayhill. Wherever the canyon widens, small farms have been developed on the soft, gray floodplain deposits of the Rio Peñasco. In many places, the stream has incised through its own deposits to depths of 20 to 30 feet. Alterations in agricultural land management, logging, and climatic change have altered the rate of stream flow and intensified channelization and erosion of gullies, or arroyos.

Cliffs that border the floodplain show interesting erosion patterns. As the stream develops sinuous curves and bends called meanders, indentations are carved in cliff walls that bound the stream. Mature rivers and streams erode most rapidly at the outer curves of their meanders, forming features called cutbanks, which are well displayed here. Meandering streams feature cutbanks on the outer sides of large bends where the stream energy is high and removes bank material. The cutbank is often paired with a location of low energy and deposition of sandbars on the inner side of the river bend. If these meanders continue to develop, they can become so sinuous that they cut off their long loops, creating an abandoned oxbow lake.

The limestone layers begin to level out near milepost 59, though the San Andres continues to floor the broad valley. The valleys are deeply gullied here. Circular sinks

SUNSPOT SCENIC BYWAY

The cream to whitish-gray travertine, a carbonate rock, continually precipitates from flowing springwater at Bluff Springs, accessed by trail from the Taylor Canyon Trailhead off Sunspot Scenic Byway. The bright-orange striping at the base of the travertine cliff is due to the presence of carotenoid-containing bacteria, which inhabit the warm waters.

The Sunspot Scenic Byway (NM 6563) is a 16-mile jaunt south from Cloudcroft to Sunspot, home of the Apache Point and National Solar Laboratories. The road number is special: 6563 angstroms is the wavelength of light scientists use to locate active areas on the sun. A visitor center and the telescope grounds are open to walkers. The highway wends over heavily forested, ridge-capping San Andres Formation, while the Yeso Formation becomes visible in areas of lower topography.

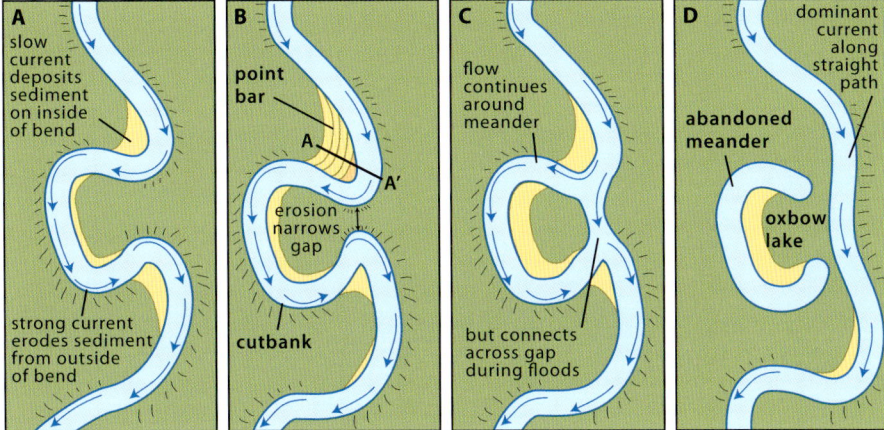

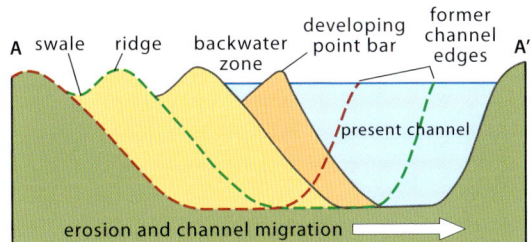

Cutbanks and cliffs form on the outside of a channel bend, with sandy point bars forming on the inside bend.

caused by the collapse of underground limestone caverns can be seen east of milepost 59. Look for a well-defined sink north of the road just west of milepost 66 and more on both sides of the road near milepost 68. Precipitation can infiltrate through some of these sinks, running through joints in the San Andres Formation that have enlarged over years of dissolution. These waters flow downdip along the underlying layers and into the Pecos River valley. Other sinks now tenuously hold water due to the collection of fine silt and mud within the cavities.

At the intersection with NM 13 (milepost 74), the highway encounters Quaternary alluvium, then runs in the nearby modern stream deposits of the Rio Peñasco. The highway crosses Eagle Draw, an ephemeral stream at milepost 91, and enters the Pecos River valley near Artesia (milepost 107), where the surface of the valley is dominated by alluvial sediment deposited at the end of the Pleistocene and in the Holocene. Small oil and gas wells between Hope and Artesia penetrate this alluvial cover and pierce the San Andres Formation below, which here serves as a reservoir unit for oil, natural gas, and water.

In downtown Artesia, the Navajo Refinery can process 100,000 barrels of crude oil per day, refining it into gasoline, kerosene, diesel fuel, lubricating oil, propane and butane gas, asphalt, and chemicals used in manufacturing synthetic fabrics, insecticides, and fertilizers. Petroleum refineries convert crude oil into its various components by heating it, a process called fractional distillation. The separation process heats crude oil in furnaces. Various liquid and vapor components of the crude oil separate and are harvested by their distinct boiling points and density. Fuel

The Navajo Refinery in Artesia.

oil and diesel are heavier and have higher boiling points, while kerosene, gasoline, butane, and propane have lower densities and lower boiling points. After the different components are separated, they are further broken down and converted into more refined products, a process called cracking. The final step in refining is the treatment and quality checking on the commercial product. Gasoline, for example, must be checked for octane level and vapor pressure. These many products are then stored near the refinery or transported for use.

Artesia is at the northern edge of the Permian Basin, an asymmetrical sedimentary basin in part of Texas and New Mexico, with several component basins within it. It is the largest petroleum-producing basin in the United States. The component basins are the Delaware Basin, Central Basin Platform, and Northwest Shelf located in southeastern New Mexico, and the Midland and Val Verde Basins and Eastern Shelf in west Texas. The Permian Basin formed as a downwarped foreland basin on the northwest side of the Ouachita-Marathon thrust belt, which bounds the basin to the south in west Texas. The foreland basin began collecting sediment during Pennsylvanian and Permian time, reaching more than 29,000 feet. Increases in clastic sedimentation reflected differential tectonic movements: Permian sedimentation occurred during the slowing and cessation of tectonism as the region became more stable and marine. The thickest accumulations of Pennsylvanian and Permian units are in the Delaware Basins.

The Central Basin Platform, an uplifted, fault-bounded structural high, separates the major Delaware and Midland Basins. This multilayered platform includes the carbonates of the Wichita-Albany Group and the Clear Fork, San Andres, Grayburg, and Yates Formations. These units were deposited as reefs and as clastic sediments in shallow seas.

Artesia developed where artesian water—groundwater that rises to the surface without being pumped—was found, which is how it got its name. This groundwater

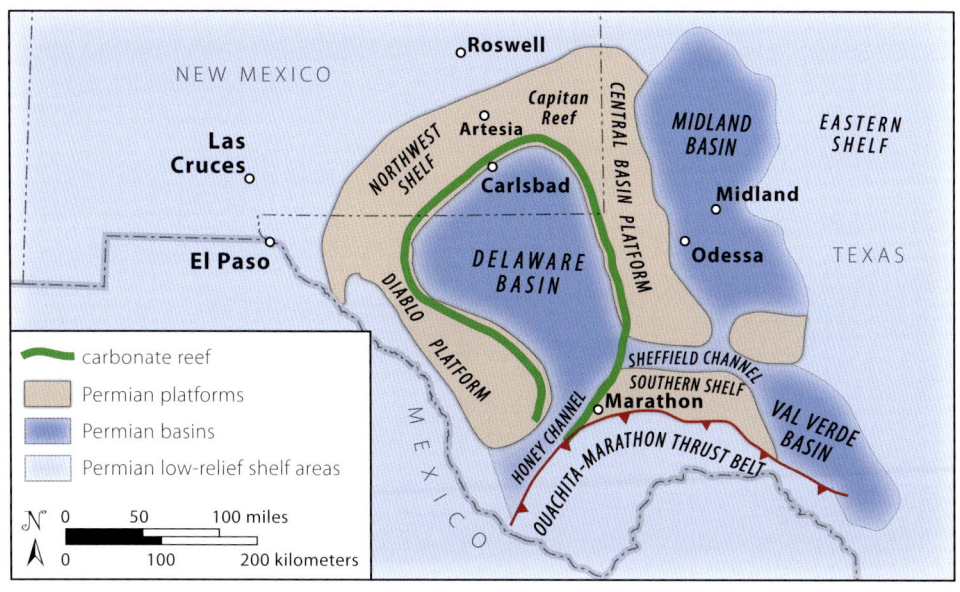

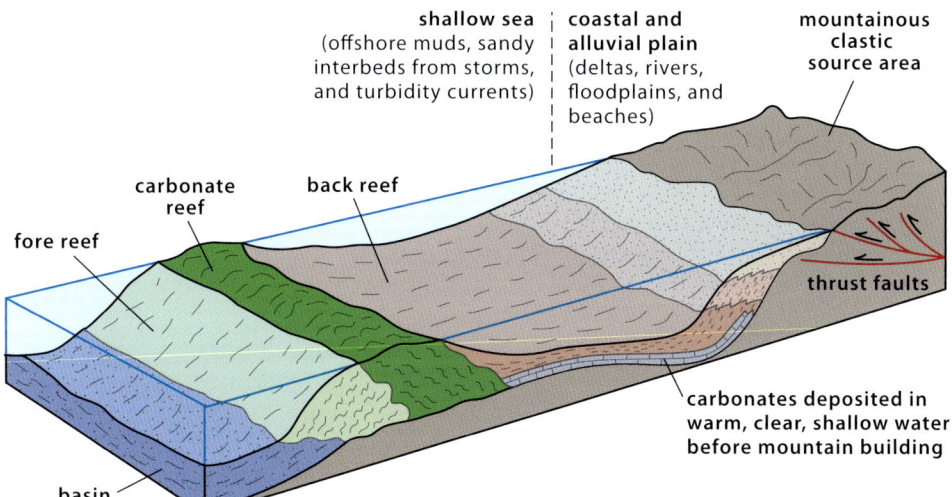

Map and cross section of the Permian Basin, a deep sedimentary basin that formed on the continental side of the Ouachita-Marathon thrust belt in Texas. The Delaware subbasin extended north into New Mexico, where shallowing waters along its edge permitted the formation of a reef.

originates from mountain precipitation that flows down and enters the porous and permeable Permian rock on the mountain slopes. These waters flow within the aquifer rock, the San Andres Formation limestone, because it is contained by less permeable layers in the Permian Grayburg, Queen, and Seven Rivers Formations above it. As groundwater flows down the dipping limestone layers, it develops hydrostatic head, the pressure that prompts the water in artesian wells to bubble up of their own accord. The weight of the confined upslope water pushes water down due to gravity. The weight of the stored waters within the aquifer pushes the water out of the ground. This same process brings water from a hilltop reservoir down sloping pipes and up into a house. Near Artesia, wells drilled through the less permeable layers provide escape routes for the water as this hydrostatic pressure forces it upward. Early settlers report that water from some wells shot up almost 100 feet above ground level when the aquifer was encountered.

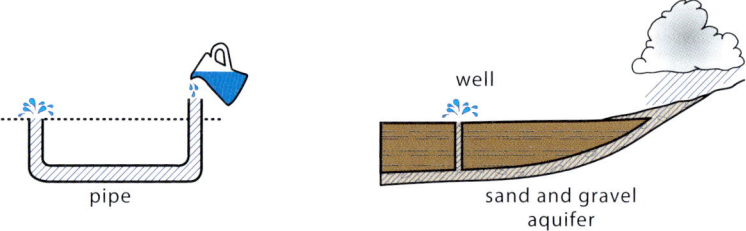

In an artesian system, the point of entry is higher than the wellhead, so well water rises without pumping. Impermeable shale or clay layers prevent the water from rising, so wells are drilled to access it.

US 285
Santa Fe—Encino
69 miles

This guide for US 285 starts in the southern end of the Sangre de Cristo Mountains, where pink hills of Proterozoic granite and metamorphic rock are the setting for some of Santa Fe's fine homes. South of the plutonic granite and granitic gneisses, the highway drops down onto the broad plains of eastern New Mexico. The transition from Proterozoic bedrock through the zone of radically deformed Paleozoic sedimentary rocks and onto the more flat-lying sedimentary units of the plains is unexpectedly scenic.

In the distance to the southwest are the Sandia Mountains, the other side of which is Albuquerque. About 300 feet of the visually striking 300-million-year-old Madera Group limestone caps the range. This ledge- and cliff-forming limestone contains abundant Pennsylvanian crinoid, brachiopod, gastropod, coral, and bryozoan fossils. The Madera Group forms a steady ramp from the mountain crest to the eastern plains.

At milepost 286, the highway drops down an escarpment of one of the faults extending south of the Sangre de Cristo Mountains, then parallels it briefly. The escarpment exposes a large roadcut of Quaternary sediment eroded from the surrounding mountains. The village of Lamy (east of the highway from milepost 284) was established as a stop along the Santa Fe Railroad and exists now as an artsy bedroom community to Santa Fe. At milepost 284, just south of the turnoff to Lamy, Cerro Colorado (literally meaning "red hill") lies just north of US 285 with its purples and reds characteristic of the Chinle Group. Deposited in the Triassic when active subduction along the west coast caused numerous volcanoes to erupt, the Chinle Group is full of bentonitic layers high in volcanic ash. Sandstone lenses in the formation form capstones; where there are none, the Chinle erodes easily into a monotonous flatland.

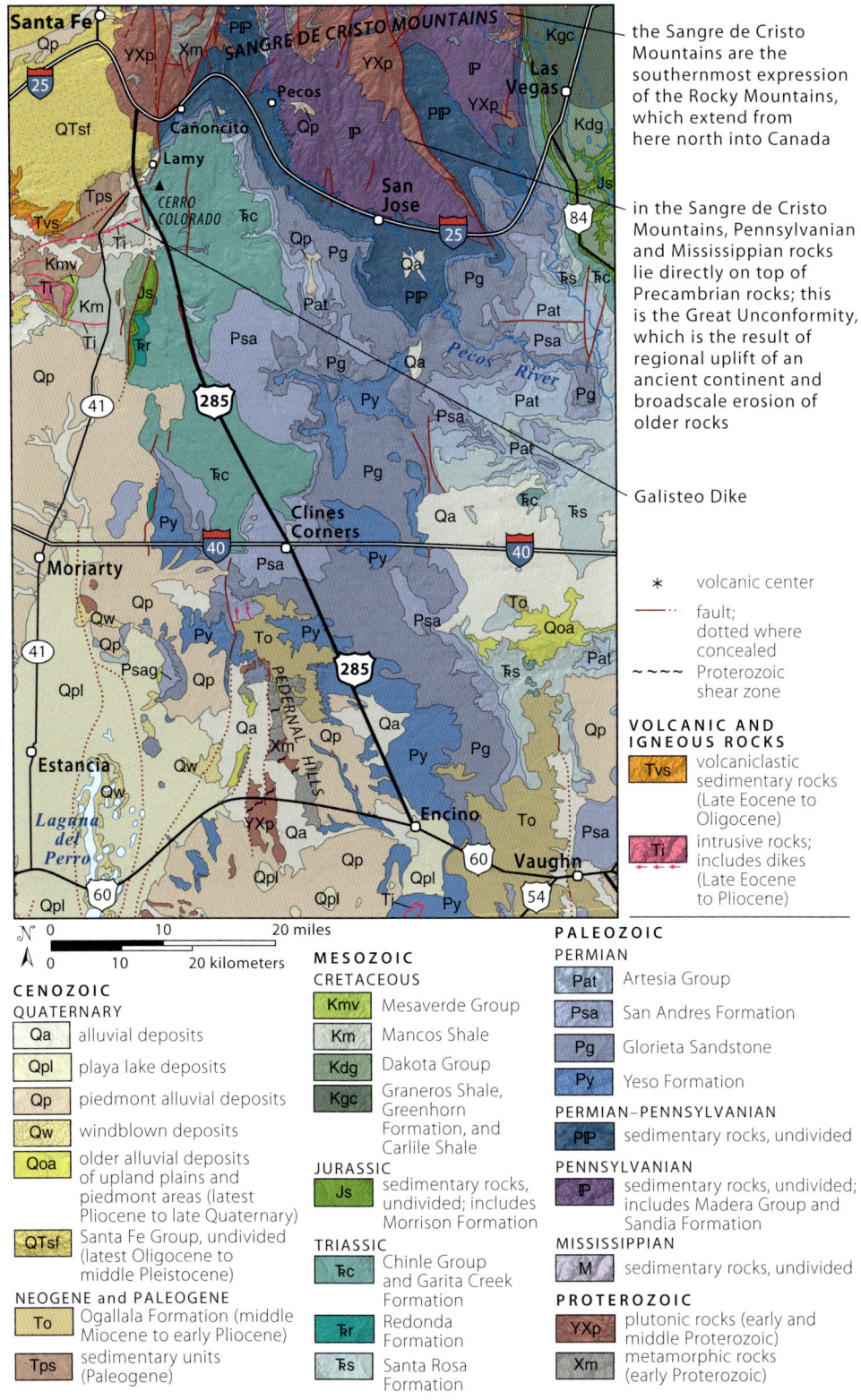

Geology along US 285 between I-25 south of Santa Fe and Encino.

The stunning colors of the Chinle Group come from differing concentrations and compositions of iron minerals in the sediments.

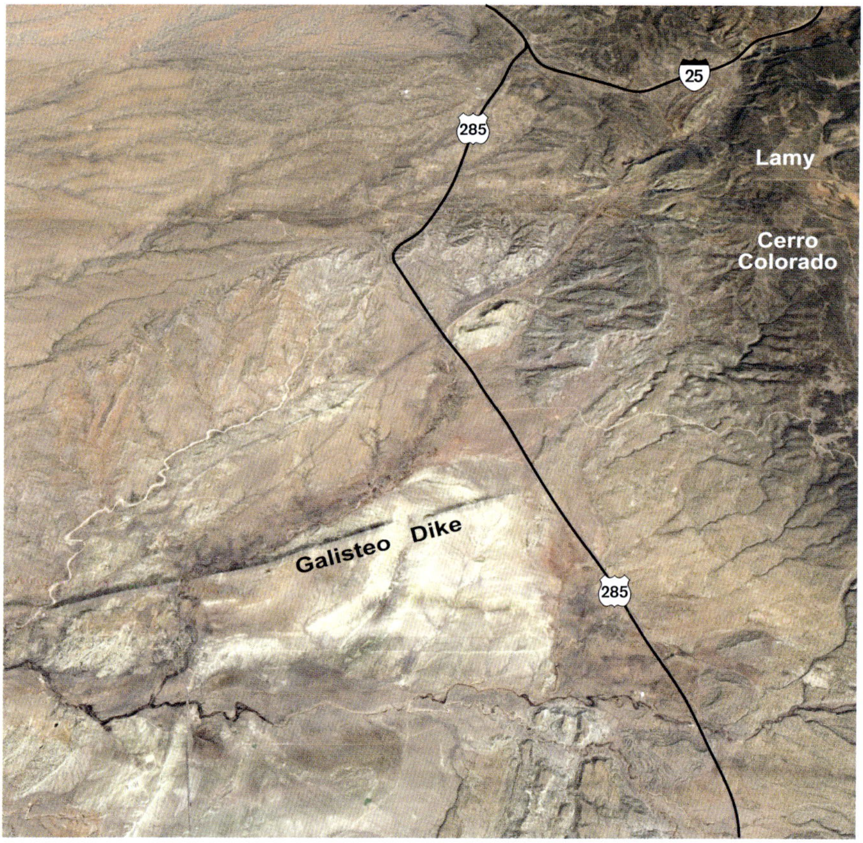

The Galisteo Dike is visible on satellite imagery. —Google Earth base image

A narrow rampart ridge, the trace of the Galisteo Dike, stands above the Mancos Shale to the southwest at milepost 282. This 26-million-year-old Oligocene dike is a series of overlapping segments that vary from 200 to 1,200 feet in length and are up to 18 feet thick. At the surface, the dike rock weathers to a rich dark brown.

At milepost 278, US 285 crosses the Arroyo San Cristobal. Here, faults bring the Triassic Chinle Group to the east up against the Jurassic Morrison Formation and Cretaceous Dakota Sandstone that form the hills to the west. US 285 follows the trace of these faults as it moves south through the canyon. Continuing south, the road remains on the Chinle until milepost 255, south of which is the San Andres Formation.

South of Clines Corners, US 285 remains on Paleozoic sedimentary units until south of milepost 241, where it climbs onto an exposed surface of the Ogallala Formation, a remnant far west of the more extensive unit that surfaces much of the southern Great Plains. Red layers overlain by white caliche characterize the Ogallala. South of milepost 239, the road drops suddenly through the white caliche layer, then continues down through the underlying reddish layers of the Yeso Formation. Look for other areas of the caliche between here and Encino.

Small knobs of Proterozoic rocks standing above the plain to the west of the highway at milepost 238 are the Pedernal Hills, composed of 1.7-billion-year-old Ortega Quartzite and 1.45- to 1.35-billion-year-old granites with large crystals in them. The Pedernal Hills are the ancient roots of the Ancestral Rocky Mountains. The Pennsylvanian and Permian sedimentary rocks surrounding the hills record pulses of mountain building, when rates of erosion were high. Siliciclastic sediments eroded from these former mountains collected in the basins surrounding these ancient uplifts. Rocks in the hills tell of a much higher and longer uplift than the present low hills reflect. Understanding the geometry of these ancient uplifts and basins is important in understanding hydrocarbon reservoirs because many of the ancient basins are now highly productive oil and gas reserves.

US 285
Vaughn—Roswell—Carlsbad
206 miles

US 285 is exceptionally flat between Vaughn and Roswell, with very few roadcuts. South of Vaughn, the road follows the eastern edge of a remnant of the Ogallala Formation and the Quaternary alluvial deposits to the east that overlie the Ogallala. The Ogallala Formation was deposited in early Pliocene to Miocene time when streams deposited clasts of eroded bedrock east across the plains in overlapping alluvial fans from the base of the mountains. A thick layer of caliche at the top of the Ogallala resists erosion, so the eroded edge of the unit often forms an escarpment, similar to the one paralleling the road just to the west but out of sight from the highway, which lies atop of the flat surface.

Quaternary and recent alluvial gravels and sediments surface the area to the east. These deposits are a thin veneer over the underlying thick mass of Permian San Andres Formation and Artesia Group limestone and gypsum. Along most of the road, the land is pockmarked with hummocks and sinkholes, reflections of the heavily eroded

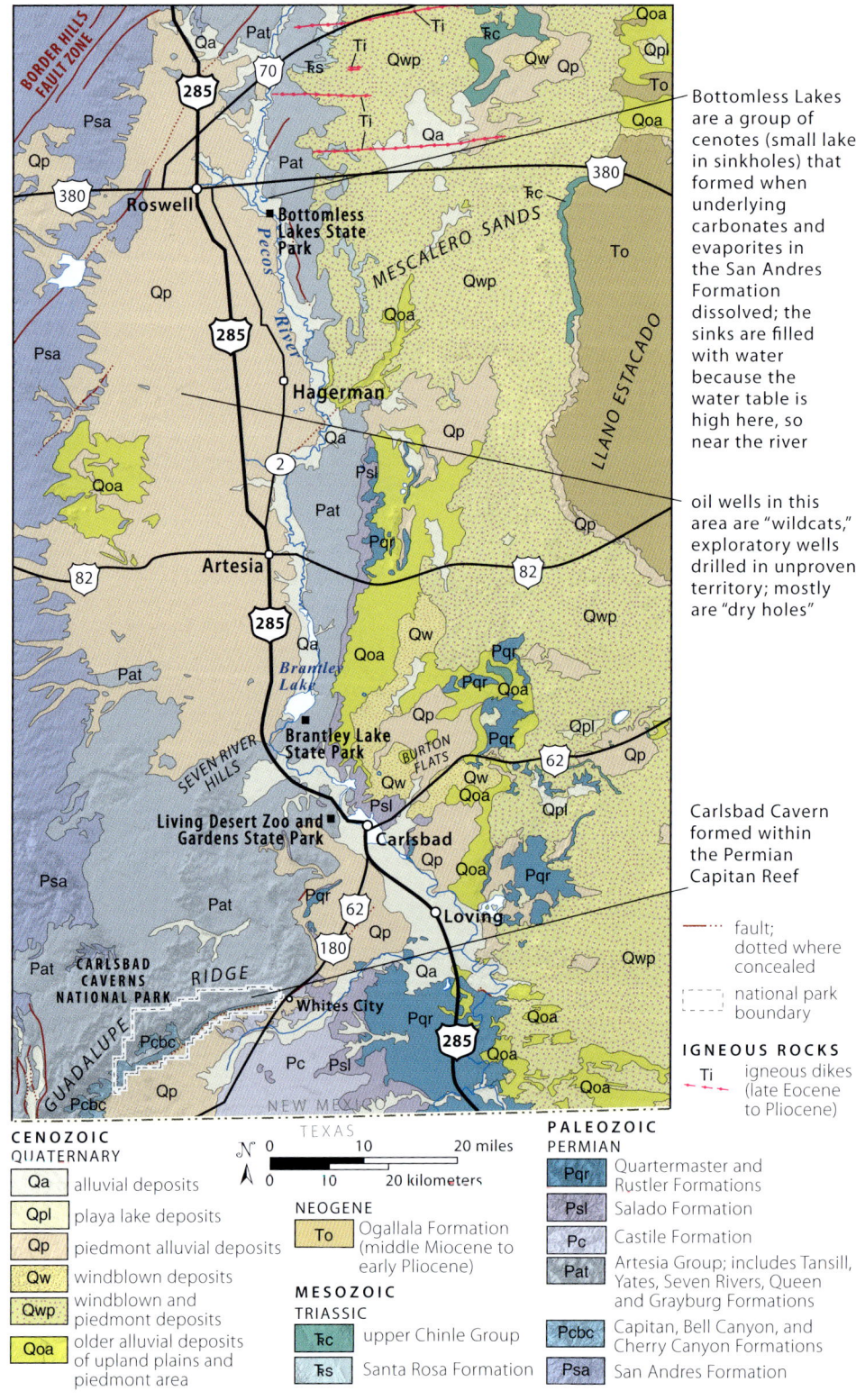

Geology along US 285 between Roswell and the Texas border. See map on page 239 for geology between Vaughn and Roswell.

limestone and gypsum bedrock below. Look for a sinkhole on the west side of the road just north of milepost 170. There are more to see between mileposts 160 and 156 where the Permian rocks are no longer veneered with alluvium. Notice how the soil has become reddish because it formed on Permian rocks high in oxidized iron.

The Artesia Group, which surfaces around milepost 150 as a broad south-trending outcrop, was deposited on a stable continental shelf. The Artesia Group records the youngest Permian deposition and includes the carbonate, evaporite, and clastic facies of five named formations: Grayburg, Queen, Seven Rivers, Yates, and Tansill Formations. It is a readily traceable unit in the Permian Basin in north-central New Mexico, western Texas, and southwestern Oklahoma. Changes in sea level are recorded in its variable shallow marine limestones, evaporites deposited in restricted basins, and shoreline siltstones. Note the large, reddish sinkhole east of the highway at milepost 138. Pecos diamonds, quartz crystals with points at both ends of the prism, are found in the Seven Rivers Formation along the Pecos River.

Roswell stands on the Quaternary alluvium of the Pecos River floodplain. In 1985, a woolly mammoth tusk was found in a gravel pit near Roswell. These ice age beasts went extinct about 10,000 years ago. See US 70 beginning on page 248 for information about the Bitter Lake National Wildlife Refuge. See page 275 in the US 380 road guide for information about Bottomless Lakes State Park.

South of Roswell, US 285 remains for some time on terraces that border the Pecos River. Lower terraces here, as well as the present floodplain, are farmlands irrigated with groundwater from the artesian aquifer in the San Andres and Grayburg Formations. Higher, unirrigated terraces are grazing land. The modern Pecos River meanders on the east edge of its floodplain, east of which the sandstone and limestone of the Artesia Group form low bluffs. Drilling and pipeline supply companies along the highway, as well as tank cars and trucks on the railroad and highway, announce that the route is in Oil Country, USA.

Pale-gray hills on either side of the river are in the Permian gypsum and mudstone of the Artesia Group. The high gypsum content and arid climate restrict plant growth on these hills. Rare precipitation percolates deep underground through the porous limestone, further limiting soil development and plant growth. The calcium carbonate rock and arid environment also result in highly alkaline soil and water. Caliche, deposits of calcium carbonate in soils, develops where calcium carbonate has precipitated out of mineral-rich groundwater and surface waters.

Artesia (milepost 70) is named for the artesian springs and developed wells in which water bubbles to the surface without pumping. This groundwater originates in the Sacramento Mountains, some 60 miles to the west, as precipitation enters the bedrock by percolating into the many sinks and small caverns that dot the surface of the Permian limestone between here and the mountains. As it flows downdip through the San Andres Formation aquifer, it develops pressure. The water is literally being pushed out of the ground by the weight of the stored waters within the aquifer.

Oil and gas wells surround Artesia. Those with large rockers are oil wells; those with intricate arrangements of pipes, valves, and gauges (known as Christmas trees) are mostly natural gas wells. Oil is stored in silo-shaped tanks, while gas, stored under pressure, requires strong round-ended silvered tanks, making it easy to tell which wells produce which. Both oil and natural gas come from the porous San Andres Formation. In this area, as well as farther south, Permian rock layers are warped and

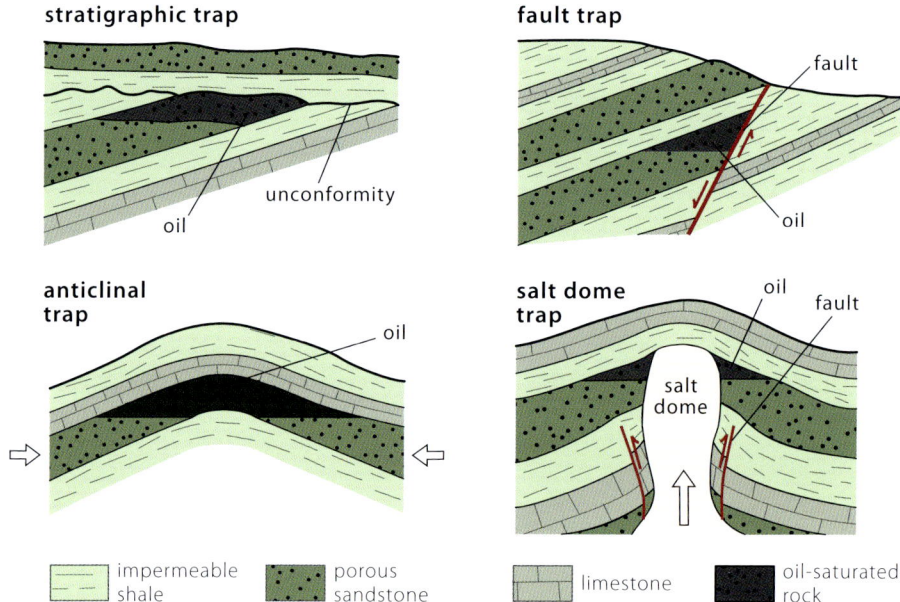

Oil and natural gas collect in petroleum traps. Buoyant petroleum rises into a porous host rock until an overlying impermeable layer traps it. Traps are often formed by natural geologic structures, such as anticlines, or along faults.

bent into small anticlines and synclines that make the area a petroleum and structural geologist's paradise. In anticlines, in which rock layers are domed up, oil and natural gas rise into the fold until impermeable layers trap the hydrocarbons.

Formed from uncounted billions of tiny marine organisms whose bodies were trapped in marine sediments, oil takes millions of years to develop. It tends to migrate from its source rock into and through permeable reservoir rock layers such as sandstone and cavernous limestone. There, it collects in various kinds of geologic traps, held by impermeable caprocks such as shale and mudstone. It is up to geologists to find the trapped oil, and finding it is a complicated business requiring surface mapping, core drilling, subsurface mapping by means of drill cuttings, and study of tiny microfossils brought up in cores and cuttings. Geophysical tools such as electrical logging of wells, study of manmade seismic waves transmitted by rocks below the surface, and studies of local variations in gravity and Earth magnetism also help.

The highway remains on the river terraces to about milepost 52, then rises onto the limestone surface. There are oil and gas wells here as well. The Ogallala Formation used to cover the limestone here when the Pecos River area was buried under hundreds of feet of sand and gravel. Since Pliocene time, however, the Pecos River has cut through these sediments, carrying rock debris downstream to the Rio Grande.

To the west at milepost 50 is the northern end of the Seven Rivers Hills. Brantley Lake, visible to the east at milepost 50, is a reservoir on the Pecos River. The dam consists of a 730-foot-long concrete section and a 3.9-mile-long earthen embayment and stands 143 feet above the valley floor. The dam sits on the Seven Rivers Formation of the Artesia Group. The soluble limestone would be a poor foundation for a dam, but the engineers constructed the dam at the juncture of permeable evaporite and less-permeable dolostone and sandstone. The less-soluble rock units provide a stable base for the dam structure. The limestone, dolostone, and gypsum of the Seven Rivers Formation outcrop on the eastern side of the lake in Brantley Lake State Park. Picnic grounds are built on and of this limestone. To get there, turn east on Capitan Reef Road between mileposts 46 and 45.

A geologist stands above a sinkhole in Permian carbonates at Burton Flats, northeast of Carlsbad. —Photo by Dave Decker

At Carlsbad, the road enters the Delaware Basin portion of the Permian Basin. This region of New Mexico and Texas was, during Permian time, a shallow embayment of the sea. Today, the one-time embayment is one of the most important and productive hydrocarbon sources in the United States. Carlsbad was first known as the town of Eddy, established by the Eddy Brothers in 1888. The name was changed to Carlsbad in 1899 because the supposedly beneficial mineral content of springs northwest of town rivaled that of the famous Carlsbad spas in western Bohemia, now the Czech Republic. The water chemistry is not similar at all.

Living Desert Zoo and Gardens State Park, accessed just south of milepost 38 in Carlsbad, highlights the flora and fauna of the Chihuahuan Desert. With a setting in the Ocotillo Hills, the park includes indoor and outdoor facilities, trails with interpretive signage, and other educational features. In the park, sedimentary units of the Tansill and Yates Formations of the Artesia Group are draped over preexisting reef mounds. The dolomites of the Tansill were deposited 260 million years ago in a shallow marine-shelf environment. The sandstone, siltstone, and limestones of the Yates Formation are exposed in arroyos. Sparkling-white gypsum, deposited as the Permian sea evaporated, is used for edging outdoor gardens.

Potash, a mixture of potassium salts and potassium carbonate, is another resource abundant in the Carlsbad region. It is valued as fertilizer and for improving water retention in soils. Potash forms as seawater evaporates and is found within the

Permian Salado Formation, a thick unit of halite salts and anhydrites. Thick layers of potash were discovered 900 to 1,800 feet below the surface near Carlsbad in 1925 during the drilling of a petroleum exploratory well. The potash is extracted from highly mechanized underground mines.

CARLSBAD CAVERNS NATIONAL PARK

Carlsbad Caverns National Park, one of the largest explored cavern complexes in the world, is developed in the complex geologic setting of the Capitan Reef. To reach the park from Carlsbad, head 20 miles south on US 62/US 180. To the west is Guadalupe Ridge, the northern finger of the Guadalupe Mountains, one of the most remarkable mountain ranges in the US. These massive limestone mountains are the remnants of an ancient reef composed entirely of myriad of small marine plants and animals of Permian time. Called the Capitan Reef after a prominent peak at the southern end of the mountains, this reef developed in shallow water along the edge of the Delaware Basin embayment. Confined shallow marine water conditions reached optimal

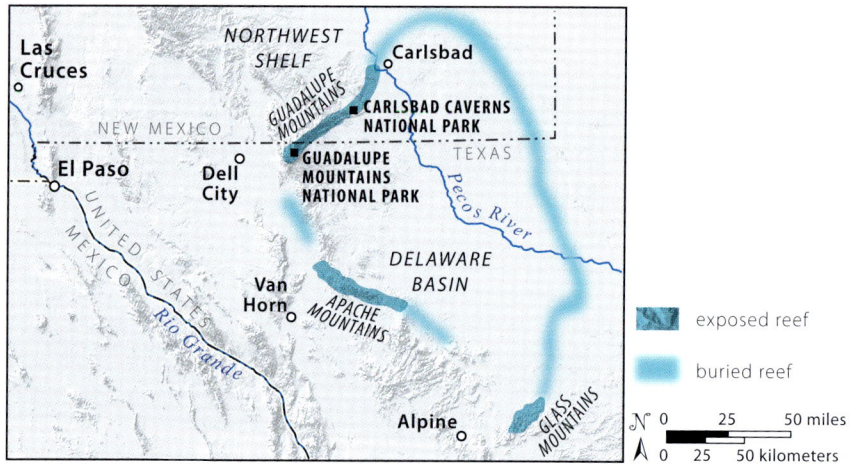

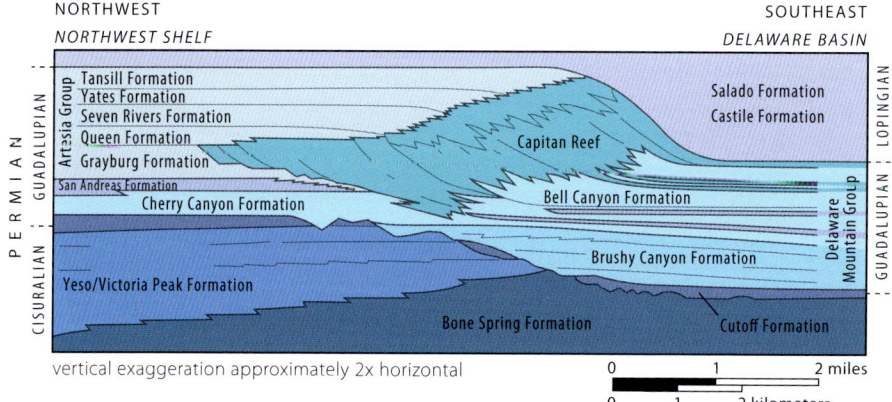

Map and cross section of the Capitan Reef. —Modified from Scholle, 2020; National Park Service

temperature, salinity, and water movement for the rapid growth of reef organisms. The Guadalupian Epoch of the Permian Period, 273 to 259 million years ago when the reef was actively growing, was named for the Guadalupe Mountains. This fossil reef shares many characteristics of today's Great Barrier Reef of Australia.

The Capitan Reef is constructed predominantly of sponges and algae, though other organisms existed, including bryozoans, corals, brachiopods, clams and snails, crinoids, and other calcium-secreting plants and animals. These together created a ridge of limestone, called the Capitan Limestone, that separated flat-lying sediments of the basin floor from equally flat-lying sediments of a protected lagoon to the west. The Goat Seep Dolomite is an underlying reef unit that also preserves algae and sponges.

Thanks to abundant subsurface information from oil and gas wells, the Capitan Reef is the best-understood fossil reef in the world. The reef is horseshoe-shaped, with the Guadalupe Mountains forming the visible northwestern curve of the horseshoe arching from west Texas through Eddy and Lea Counties in New Mexico. Where the reef bends back south into Texas, it is underground in the subsurface. The exposed portions of this backreef complex, notably Guadalupe Ridge and the Guadalupe, Glass, and Apache Mountains of Texas, have been elevated by faulting above the regional surface. This complex reaches maximum elevation of 8,750 feet at Guadalupe Peak in Guadalupe Mountains National Park, east of El Paso, Texas. The reef is present in the shallow subsurface beneath Carlsbad, where it forms a limestone aquifer that supplies drinking water to the town.

Eventually, the outlet between the restricted, saline section of the Delaware Basin and the ocean became so restricted that the waters of the Delaware Basin were evaporating faster than they were receiving new water. These increasingly saline waters left varves, seasonal deposits of gypsum and calcite, that form the Castile Formation. Nearly 2,000 feet of limestone rock, which was eventually buried under thousands of feet of sediment as the basin filled and subsided, formed the Capitan Formation.

This region underwent tectonic uplift, and the reef and neighboring host rocks were intensely fractured. Regional uplift and more localized Cenozoic block faulting have resulted in the exposure of the dramatic reef mountain ranges. This largely east-west faulting and associated fractures act as water conduits. Over millions of years, moving waters have dissolved and hollowed out the porous limestone of the Capitan Formation, forming caves.

A unique process formed the Carlsbad Caverns. Most limestone caves dissolve because carbon dioxide forms a weak carbonic acid. At Carlsbad, hydrogen sulfide found in the abundant hydrocarbon reservoirs present in the region interacts with groundwater to form sulfuric acid. This acid is much stronger than carbonic acid. Sulfuric acid interacted aggressively with the limestone, forming massive caves within the reef complex as the mountains were uplifted and the water table shifted and flowed along local faults. This sulfuric acid–limestone interaction, which began between 6 and 4 million years ago, also resulted in the deposition of large gypsum deposits on many cave floors.

Rocks close to US 62/US 180 both north and south of the park entrance road are dolomites of the Tansill Formation, once muddy sediments in the lagoon behind the Permian reef. Interbedded originally with salt, gypsum, and potash, they have bent, broken, and collapsed as these soluble substances—particularly the salt—were leached out by rainwater and groundwater. Because they are still highly mineralized, they support little vegetation, enhancing the desert aspects of this part of New Mexico.

Uplift of Guadalupe Ridge and the Guadalupe Mountains to the south occurred in several installments starting about 12 million years ago. Every prominent level in the caverns represents a period of time when the water table remained fairly constant. Submersion depended on the rate of uplift and the climate and, at times, once-dry portions of the caverns were resubmerged. The result of this interplay of strong acid and highly dissolvable reef rock is more than three hundred caves in the Guadalupe Mountains, many within the Carlsbad Caverns National Park. Older passages, at the top, have been tilted by continuing uplift of Guadalupe Ridge. The main part of Carlsbad Caverns is formed in the Capitan Formation. The cavern entrance, created by collapse of part of the cavern roof, is framed with lagoon dolomite from the Tansill Formation deposited behind Capitan Reef.

The floor plan of the cavern follows joints in the massive limestone. Most of them are vertical, and they are arranged in parallel sets that trend either northeast—parallel to the reef escarpment—or southeast. As expected, these are the directions of most of the passageways and rooms within both Carlsbad Cavern and other park caverns. As water drained from any one level, ceilings and partitions must certainly have collapsed, enlarging some parts of the caverns and reducing others.

Fairly late in the history of the caverns, slightly acidic surface water dissolved some of the limestone as it flowed downward through it. When this water reached the cave

Gypsum, a byproduct of the interaction of hydrogen sulfide and limestone, forms large crystals in Lechuguilla Cave, a cave in the national park discovered in 1986. The large amount of gypsum, and its composition, was one of the clues geologists used to determine that hydrogen sulfide was the agent of solution. —Photo by Gavin Newsom, National Park Service

Stalactites on the ceiling at Carlsbad Caverns.

Cave popcorn forms when warm, humid air currents laden with minerals cooled, dried, and dropped their loads of minerals on walls and other cave formations.

Stalactites and stalagmites, which, still growing by tiny additions of calcium carbonate, eventually join to form columns.

passages, the water released the calcium carbonate. The water, some in tiny droplets, some in thin sheets—began to deposit its dissolved minerals on cave surfaces. Spelethems, or cave decorations, begin forming very slowly. Calcium-carbonate-laden water dripping from the ceiling first forms straws, thin tubes with a sheathing of calcium carbonate. When the straws become plugged, the water flows down the outside of the tube instead, thickening it into a stalactite. Water dripping from the tips of a stalactite builds up a stalagmite on the floor. Today, the caverns are drier than they were in the past, and the cave formations are not growing as fast.

An excellent side trip through McKittrick Canyon, in Guadalupe Mountains National Park in Texas, exposes the entire reef in cross section from back reef to basin and is a world-famous exposure. To reach the turnoff to the canyon, continue on US 62/US 180 for 27 miles south of the Carlsbad Caverns turnoff.

US 380
Texas—Roswell
82 miles

The vast plains of eastern New Mexico and the panhandle of Texas are surfaced with gravel and sand of the Ogallala Formation. In Miocene and Pliocene time, streams carried gravels eastward from the southern Rocky Mountains and newly uplifted fault-block mountains east of the Rio Grande rift. Once a continuous surface all the way to the mountains, these gravels have been eroded and removed from the eastern slopes of the mountains and from the broad valley of the Pecos River. In New Mexico, only localized patches of Ogallala Formation remain along the eastern border of the state. US 380 crosses the largest remnant, a broad plateau called the Llano Estacado. The Ogallala is an incredibly important aquifer here in the dry southern Great Plains, but the flat gravel surface doesn't provide any interesting roadcuts.

Between the state line and Roswell, the highway goes through the northern fringes of one of New Mexico's two major oil-producing provinces. (The other is at the opposite corner of the state, in the San Juan Basin.) Oil and gas in this part of New Mexico originated in Paleozoic sedimentary rocks and occur in a number of pools—not pools in the usual sense of the word but underground masses of rock saturated with oil and gas. Wells vary in depth from about 1,000 to more than 17,000 feet. Nowadays, every effort is made to prevent "gushers," the dangerous and costly fountains of oil common in the early days of the oil industry.

Old windmills pump water from gravel layers of the Ogallala Formation, or at least they used to when the wells reached the water table. The water level has dropped because of overpumping, and many old wells are now dry.

In addition to oil wells and windmills, the Llano Estacado is marked by low, circular basins often containing water or mud, characteristic of parts of the Great Plains. These hollows are shallow sinks caused by solution of gypsum and salt in underlying Permian sedimentary rocks. There are dozens of such sinks between the Texas border and Roswell. Look for one north of the highway just east of milepost 204. They are easiest to pick out in summer when they contain the only green grass around and everything else is dry and brown.

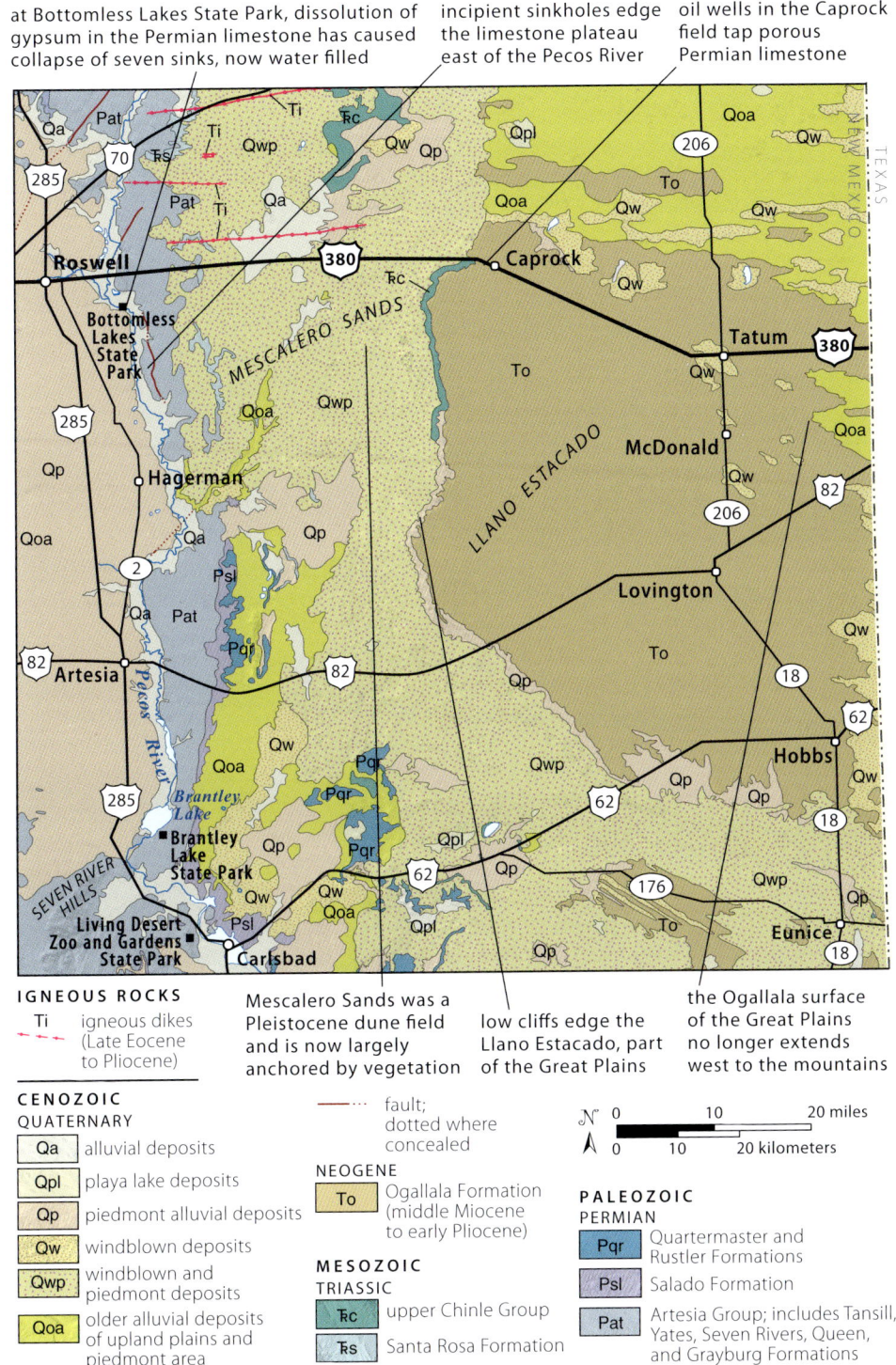

Geology along US 380 between the Texas border and Roswell.

The edge of the Llano Estacado is clearly marked by a line of bluffs capped with Ogallala Formation gravel cemented with caliche. In the foreground are dunes of the Mescalero Sands Recreation Area.

At milepost 200, 2 miles west of the small outpost of Caprock, the highway begins to drop off the surface of the Llano Estacado. The humpy topography visible ahead from milepost 196 is a Pleistocene sand dune field—the Mescalero Sands. The dunes are mostly stabilized by vegetation now, though still active sand surfaces can be seen to the south from milepost 194, where a road provides access to the Mescalero Sands Recreation Area. Prevailing winds in this area are from the west, and the dunes show a classic dune profile—a long, gentle windward slope, and a shorter, steeper leeward one. This region of sand dunes formed in two episodes: during the late Pleistocene (90,000 to 75,000 years ago) and then during the early Holocene (9,000 to 5,000 years ago). This sandy area is approximately 25 miles wide and home to large parabolic and coppice dunes stabilized by mesquite trees. These sand-sheet deposits rest on Permian limestone and evaporites of the Artesia Group, with some remnants of Triassic Moenkopi Formation that once covered the Artesia Group here.

Between milepost 175 and 174, the highway passes through another oil field where pumps are smaller because reservoir rocks are nearer to the surface. Some dark-red Triassic sedimentary rocks and light-tan and pinkish-tan Permian rocks belonging to the Artesia Group can be seen.

The Seven Rivers Formation of the Artesia Group is well exposed along the edge of the Pecos River valley. Look for low roadcuts near milepost 164, where US 380 crosses an arroyo that drains into the Pecos River. Thin, wavy, white stringers are layers of gypsum in the otherwise reddish rock. At milepost 163, as the road drops down off the bluffs through more red-and-white roadcuts of the Seven Rivers Formation, is a good view of the expansive Pecos River valley. It used to be spring fed, but with irrigation wells intercepting the groundwater, the Pecos is now a shadow of its former self. The floodplain on either side of the river is not used for farming because the

BOTTOMLESS LAKES STATE PARK

Bottomless Lakes State Park, reached by turning south on NM 409 at milepost 165, contains a chain of lakes. Fishing, swimming (only at Lea Lake), and picnicking are popular here. The lakes are technically cenotes, sinkholes, that vary in size, many with near-vertical walls that tower above their perennially present water. Cenotes form in a multistep process when carbonate material within bedrock dissolves and is removed by flowing groundwater, forming a cave. Either the overlying, unsupported ground material collapses into the underlying cave, or surrounding bedrock collapses when groundwater table changes elevation, leaving the rock dry, weak, and unsupported. The bedrock here is the Artesia Group, a sedimentary unit of gypsum, limestone, sandstone, siltstone, and shale deposited in the increasingly saline waters of the shelf setting of the Delaware Basin. The Artesia Group easily flows, bends, and warps as pockets of gypsum are dissolved and removed from within underlying limestone of the San Andres Formation. Some of this warping is visible in the lakeside exposures at Lea, Mirror, and Cottonwood Lakes. A few of the lakes are bordered by crusts of gypsum and other salts—deceptive crusts not strong enough to bear the weight of humans or automobiles. The lakes are not really bottomless, but they are as deep as 90 feet. They are fed by upward flow of groundwater from the underlying artesian aquifer, the same aquifer that feeds irrigation wells around Roswell. Evaporation rates are high.

Bottomless Lakes near Roswell occupy sinks formed by the collapse of underground caverns, which were produced by the solution of gypsum in the Permian rocks.

White stringers of gypsum cut through the reddish Artesia Group along the entrance road to Bottomless Lakes State Park.

groundwater becomes increasingly saline to the east. It is used for grazing, however. The terraces on the west side of the river are irrigated.

Originally a ranching and cattle-shipping center, Roswell is now a petroleum town—complete with the sulfurous fragrance of natural gas. Its water comes not from the Pecos River but from the Capitan, Sierra Blanca, and Sacramento Mountains to the west. Flowing through cavernous, gently east-dipping Permian limestone, the water reaches the Pecos Valley with a buildup of hydrostatic pressure, so that it rises in wells without pumping. The discovery of this artesian water in 1891 led to the explosive growth of agriculture in the Pecos lowlands.

US 380
Roswell—Carrizozo
90 miles

Roswell receives only 13 inches of rainfall a year—not enough to nourish the fields and orchards seen around the city. Artesian water was discovered here in 1891. Rain falling on the broad slopes of the Sacramento Mountains to the west percolates quickly into fractures and underground caverns dissolved in limestone of the San Andres Formation and overlying Grayburg Formation. This water takes ten to fifty years to journey eastward to the Pecos Valley. Pressure builds as it flows to lower elevations, so near Roswell, the waters in wells that tap the limestone aquifer flow without pumping.

Climbing west out of the valley, US 380 crosses several poorly defined Quaternary-age stream terraces of the Pecos River. These well-drained, gravelly surfaces provide excellent soils for agricultural fields. The final rise around milepost 315 is to a stony

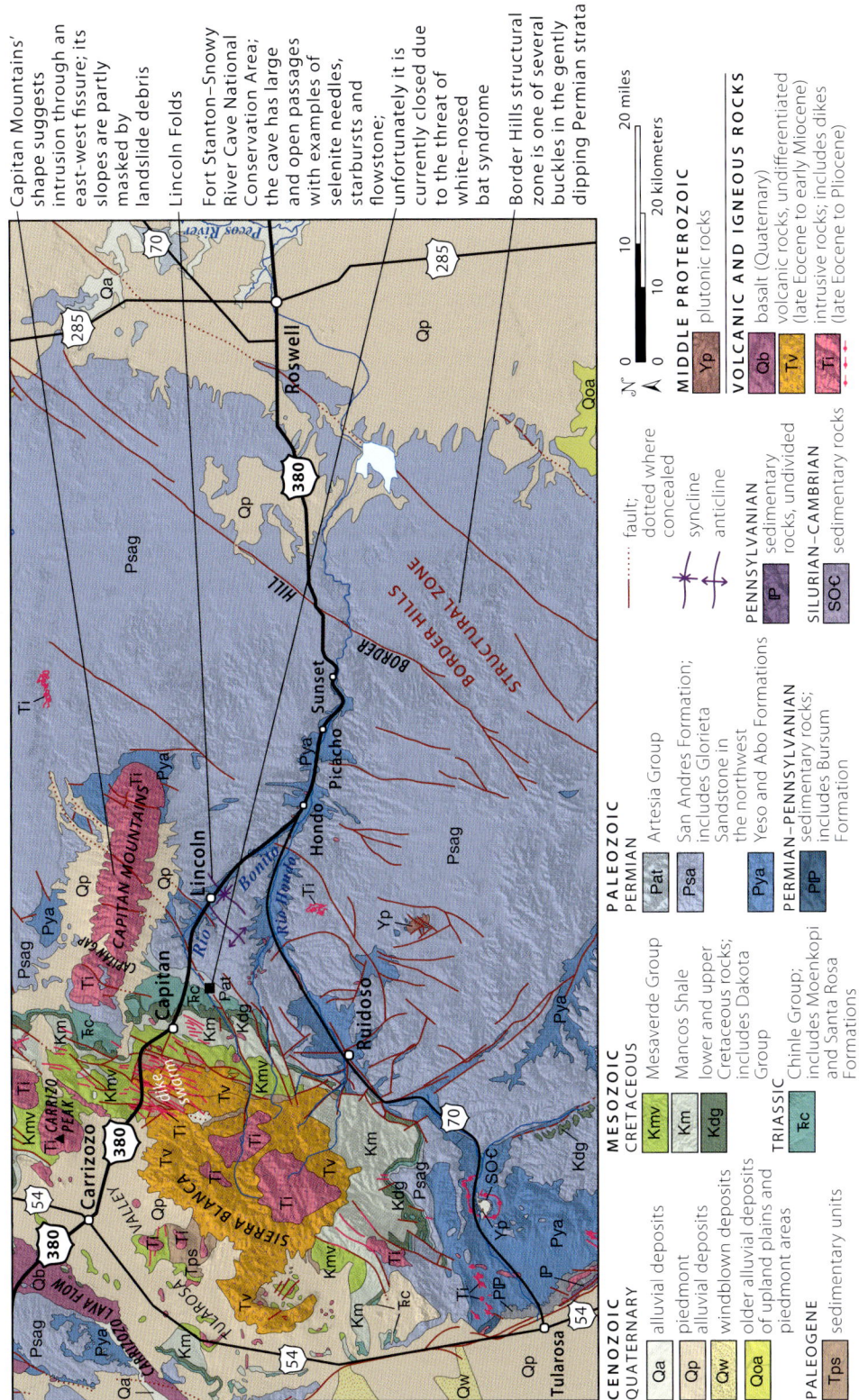

Geology along US 380 between Roswell and Carrizozo.

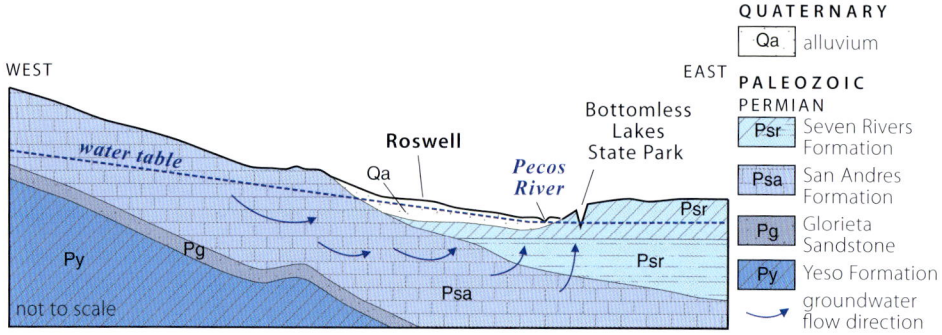

Water flowing through the San Andres Formation reaches Roswell and the Pecos River under pressure. —From Brandes, 2021

limestone surface of San Andres Formation, which is thinly covered with gravel. This whitish or yellowish Permian limestone is exposed in small roadcuts. Originally deposited in flat-lying layers on the seafloor in an open marine setting, the unit now tilts at approximately 1 degree to the east-northeast as part of the Pecos Slope, a broad structural feature in southeast New Mexico. Evidence of underground passages that honeycomb the San Andres Formation limestone can be seen from the highway; sinks form as cavern roofs collapse. The formation contains fossil gastropods, scaphopods, and bivalves, as well as evidence of burrows and bioturbation.

This region is part of the Northwest Shelf of the Permian Basin and marks the transition between the extensional Rio Grande rift and the stable Great Plains. Near milepost 326 and again near milepost 307, the road crosses long, northeast-trending faults, defined by long ridges and arroyos. These faults are part of the Border Hills structural zone.

The cone-shaped end of the Capitan Mountains dominates the view northwest of the weigh station at milepost 315. The east-west-trending mountains are an Oligocene pluton that was emplaced in the crust approximately 29 million years ago. The Capitan Mountains are the largest exposed Cenozoic intrusion in New Mexico, with more than 100 square miles of fine-grained granite outcrop. Farther west is the towering summit of Sierra Blanca, an eroded composite volcano. The Sierra Blanca massif is the boundary between the stable Great Plains to the east and the active Rio Grande rift and the collection of actively faulting mountain ranges of western North America. The Sierra Blanca was active 38 to 28 million years ago and produced lava flows, volcaniclastic sedimentary deposits, and minor welded ash-flow tuffs.

Much of the ground surface along the highway west of the weigh station is caliche, precipitated deposits of calcium carbonate and associated minerals in soil. Used for road material here, caliche forms a good aggregate that tends to harden with time.

The hill crested at milepost 307 is called Border Hill, one of the fault-created ridgelines in the Border Hills structural zone. At milepost 300, the highway drops into the valley of the Rio Hondo, with good exposures of the San Andres Formation limestone and the underlying Permian-age Glorieta Sandstone and Yeso Formation. Individual limestone layers in the San Andres Formation alternate with thin beds of shale, deposited as the marine depositional setting deepened and shallowed.

Roadcut through the San Andres Formation near milepost 301.

Sandstone and siltstones dominate the reddish-pink layers of the Yeso Formation, though the unit also contains limestone and gypsum. It was deposited in a shallow marine shelf and tidal setting where tides and low sea level events exposed the surface, resulting in frequent precipitation of gypsum and salt. Being very soluble, much of the salt and some of the gypsum have largely been removed through dissolution. The remaining gypsum is geologically incompetent, meaning it is weak, soft, and prone to deformation. As such units deform, the overlying layers tend to buckle and crumple, which is expressed in the common twisted layers and slumps seen in exposures. The Yeso Formation is separated from the San Andres Formation above by the yellowish cliffy slopes of the Glorieta Sandstone, visible west of milepost 297.

Layered yellow-to-tan siltstone, fine sandstone, limestone, dolomite, and gypsum of the Yeso Formation form the high roadcut on the eastern side of the highway just north of Picacho near milepost 292. Bedding is chaotic in places where gypsum has been partially removed through dissolution.

The Rio Hondo's floodplain, with small farms and orchards, is gullied by the present river. Gullying started during the 1860s, possibly as a result of overgrazing. The gullies are filled with cottonwood and tamarisk trees. Low, abandoned river terraces are cut in the San Andres Formation and record the incision of the Rio Hondo as it eroded its current channel. These terraces are capped with gravels deposited by the river before it started its most recent cycle of downcutting.

West of Hondo junction (milepost 285), US 380 follows the valley of Rio Bonito, and more exposures of the Yeso Formation, Glorieta Sandstone, and San Andres Formation are seen. The San Andres Formation is essentially flat lying, but the Yeso beds beneath the flat-lying limestone are buckled into the Lincoln Folds, spectacular features exposed for about 7 miles in bluffs north of the river. The absolute origin of the Lincoln Folds is debated, but two main hypotheses exist. The first is that the folds resulted from gravitational sliding of the unit during the late Eocene to Oligocene when the intrusions of the Sierra Blanca igneous complex bowed up the rock layers. The more recently adopted hypothesis for the folds is that they are slightly older and formed during the Eocene Epoch when Laramide compression and deformation resulted in elevated fluid pressures within the Yeso Formation layers. This high amount of fluid for millions of years resulted in the dissolution and removal of soluble layers and the buckling of the remaining, incompetent layers. Look east from milepost 97 across the valley at the base of the hills to see tilted rock. A large roadcut halfway between milepost 97 and 96 exposes similarly contorted Yeso Formation.

Side streams cut deeply into the soft Yeso Formation. Most are ephemeral, flowing only during and after heavy rains. When they enter the main river channel of the Rio Bonito, the streams rapidly lose energy and deposit gravels and cobbles in bouldery

Look north from around milepost 100 to see the Lincoln Folds beneath the flat-lying units that form the ridge.

alluvial fans that are visible along the valley margins. Where the highway crosses these fans, the poorly sorted nature of alluvial fans is visible in the jumbled pebbles, cobbles, and sands.

East and west of the historic town of Lincoln (milepost 98), large blocks of San Andres Formation have tumbled down across the Yeso Formation. Both formations rise westward to the summit of an anticline at milepost 92. Continuing its downward dip, the Yeso Formation disappears underground. Slope-forming reddish limestone, siltstone, and gypsum exposed on the western slope of the anticline are part of the Artesia Group, which overlies the San Andres. By milepost 90, near the Fort Stanton junction, these Permian rock units, too, disappear underground. Overlying them is the Triassic Santa Rosa Formation of the Chinle Group, with a silica pebble conglomerate layer about 6 feet thick as its base. The conglomerate contains pebbles of chert, quartzite, and igneous rocks not known in this region—smoothly rounded pebbles that were carried many miles by Triassic streams and rivers, primarily from the southeast, where the Ouachita fold belt was actively uplifting. The rest of the Santa Rosa Formation is a beach deposit of clean quartz sand with the uniform grain size typical of beach sand. The basal conglomerate of the Santa Rosa Formation is considered to be equivalent to the basal conglomerate of the Shinarump Formation of the Colorado Plateau, also a high-energy conglomerate. Together, these formations indicate an active river system extending from the Ouachita fold belt clear across New Mexico and into Nevada.

In this area, some of the many caves in the San Andres Formation have yielded prehistoric Native American artifacts. The Fort Stanton–Snowy River Cave National Conservation Area, both north and south of the highway here, protects the caverns and their contents. Fort Stanton Cave is 31 miles long, the fourteenth-longest cave in the United States. Another nearby protected site, Feather Cave, has evidence for AD 1350 ritual activity, including pictographs, feathers, prayer sticks, and painted mask fragments.

Along the straight section of the highway between mileposts 89 and 87, look north to see the long, east-west-trending ridge of the Capitan Mountains. The 29-million-year-old intrusive rock that forms it pushed its way between the San Andres Formation and underlying rock layers, doming up the limestone and younger rock units that lie above it. Erosion has removed most of the domed-up rocks, but their cut edges appear around the mountain base wherever they are not covered with landslide and rockfall debris. The magma rose up along the Capitan lineament, a line of intrusions and mineralization. Heated groundwater and magmatic fluids, or brines, concentrated and transported metal-rich waters along fractures and veins, and upon cooling, these concentrated waters precipitated and deposited iron and other metals and minerals.

As the highway climbs a low summit between mileposts 87 and 86, younger sedimentary rocks are exposed. The cliff-forming Late Cretaceous Dakota Sandstone overlies red beds of the Triassic Chinle Group. The band of sandstone to the north of the summit is the Dakota Sandstone, known for its widespread cliff-forming sandstones, mudstones, and shales that were deposited in nearshore and beach environments. The unit records the gradual encroachment of the Western Interior Seaway in Cretaceous time. Though the sandstone is light tan in color, exposed surfaces are usually dark due to lichens and desert varnish. Large blocks of resistant Dakota

Sandstone have fallen down near the road and rest on the slope-forming Chinle. Farther downhill, a dike cuts the sandstone in a roadcut (uphill from milepost 86).

Capitan (milepost 85) lies in a valley of the dark-gray Mancos Shale, which was deposited across the bottom of large portions of the Western Interior Seaway. The Mancos Shale contains Late Cretaceous fossil clams and ammonites, shelled relatives of today's octopus and squid. Capitan was originally a coal- and iron-mining center. The largest deposits of this region are about 6 miles north of town and are associated with the Capitan pluton. The iron is found in the zone of contact between the intrusion and the Paleozoic sediments it intruded, where it was concentrated by hot fluids during and after intrusion. Coal mines are just west of town, in rocks of the Mesaverde Group. These mining sites were called the Salado coal fields, and the sandstones, siltstones, and coal deposits reflect deposition during rising and falling sea levels in the Western Interior Seaway.

More Eocene-age andesitic dikes become visible west of milepost 85. More than two hundred dikes, a collection known as a dike swarm, cross the highway in the next 12 miles, many forming distinct ridges. These dikes trend northeast-southwest and are steeply inclined, ranging from 3 to 10 feet in width. They appear to have been injected into host sedimentary units along joints of the same directionality, and some of these features are up to 4 miles in length. Look for a dark dike in a roadcut of Mesaverde Group on the north side of the highway just east of milepost 83.

West of Capitan's coal mines, the highway crosses an unconformity into the sugary sandstones and dark-red and purple mudstones of the Eocene-age Cub Mountain Formation. This unit features prominently in a large roadcut on the south side of the road between mileposts 82 and 81. Braided streams deposited it. The highway continues into the Eocene to Oligocene volcanics of the Sierra Blanca igneous complex, which holds up the upland here. The dikes cut these units, too.

Basalt dikes cut the rocks on the north side of the road near milepost 76.

At Indian Divide (milepost 79), the highway crosses another northeast-trending fault, this one bringing the Mesaverde Group to the surface again. As the highway begins its descent toward the Tularosa Valley, it sweeps around multiple corners with dramatic, colorful roadcuts of Mesaverde sandstone and coal, again followed by the purple-and-white Cub Mountain Formation. Numerous dikes cut both. A particularly large dike crosses the highway just uphill from milepost 76 and takes the form of a prominent northeast-trending ridge. To the west of milepost 76, the highway descends a broad alluvial apron to Carrizozo. Small peaks north of the highway, including Carrizo Peak, are Oligocene intrusions. Alluvial fans cover the base of these peaks. Gold and silver were mined in these hills and in the White Oaks mining district (now in Lincoln National Forest) from the 1880s to the 1940s. To the west, across the Tularosa Valley, the black flows of the Valley of Fires lavas can be seen. See page 209 for more information about these flows and the Tularosa Valley, a basin of the Rio Grande rift.

THE VOLCANIC SOUTHWEST

The landscape of southwest New Mexico is typical of the vast Basin and Range Province, a distinctive topography of north-south-oriented mountain ranges separated by valleys. The Basin and Range extends from western New Mexico west through Arizona to southern California, and north through Nevada to southeast Oregon and southern Idaho. This topography resulted from the clockwise rotation and extension of the Southwest as motion of the Pacific Plate caused east-west tension on the continent, beginning in New Mexico about 20 million years ago. Normal faults developed to accommodate the tension, or pulling apart, of the crust. The mountain ranges are uplifted fault blocks, called horsts, and the valleys are down-dropped fault blocks, called grabens. On the east, the Basin and Range merges with the Rio Grande rift near Deming, New Mexico. To the north, the Basin and Range merges into mesas of the Colorado Plateau.

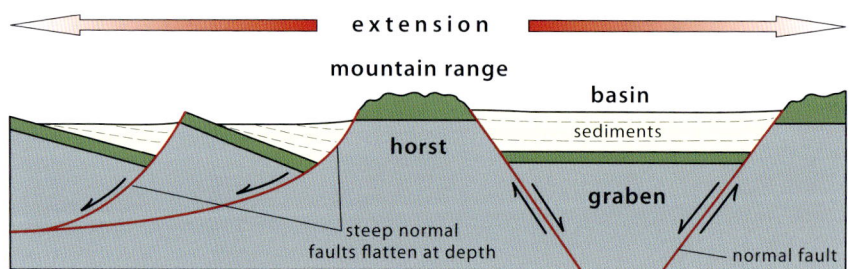

The stretching of the Basin and Range province in the last 20 million years broke the landscape into blocks uplifted and downdropped along normal faults.

The down-dropped valleys are filled with debris that has eroded from the mountains for millennia. The widespread basin fill of Miocene to Quaternary age is lumped into the Gila Group (also called the Gila Conglomerate). It is the time equivalent to the Santa Fe Group, the basin fill along the Rio Grande. The Gila Group is generally present on the western side of the Continental Divide in New Mexico as basin fill along tributaries of the Colorado River such as the Gila and San Francisco Rivers. It is also present in closed basins that don't have an outlet to a through-flowing river. The Gila Group records largely alluvial deposition but also includes lake deposits, volcaniclastics, windblown sands, and river gravels. It is notable for including clasts of the Bloodgood Canyon Tuff, which have a striking blue reflection, as well as locally preserving fossils of turtle, heron, and many mammalian species.

Much of the Gila Group is conglomerate, a rock unit solidified from the pebbly to cobbly deposits of alluvial fans, which are triangular-shaped deposits that angle down from the mountain front to the valley. Where fast-flowing, sediment-filled water exits the steep confines of a narrow mountain canyon and reaches a lower-gradient valley, the stream energy drops suddenly and sediment piles up in the stream path. As the channel fills with sands, pebbles, and larger clasts, the stream shifts to a lower area, and this constant shifting of stream orientation results in a triangular-shaped fan deposit. The coarsest and largest material is found at the apex of the fan, while smaller sediments can be transported farther out onto the fan toward the valley floor.

The mountain ranges in southwestern New Mexico are primarily composed of volcanic rocks of the Mogollon-Datil volcanic field, where volcanoes poured out vast quantities of andesitic-to-silicic lavas 40 to 24 million years ago in the Eocene to Oligocene Epochs. This major period of volcanism is known as the Oligocene ignimbrite flare-up. Volcanism stretches from the San Juan Mountain calderas of southwestern Colorado south to the Trans-Pecos volcanic field in west Texas and into the massive Sierra Madre Occidental of western Mexico. The Mogollon-Datil volcanic field is composed of two major caldera complexes: the Mogollon and the Datil. In the southwestern part of New Mexico is another large field, the aptly named

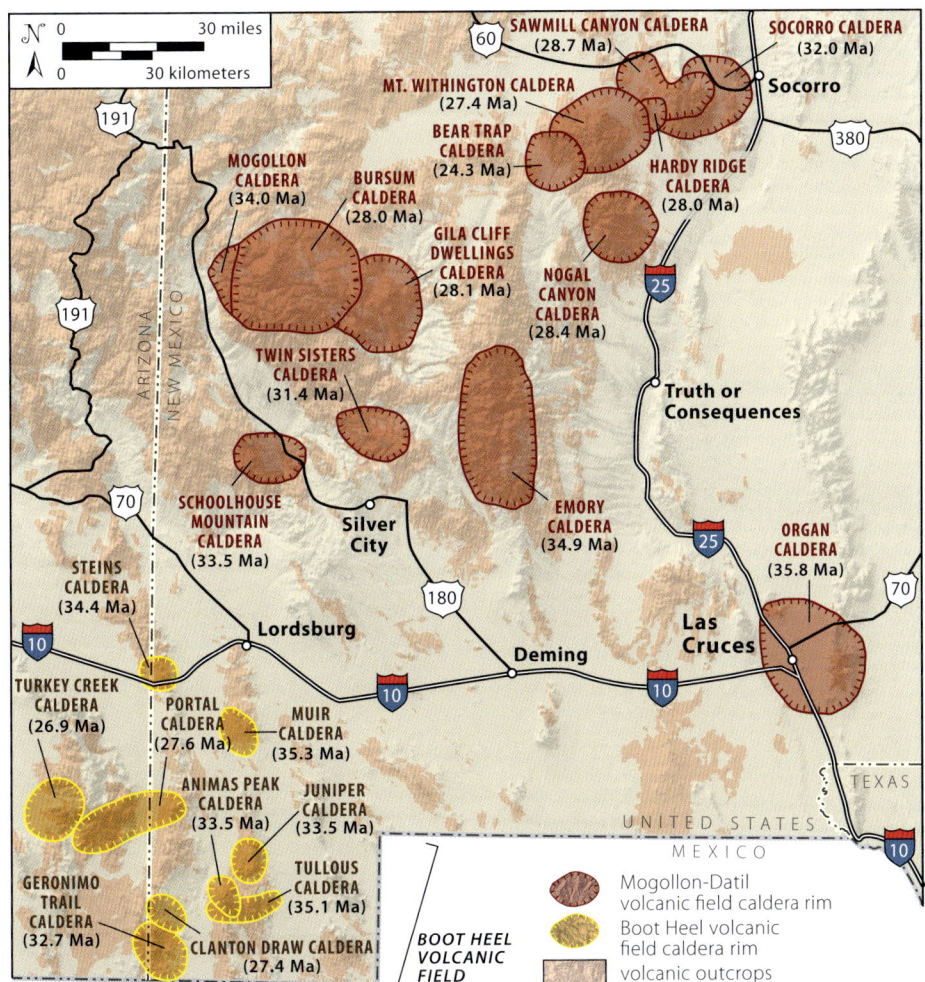

The Mogollon-Datil volcanic field consists of about a dozen large calderas that erupted 36 to 24 million years ago. A smaller volcanic field dominated by calderas occupies the Boot Heel of New Mexico. The tuff and volcanic rock erupted from these calderas outcrop throughout much of southwestern New Mexico. —Modified from McLemore, 2010

Boot Heel volcanic field. It was active from 35.3 and 26.9 million years ago with nine calderas, some of which extend into Arizona.

Lava composition varied from basaltic andesites, andesites, and dacites to rhyolites and associated tuffs. Silicic lava (dacitic and rhyolitic) can be highly viscous, meaning it is thick and sticky and does not flow as readily as the more mafic basalt. Individual silicic lava flows are short and stubby. They commonly alternate with layers of tuff and volcanic breccia, products of explosive eruptions. Where enormous volumes of

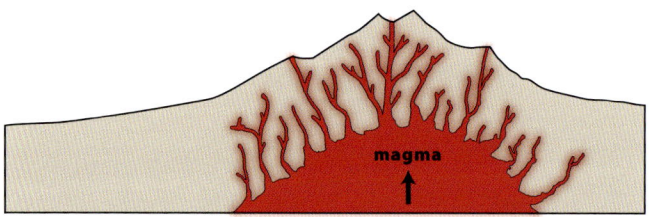

A. precursor stage; magma rises, inflating the surface

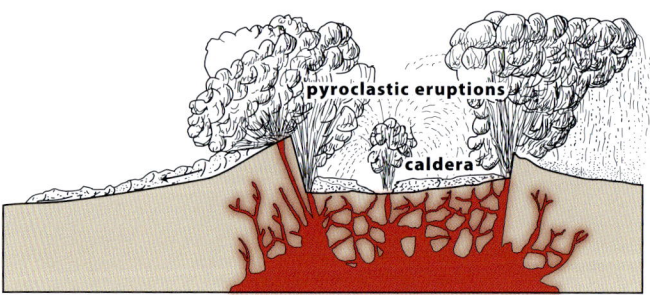

B. explosion, eruption, and caldera-forming stage

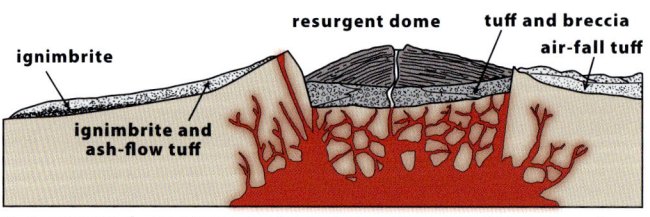

C. resurgent dome stage; continued rise of magma forms lava dome

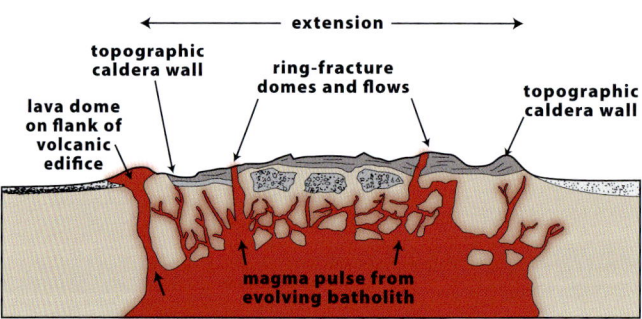

D. ring-fracture dome stage; erosion widens caldera

Calderas form when the pressure of rising magma (A) gets too high and the volcano explodes catastrophically (B). A depression, or caldera, forms above the emptied magma chamber. The pyroclastic material cools to become ash-flow tuff and welded tuff (ignimbrite), while lava that flows within the caldera forms resurgent domes (C). Erosion widens the calera, rejuvenated magma leaks through ring fractures and forms ring-fracture domes within the caldera wall (D). —Modified from Elston, 2008; Decourten and Biggar, 2017 (RGNV)

silicic magma erupted, the underlying magma chamber emptied and the overlying and nearby rock collapsed into it, forming a caldera. Some of these calderas are more than 30 miles across.

In the Mogollon-Datil volcanic field, lavas were initially andesitic from 40 to 36 million years ago, followed by basaltic and andesitic volcanoes and silicic calderas that dominated from 36 to 24 million years ago. Caldera activity began near the Organ Mountains, then migrated eastward toward the Bursum caldera and eventually northward toward Socorro. Many of the lavas of the Mogollon Group are quite different from those of the Rio Grande rift. They are lighter in color (gray or purplish gray) and contain a higher percentage of silica. Older plutons along the edges of the Mogollon-Datil volcanic field were highly mineralized during this volcanic history and now host lucrative copper and silver mines.

I-10
Arizona—Lordsburg—Deming
82 miles

Much of the outcropping rock between the Arizona border and milepost 50 is part of the Boot Heel volcanic field that erupted between 35.3 and 26.9 million years ago. The Peloncillo Mountains straddle the Arizona–New Mexico state line and extend south to the US–Mexico border. To the south of the highway, the Peloncillo Mountains are geologically complex, with 1.4-billion-year-old granites in the core of a faulted anticline. Faulting has juxtaposed the granite with Pennsylvanian, Permian, and Cretaceous rocks, along with various intrusions and rhyolitic, andesitic, and dacitic extrusive lavas, all Cenozoic in age.

In the ghost town of Steins (exit 3), I-10 summits the range at 4,400 feet elevation. Volcanic material from this region was quarried and crushed for use as ballast for the Southern Pacific Railway, which was completed in this area in 1880. Foundations of the old mill still stand in Steins. Outcrops of Eocene volcanics occur along the interstate. Look on the north side of the road for vertical columns between mileposts 2 and 3, then cliffy outcroppings on the south side just west of milepost 3.

Vertical joints create columns in the Eocene volcanics in the Peloncillo Mountains north of milepost 2.

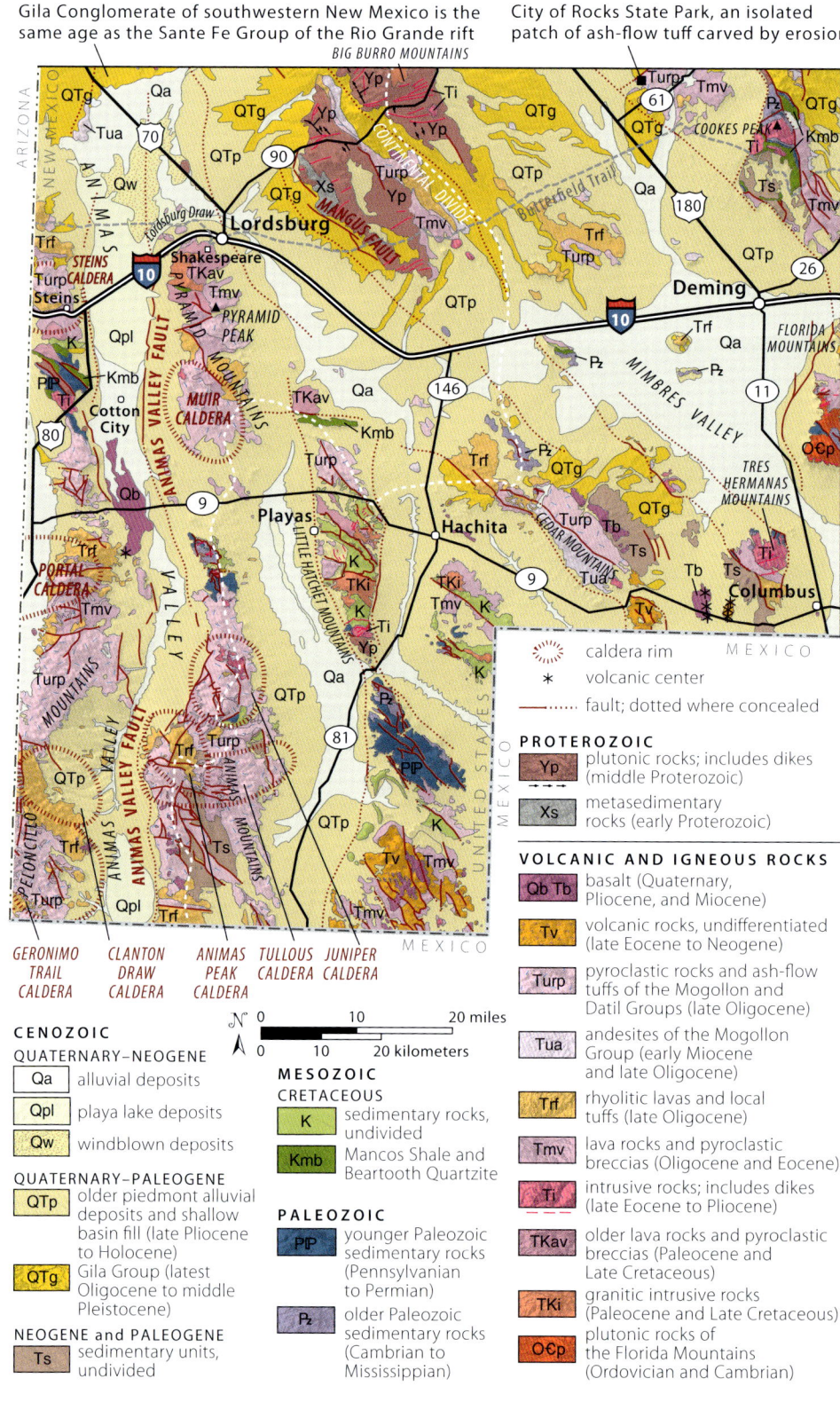

Geology along I-10 between the Arizona border and Deming.

At milepost 3, charcoal-gray rhyolite dikes cut the pink andesite lavas and pyroclastic flow breccias of the Cenozoic-age Steins Mountain quartz latite porphyry immediately north of the interstate. Abandoned ruins of the mining town lie just north of the highway. East of the Peloncillo Mountains, the highway closely follows the route of the Butterfield Trail, a pre-railroad stagecoach route to California. It descends the broad bajada, or alluvial apron, that surrounds the mountains.

Near milepost 5, the route crosses three beach ridges of the Pleistocene Lake Animas but, unfortunately, these shoreline features are obscured near the highway. Lake Animas was once 12 miles wide and up to 65 feet deep, occupying a fault-bordered graben bordered by the Peloncillo Mountains to the west and the Pyramid and Animas Mountains to the east. At its greatest extent 12,000 years ago, Lake Animas was probably freshwater, and it is thought to have drained northward toward the Gila River Valley. As its waters evaporated, bringing lake level below the outlet level, the lake became increasingly salty. A broad modern playa, with deposits of salt and other evaporite minerals, now occupies part of the old lake floor. Like many other valleys in the Basin and Range region, it is internally drained. Water entering it flows toward its lowest spot and remains trapped, slowly evaporating and leaving any dissolved minerals—salt, gypsum, sodium carbonate, and other compounds—behind on the playa surface. Large travertine deposits and near-surface vein deposits of calcite, fluorite, and psilomenane (black manganese oxide) are also present within the Animas Valley.

When dry, the playa is marked with a network of mud cracks, the result of clay-rich sediments shrinking. Over much of the lake surface, the mud-crack pattern is quite small scale, but out near the center of the playa huge cracks as much as 1 foot wide delineate 10-foot or larger polygons arranged in long, parallel rows. They are visible on satellite imagery and Google Earth.

South of I-10 near Cotton City in the southern part of the Animas Valley lies a geothermal area where naturally heated water is close to the surface, warming the ground enough to cause rapid snowmelt (if it snows) and earlier-than-normal springtime greening of the desert and crops. The geothermal area hosts a power plant, greenhouses, and an aquaculture farm. The power plant pumps out hot water, runs it through a heat exchanger to turn turbines that produce the power, then pumps the now-cooled-but-otherwise-unchanged water back into the ground. The hot-water system at Cotton City lies near the intersection of the 35.3-million-year-old Muir caldera and the Animas Valley fault. The caldera is one of nine located in the Boot Heel volcanic field.

At milepost 11, views to the south show the Animas Mountains, just north of the US–Mexico border. As trapped waters evaporate, they leave behind their mineral contents in the form a of pale caliche crust, which coats the valley floor here. About halfway between mileposts 12 and 13, the road cuts through two Pleistocene beach ridges, but they are difficult to discern. Look for yuccas, which seem to prefer the sandy soil of the beach ridges.

The highway then climbs gradually across the bajada surrounding the Pyramid Mountains, which get their name from the dark, conical peak—a volcanic neck—at the north end of the range. Other than intrusive rocks of the same age exposed near the north end of the range, these mountains are composed almost entirely of volcanic rocks of Eocene to latest Oligocene age, the remnants of caldera eruptions including andesitic, rhyolitic, and dacitic lava flows of the Datil and Spears Groups and others. Hydrothermal alteration of rocks is widespread throughout the Pyramid Mountains.

Silver was discovered in the Paleocene intrusion near the north end of the Pyramid Range, triggering a local mining boom. The first mining claim in what is now the Lordsburg mining district was initiated on April 7, 1870. The initial boom was a hoax, though. Ores were not nearly as rich as claimed in this announcement about it in the *Tucson Weekly Arizonian* in April 1870: "We have found it! The greatest treasures ever discovered on the continent, and doubtless the greatest treasures ever witnessed by the eyes of man." The original claim was named the Mountains of Silver Mine.

In the end, there was some silver in the district and it was mined at Shakespeare, now a ghost town, until 1885. Copper, gold, lead, zinc, and uranium have also been mined here, largely from three mines: the Eighty-Five, the Bonney-Miser's Chest, and the Henry Clay–Atwood Mines. Mines within the Lordsburg district reach more than 2,000 feet deep, some of the deepest in New Mexico. Material from these mines has produced an estimated $60 million between 1870 and 1975. Ore occurs in copper-tourmaline veins that are on average 6.5 feet wide and more than 4,000 feet long. Northeast-trending faults control these veins and their precious deposits. Mining continues to be of interest as the global economics of minerals fluctuate.

Immediately east of Lordsburg (milepost 20), the interstate crosses Lordsburg Draw, once part of Lake Animas and a place of geologic interest because of the discovery and possible excavation of mammoth skeletons during the 1960s. Unfortunately, records of the excavation and the bones recovered are incomplete.

Northeast of Lordsburg are fault slices of the Burro Mountain batholith and metasedimentary rocks of Proterozoic age, flanked with Eocene and Oligocene pyroclastic rocks and ash-flow tuff rocks of the Datil Group that are, in turn, flanked by the basin-filling Gila Group of mostly Miocene to Quaternary age. The Big Burro Mountains were uplifted during the Laramide orogeny, then experienced Basin and Range extension, and finally were buried by volcanics of the Mogollon-Datil volcanic field. The Great Unconformity is present here, where the Datil Group volcanic rocks lie directly on a weathered surface of Proterozoic basement rocks. Paleozoic and Mesozoic sediments were either eroded or never deposited on the Burro uplift.

The Burro uplift, to the north, is adjacent to what is believed to be a failed rift zone, the Late Jurassic to mid Cretaceous Bisbee Basin, which extends to the northwest across Arizona and to the southeast, where it joins the Rio Grande rift. The Big Burro Mountains are also home to the Tyrone mining district, a major source of copper. See US 180 on page 309 for more information about the Tyrone Mine.

Small ranges and clustered hills south of the highway are fault blocks of Oligocene volcanic rocks of the Mogollon-Datil volcanic field, Cretaceous sedimentary units, and small exposures of Proterozoic basement rocks. These hills are the tops of Basin and Range horsts, now mostly buried by the thick valley fill.

The smelter town of Playas, about 20 miles south of the interstate, is where ores mined at Tyrone are processed. When built in the 1970s, the smelters were among the most modern in the world and processed some 2,000 tons of concentrated ore per day. The town and smelters were abandoned when the price of copper fell late in the 1990s. The facilities were acquired by New Mexico Tech and are now used as a training and research facility for first responders and counter-terrorism trainees.

At milepost 55, I-10 crosses the Continental Divide at an elevation of 5,845 feet, but the divide is scarcely discernable here and there are no streams to separate. Ephemeral drainages west of the divide course into the Animas Valley, which has no outlet, typical of many basins in the Basin and Range. Runoff that flows in gullies east of the divide similarly reaches a closed basin and sinks into the alluvium. East of the divide, the highway descends an imperceptible slope among scattered hills and small ranges that are the eroded summits of deeply buried ranges.

In the small ranges nearby, for example the Cedar Mountains 10 miles south of I-10 between milepost 50 and 60, Cenozoic granite intrudes Paleozoic limestones. Along the contact, the calcium carbonate of the limestone catalyzed mineral enrichment. Silver, lead, zinc, and gold were produced from these ranges, mostly in the late nineteenth century.

The Tres Hermanas Mountains, about 25 miles south of Deming, are another intrusion that generated zinc, lead, and minor silver, gold, and copper in the form of both skarn and vein deposits of sphalerite, galena, chalcopyrite, and others. Onyx, fluorescent calcite, and other minerals have been found in these hills. The mountains have crystalline quartz monzonite cores that intruded into the surrounding Cretaceous limestone and other sedimentary units during the end of the Laramide orogeny in Cenozoic time. The mineralization occurs along fractures and faults associated with intrusion. Some garnet-rich marble deposits exist. Multiple periods of mineralization resulted in a suite of replacement mineral deposits. Economic materials were pulled from the Mississippian-age Escabosa Limestone and other Pennsylvanian-age sedimentary units. Mining here began in the 1880s, and there are many prospect pits visible to visitors.

The Mimbres Valley near Deming was settled in the 1880s by farmers brought here by mining companies to provide groceries for their miners. At that time, all irrigation was from the Mimbres River, which begins in the mountains to the north and then flows into the closed basin and sinks into the ground. Today, groundwater-tapping wells nourish farms. The basin is filled with more than 4,000 feet of chaotic basin-fill alluvium that includes gravels, sands, and clay lenses. The preserved river channels of the ancient Mimbres River have been mapped in the stratigraphy. Water levels in the

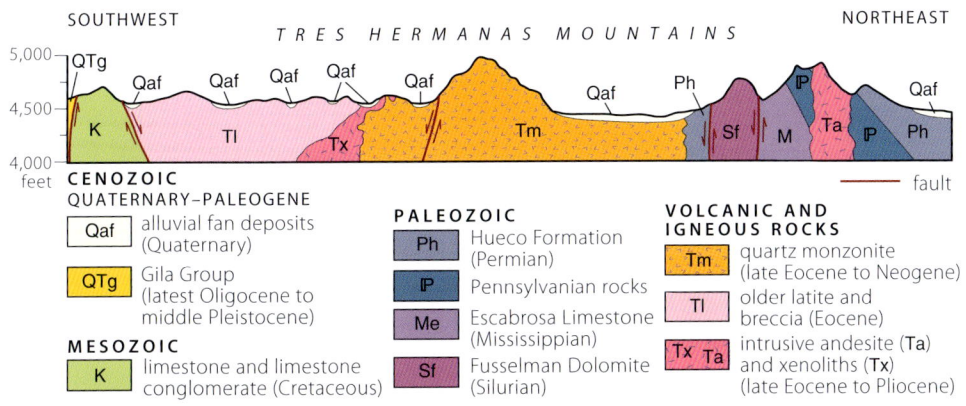

Cross section of the Tres Hermanas Mountains, a Cenozoic intrusion into Mesozoic and Paleozoic rocks. —Modified from Balk, 1962

main unconsolidated aquifer began to decline in 1913 and have continued to drop because pumping for domestic use, industry, and irrigation exceeds natural recharge. Pumping of groundwater has caused local subsidence of the desert floor and cracking or fissuring of its surface. Large linear fissures more than 2,000 feet long and ranging in depth from 1 to more than 40 feet deep have been observed, as well as polygonal-shaped fissures more than 1,200 feet across. The ground surface has subsided nearly 1 foot near areas of high water withdrawal.

For information about City of Rocks State Park north of Deming, see US 180: Deming—Arizona on page 308.

I-10
Deming—Las Cruces
60 miles

On the eastern outskirts of Deming, I-10 crosses the Mimbres River (milepost 86), an ephemeral stream with headwaters to the north in the Mimbres Mountains, part of the Black Range. When it does flow, the river carries water eastward from Deming and then sinks into the porous gravels of the desert.

To the north, between mileposts 90 and 95, are more Eocene to Oligocene volcanic rocks. Cookes Peak (8,404 feet) is a gray granodiorite intrusion that was emplaced 38.8 million years ago as volcanism associated with the Mogollon-Datil volcanic field began. Cookes Peak is bordered on three sides by normal faults and is an uplifted block (or horst) of the Basin and Range extension. Mississippian- to Devonian-age sedimentary rocks flank the inner volcanic core. Cookes Peak has long been a landmark, in part because of the perennial spring at the southeastern base of the mountain range. Fort Cummings was established at Cooke's Springs in 1863.

The very northern tip of the Cookes Range is Proterozoic basement rock, and like many ranges in New Mexico, the Great Unconformity is exposed. Here, at least 650 million years of rock record are missing—less than is missing in the Sandias near Albuquerque but still a notable lack of stratigraphy, with Cambrian rock lying directly on the basement. Paleocene to Eocene conglomerates, sandstones, and mudstones are preserved in a Laramide basin associated with the Cookes Range. Younger ash-flow tuffs from the Mogollon-Datil calderas flowed over the region before the Cookes Range was uplifted during Basin and Range extension.

Mines near Cookes Peak, most of them now closed, produced galena (lead), silver, sphalerite (zinc), copper, gold, and fluorite following the early discovery of silver in 1876. The lead, zinc, and silver deposits are found in replacement ore bodies along fractures in the Cookes Peak mining district. Here, hot fluids dissolve carbonate rocks (limestone and dolomite of the Fusselman Dolomite), and the cavities left behind are filled in with new mineral deposits. Galena and fluorite are found in these fractures, as well as jasperoid, which can contain fluorite, calcite, quartz, pyrite, and cerrusite.

The West Potrillo Mountains, the conical peaks south of I-10 between mileposts 112 and 120, are part of the Potrillo volcanic field. This young volcanic field, with 916,000- to 20,000-year-old basaltic flows, extends from New Mexico into northern Chihuahua, Mexico. The volcanic field is part of the Rio Grande rift, and the magma that erupted here came up through deep structural fractures that extend through the

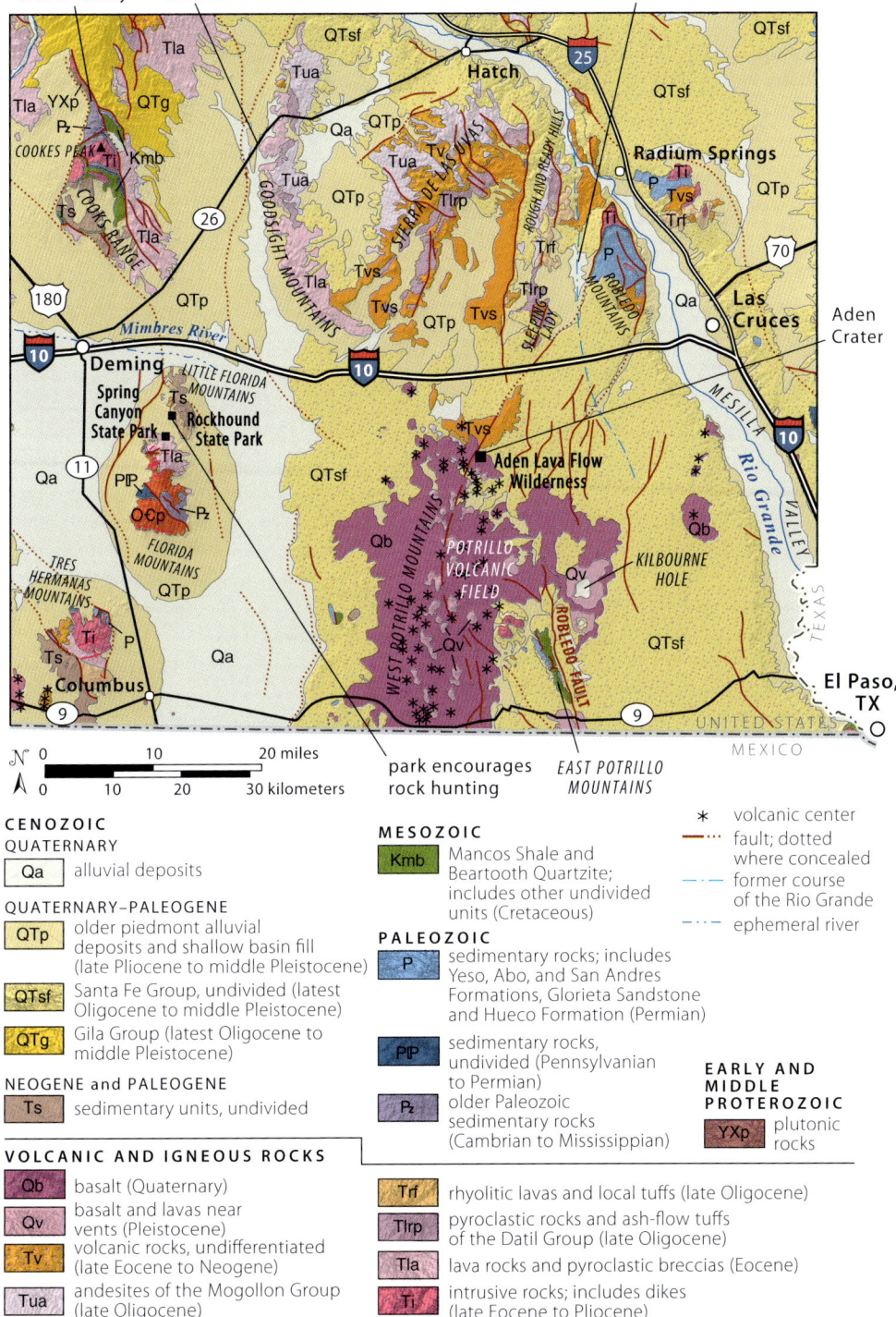

Geology along I-10 between Deming and Las Cruces.

crust. Three main basalt flows exist, including the West Potrillo Basalt, Aden-Afton Basalt, and Black Mountain–Santo Tomas Basalt. There are about 150 cinder cones, and lava flows cover hundreds of square miles. Several low craters, called maars, resulted from steam explosions that occurred when hot lava erupted into water.

The Aden Lava Flow Wilderness preserves the Aden Crater, a shield volcano of the Potrillo volcanic field, last active approximately 16,000 years ago. Aden Crater is known for the juvenile Shasta ground sloth (*Northrotheriops shastensis*) found in a lava tube. The sloth fell to its death 11,000 years ago and was well enough preserved to still be articulated and have skin and hair. The discovery provided information about what this now-extinct species looked like.

ROCKHOUND STATE PARK AND SPRING CANYON RECREATION AREA

Rockhound State Park, 12 miles southeast of Deming near the north end of the Little Florida Mountains, encourages rock collecting. Colorful agates and geodes can be found among the volcanic rocks. The campground and trailheads sit on a bajada, or alluvial fan, formed by the deposition of material eroded from the mountaintops. Mineral deposits in the Florida Mountains were first noted in 1876. Mining occurred in the region from 1880 to 1956, producing copper, gold, silver, lead, and manganese from vein deposits. From 1918 to 1959, mines on the northeast slope of the Little Florida Mountains produced manganese, which was used to harden steel and aluminum. While mine adits may be visible throughout the landscape, they are abandoned and unmaintained and, thus, very dangerous.

Volcanic activity from 34 to 24 million years ago is recorded in the layers of andesite, dacite, and rhyolite ash-flow tuffs and lavas. Rhyolite domes and dikes later intruded these original

The red cliffs to the east of the Little Florida Mountains, above the campground at Rockhound State Park, are composed of rhyolite tuff, which erupted 25.7 million years ago.

volcanic materials. The ash-flow tuffs erupted from shield or stratovolcanoes and are welded, meaning they erupted while very hot and fused into a very dense, hard rock. Later eruption of rhyolite lavas and the emplacement of rhyolite domes confuse the volcanic history.

Extensive hydrothermal activity during and after the eruptions resulted in the cementation of fanglomerates and deposition of manganese- and iron-oxide minerals and fluorite. Fluids containing dissolved minerals from original magma and country rock move through fractures in the rock, cooling and depositing their rich mineral wealth as agate, chalcedony (a cryptocrystalline variety of quartz), and quartz. The brilliant colors of minerals found in the park tell about their chemistry: reds indicate iron, black and pinks indicate manganese, blue and violets indicate cobalt, greens and blues indicate copper, orange-reds indicate chromium, and greens indicate nickel.

Both geodes and thunder eggs can be found at Rockhound State Park. Geodes, hollow rocks filled with crystals, form when water seeps into vesicles, or cavities, in igneous or sedimentary rocks, precipitating crystals from the outer edges inward. Thunder eggs, nodule-like rocks similar to geodes but solid or mostly solid, likely formed during the cooling of the rhyolite lava that hosts them. Thunder eggs vary in color and pattern depending on the chemistry and temperatures in which they formed.

The Florida Mountains just to the south are a small but extremely rugged, extensively faulted mountain range with peaks reaching just over 7,000 feet in elevation. Proterozoic gneisses intruded by a Cambrian-age pluton are at the core of these mountains. More than 4,000 feet of Paleozoic sedimentary units, the oldest being the Early Ordovician Bliss Sandstone, overlie the older crystalline rocks. Volcanic tuffs, rhyolite, and basaltic andesites and dacites form most of the range, with clasts of these volcanics dominating the alluvial fan conglomerates that surround the mountains. Spring Canyon State Park centers on a canyon walled by volcanic breccia.

Spring Canyon State Park, in the Florida Mountains, is in Eocene volcanic flows of the Rubio Peak Formation cut by younger (about 25-million-year-old) rhyolitic dikes. In the foreground is volcanic breccia.

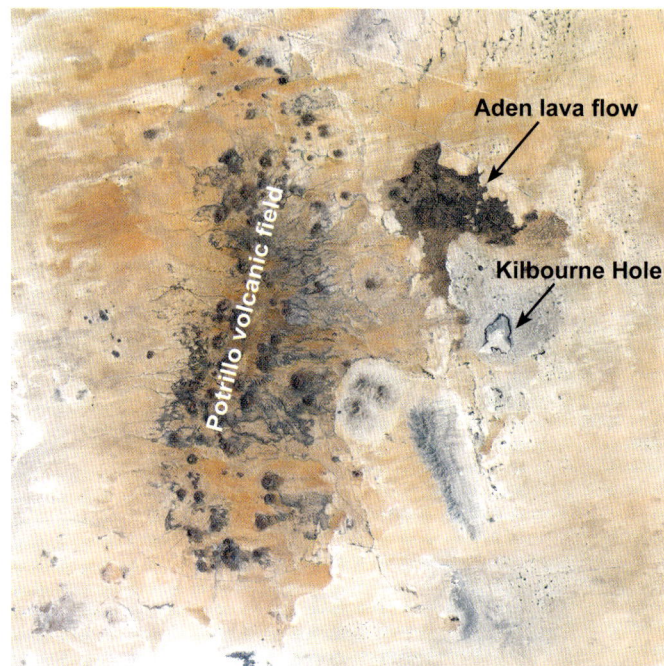

Satellite view of the Potrillo volcanic field, Aden lava flow, and Kilbourne Hole. —NASA image

Kilbourne Hole, designated a National Natural Landmark in 1975, is a maar crater within the Potrillo volcanic field. The rim is managed as part of the Desert Peaks–Organ Mountains National Monument and is accessible via dirt roads, but the valley floor is privately owned. The crater of Kilbourne Hole is approximately 1 by 2 miles in size and more than 100 feet deep. The crater formed less than 70,000 years ago when basaltic magma erupted in contact with water during the final stages of basaltic magmatism in the area.

Volcanic bombs, which form when blobs of hot magma are forcefully ejected during eruption, can be found here. Many of the bombs contain xenoliths, pieces of the lower crust or deeper mantle carried to the surface with the erupting magmas. Xenoliths include granulite containing garnet and sillimanite, pyroxene, or the mantle minerals lherzolite, harzburgite, and clinopyroxene. Sometimes, though rare, breaking open a dull-looking bomb can reveal stunning gem-quality peridot, the green-yellow olivine mineral that is the August birthstone.

The East Potrillo Mountains, 20 miles south of the interstate between mileposts 110 and 115 and out of sight, are a tilted fault block that marks the western edge of the Rio Grande rift and the transition to the Basin and Range. The still-active Robledo fault marks the eastern edge of the range.

To the north of I-10 at milepost 25, the Sleeping Lady Hills, as well as the Rough and Ready Hills farther north, owe their origin to earlier Oligocene to Miocene volcanism, part of the Mogollon-Datil volcanic field. Between mileposts 120 and 129, the highway crosses a shallow, wind-scoured depression that contains pebbles not derived from adjacent ranges. The Rio Grande transported these pebbles when it flowed south here, west of its modern course.

As the route approaches the Rio Grande Valley near Las Cruces, it is evident that most of the city is built on the floodplain and terraces of the Rio Grande. The highway drops down over several terrace levels that mark higher former river floodplain levels. After crossing the modern floodplain, the highway climbs through the corresponding terraces on the east side of the river. Man-made revetments near the highway prevent storm runoff from flooding the city and valuable bottomlands along the Rio Grande.

For more information about Las Cruces and the Mesilla Valley, including Prehistoric Trackways National Monument in the Robledo Mountains and the Organ Mountains–Desert Peaks National Monument, see I-25: Truth or Consequences—Las Cruces on page 176.

US 60
Socorro—Datil
63 miles

Socorro, first established by Spanish settlers in 1598, further developed as a mining and smelter town in the 1880s. The New Mexico School of Mines, now the New Mexico Institute of Mining and Technology, was established in 1889. The New Mexico Bureau of Geology and Mineral Resources and its public mineral museum is located on campus.

Gold, silver, copper, lead, zinc, arsenic, and other minor ores were first mined prior to 1880. By 1889, the main smelter in Socorro was producing close to 2 million ounces of silver per year, with some of the ore shipped in from Arizona, Colorado, and Mexico. Today, mining focuses on the extraction of perlite, a volcanic siliceous glass with the same composition as rhyolite. Used in horticulture and construction and as a filter aid and filler, perlite is mined here from one of the deepest and largest perlite mines in the world. Visible to the south of Socorro Peak, the mine extracts perlite from nearly 400 feet belowground. Before distribution, the rock is removed, crushed, and heated until it expands.

The Socorro Mountains to the north of US 60 and the Magdalena and San Mateo Mountains to the southwest are parts of the vast Mogollon-Datil volcanic field that extends westward to Arizona. Socorro Peak—the one with the "M"—is formed by the 32-million-year-old Socorro caldera and is the northeasternmost caldera of the Mogollon-Datil volcanic field. On its slopes, tilted, light-colored Pennsylvanian limestone and shale are overlain by volcanic breccia and volcanic ash deposited in the caldera. These deposits are overlain by Miocene volcanic rock that forms the summit of the peak. Other caldera rocks are visible as the highway continues into the mountains.

Rounding the southern end of the Socorro Mountains at milepost 136, US 60 passes a perlite mine to the north. The mine sits on a rhyolite lava dome that overlies older pyroclastic flows. The lava dome is about 60 to 100 feet thick. At milepost 132, US 60 passes Box Canyon, a slot in the volcanic rocks of the Socorro Caldera.

West of the Socorro Mountains, US 60 cuts northwest along a long valley floored in alluvial deposits. To the southwest lie the Magdalena Mountains, a north-south-trending fault-block range. The northern part of the range is primarily Proterozoic basement rocks, including 1.7-billion-year-old granodiorite, diorite, and gabbros,

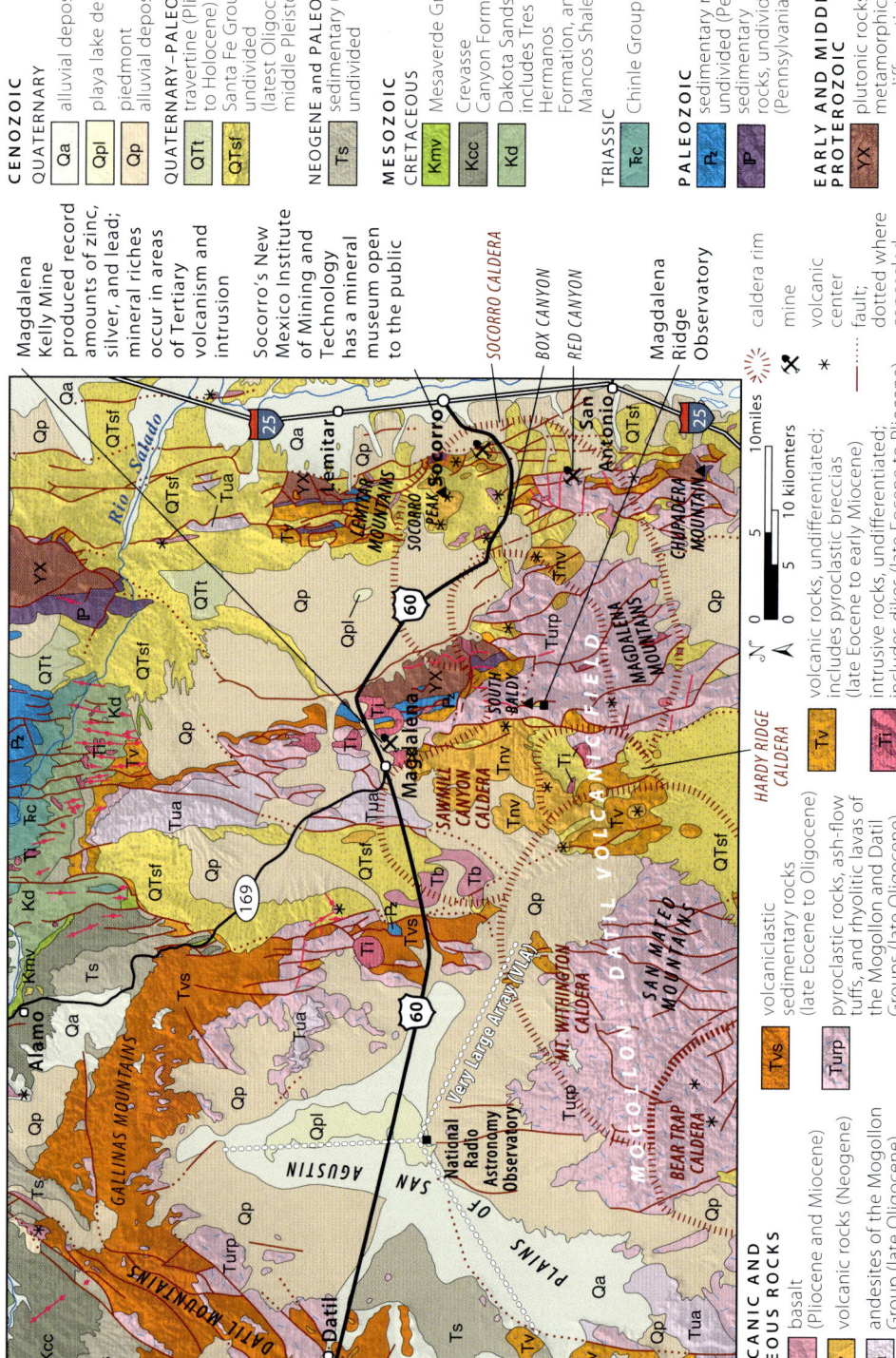

Geology along US 60 between Socorro and Datil.

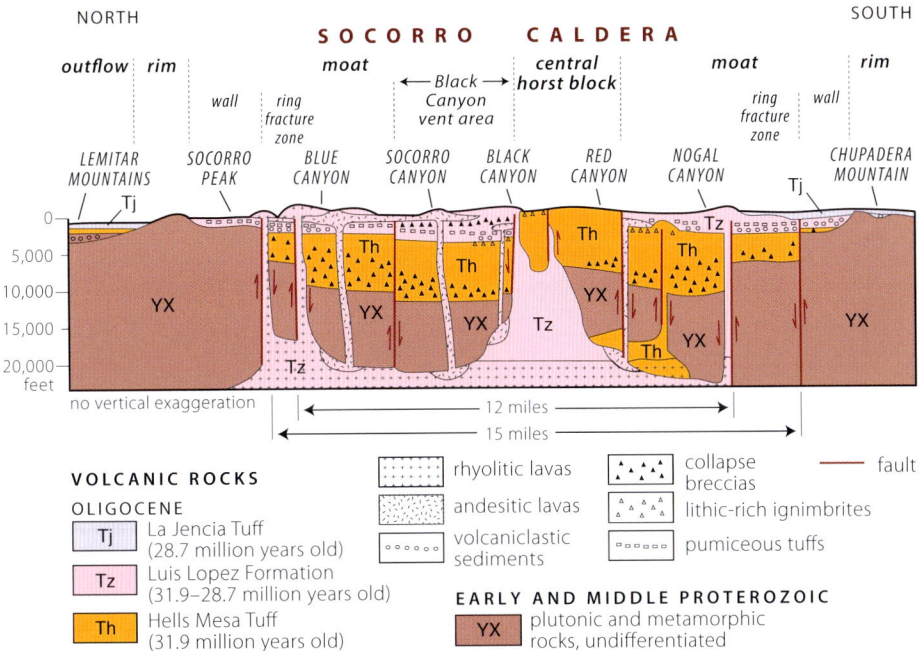

Magma that rose along multiple faults fed eruptions of lava in between the initial Hells Mesa Tuff eruption 31.9 million years ago that brought about the collapse of the Socorro caldera and the younger, 28.7-million-year-old La Jencia Tuff. The present profile of the Socorro and Chupadera Mountains, where the caldera occurs, is not shown in this reconstruction. —Modified from Chamberlain, McIntosh, and Eggleston, 2004

Box Canyon is a dry gulch, a gulch formed when its intermittent stream is suddenly swollen by flash flooding. It lies in the rugged country west of Socorro.

1.4-billion-year-old granitic plutonic rocks, and metasedimentary rocks (milepost 120). The range dips eastward, and more than 1,500 feet of movement has occurred on the faults that bound it. Much younger volcanic rocks form the high points. The Magdalena Ridge Observatory sits on a long ridge just to the south of South Baldy, the high peak (10,783 feet), on a bastion of 30- to 24-million-year-old Mogollon Group pyroclastic flows and ash-flow tuffs.

Magdalena (milepost 113) grew when silver was discovered in the Magdalena Mountains in the 1880s, but the town of Kelly, about 3 miles south of Magdalena, was the site of a more intense lead-silver boom. Kelly silver is associated with a large intrusion and occurs in linear zones along faulted sedimentary units adjacent to the intrusion. The Magdalena mining district, including the Kelly mines, shipped between $7 and $9 million worth of lead-silver ore between 1880 and 1902. The Mississippian-age Kelly Formation is the most productive ore-bearing unit. Both Magdalena and Kelly offer good mineral collecting, but avoid old mine workings, which can be dangerous. Azurite, barite, malachite, cerussite, pyrite, and other minerals can be found in old mine dumps. Most of the dumps are private property, however, and permission must be obtained to search for them.

Hydrothermal mineralization took place in three main stages. First, monzonite-granite stocks intruded into the host rock. Silicate minerals such as tourmaline, wollastonite, and diopside were deposited within the Kelly Formation as heat and magmatic fluids altered the limestone. The second stage of mineralization began once the stocks had fully cooled and solidified, and in places had been fractured along north-trending fault lines. As the circulating fluids began to cool, they deposited silica in the form of jasperoid and crystalline quartz. This mineralization is seen in a much more dispersed manner within the Kelly Limestone. The third stage was a period of replacement mineralization in which fluids followed the same fractures and mineral veins through which fluids moved in the first two stages, but this time, sulfide minerals replaced existing minerals. For example, the zinc- and copper-rich fluids dissolved preexisting silica-rich minerals—such as quartz—and deposited pyrite, sphalerite, galena, and chalcopyrite.

Limestones of critical economic importance outcrop minimally along US 60. One example of a small outcropping can be seen as the highway curves near milepost 116. Iron ores were found a few miles west and north of Magdalena, in dark-brown nodules in surface deposits. The surface-nodule supply was soon exhausted, however, and mines were opened nearby.

At milepost 103, the red beds of the Baca Formation form sloping hills of low outcrops. This Eocene unit, a sequence of mudstone, sandstone, and conglomerate, is an aggregate of rocks shed from the Defiance, Zuni, Lucero, Sierra, and Mogollon uplifts. The sediment collected in alluvial braided river systems in the Laramide-age Baca Basin, which extended from here to Arizona. The Spears Formation, an Oligocene volcaniclastic unit, conformably or gradationally overlies the Baca Formation.

The broad valley west of the Magdalena Mountains is called the Plains of San Agustin, a beautiful, flat-floored, internally drained valley. The plains are a graben that subsided between parallel Rio Grande rift faults. The graben slowly filled with sediments brought in from surrounding high ranges in long, sloping piedmont and alluvial fan deposits. It is bordered by 38- to 24-million-year-old volcanics, including tuffs, volcaniclastic rocks, and other lavas from the Mogollon-Datil volcanic field.

The western Plains of San Agustin filled with a lake during the last ice age. The lake reached its maximum extent at about 20,000 to 18,000 years ago and disappeared by 6,000 years ago. Intermittent lakes continued to fill the basin when water was abundant, though the modern area is an arid, dry lakebed.

At milepost 88 is the spectacular Karl G. Jansky Very Large Array (VLA), a collection of twenty-eight 25-meter-diameter radio telescopes arrayed in a 13-mile-long Y shape across the valley floor at 6,970 feet elevation. The many telescopes are arranged on railroad tracks and work together as a single telescope to glean a higher-resolution image than a single telescope could. One of the arms of the VLA intersects US 60 just east of milepost 90, where the highway crosses two sets of railroad tracks. The VLA gathers information on black holes, protoplanetary discs, and young stars. It also looks at the complex ways gases move within the Milky Way and helps to better understand the early solar system. Most of the array is on the even pan of lake and playa deposits. This flat surface facilitates the dynamic nature of the VLA. The dishes and the telescope array can contract and expand by moving the towers along intersecting railroad tracks, helping to "tune" the array for specific observations. This ability to rearrange the shape of the telescopes allows for the density and radius of the cumulative telescope to optimize for different imaging projects.

The National Radio Astronomy Observatory visitor center lies along old US 60, accessed via NM 52 halfway between mileposts 93 and 92. A slide show and walking tour explain the various areas of study—among them the births and deaths of stars, the properties of galaxies like the Milky Way, and the history of the origin of the universe.

A saucer-shaped receiver of the National Radio Astronomy Observatory towers above the Plains of San Agustin.

Datil (milepost 77), a former mining and smelter town, lies at the northwest edge of the Plains of San Agustin upon an alluvial apron of volcanic clasts shed from the mountains to the west.

US 60
DATIL—ARIZONA
78 miles

As US 60 heads west into the mountains west of Datil, look for dramatic boulders of pinkish-gray and lavender volcanic ash near milepost 72. This rock is welded ash-flow tuff of the Datil Group, widespread here near its source in the Eocene to Oligocene Mogollon-Datil volcanic field. Cut by vertical shrinkage joints that formed as it cooled, the rock in places erodes into free-standing pinnacles. Some unwelded ash-fall tuff occurs here as well. Where highway cuts are nearly vertical, welded or ash-flow tuff is present. Where they are slanted way back, it is softer, slide-prone ash-fall tuff.

The colossal volcanism and caldera growth that occurred during the formation of the Mogollon-Datil volcanic field resulted in large amounts of erosion and transportation of volcanic materials. Rivers deposited sandstones, and debris flows carried andesitic and dacitic clasts. Locally, these flows even contain the limestones they flowed over and picked up. Together, these form a large volcaniclastic apron that surrounds the Mogollon-Datil volcanic field. Near Datil, these volcaniclastic sedimentary rocks are called the Dog Springs Formation. Monument Rock (exit north at milepost 65), a prominent erosional pillar, is made of this 37-million-year-old

Ash-flow tuff sometimes erodes into pinnacles and residual boulders. Photo taken between mileposts 72 and 71.

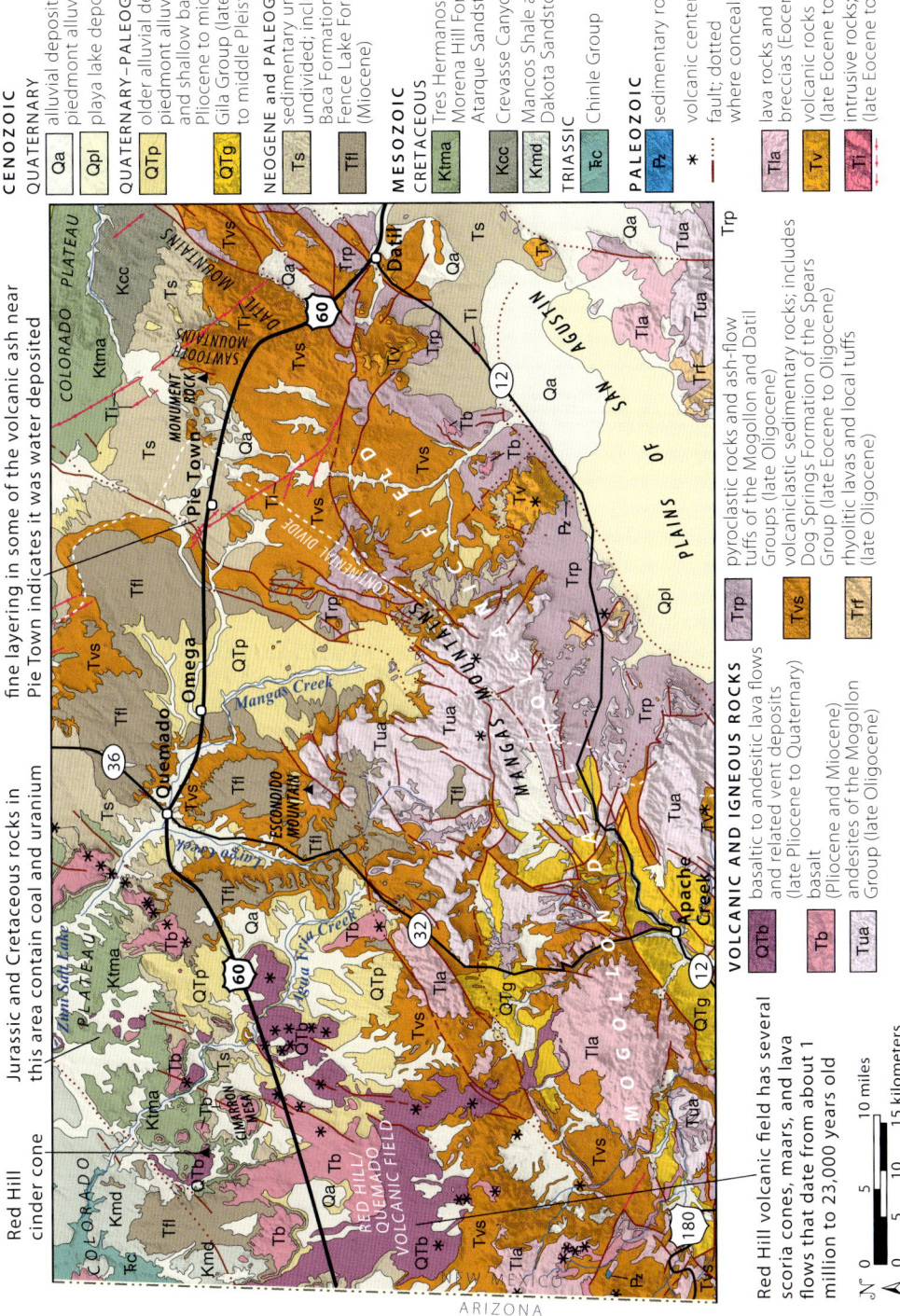

Geology along US 60 between Datil and the Arizona border.

The Sawtooth Mountains just east of Pie Town, between mileposts 65 and 60. This photo was taken a short drive into the mountains on Criswell Road, which intersects US 60 at milepost 65.

volcaniclastic conglomeratic sandstone. To the northeast of Monument Rock, fabulous spires and pinnacles are weathered out of more volcaniclastic deposits of the Dog Springs Formation.

At milepost 63, the highway crosses onto units of basin fill and volcaniclastic sourced layers, including the Baca Formation, which preserves the transition from basin filling in Eocene time to the beginning of the Mogollon-Datil volcanism.

At milepost 56, just east of Pie Town, the highway crosses the Continental Divide at an elevation of 7,780 feet. The divide here runs northeastward along the crest of the Mangas Mountains. The Continental Divide Trail, a hiking trail that roughly follows the geographic divide, joins US 60 at Pie Town from the north. The "trail" follows the highway until it moves south at milepost 43, just east of Omega. In the 1920s, a miner with a taste for pies established Pie Town, which continues to draw in tourists with an annual pie festival.

Escondido Mountain, the peak to the southwest at Omega (milepost 42), consists of a series of 29- to 26-million-year-old basaltic andesites and andesites that caps a broader apron of volcaniclastic sedimentary units. A Miocene-age fan system of large, coalesced alluvial fans surrounds Escondido Mountain. Collectively known as the Fence Lake Formation, the system may have been much larger and extended into Arizona, flanking the assorted calderas of the Mogollon-Datil volcanic field. The formation, which includes conglomerates, conglomeratic sandstone, stream-deposited volcaniclastic sediments, and windblown units, is exposed at milepost 40 and much more extensively exposed at milepost 30.

Quemado sits on the modern alluvium of ephemeral streams. To the northwest, mountain slopes are more basaltic to andesitic lavas, flanked here with the Paleogene basin-filling sedimentary units. The road stays on these mixed young sedimentary and volcaniclastic units as it continues to the southwest. Between mileposts 30 and

29, US 60 cuts through a large swath of the Fence Lake Formation, composed of Miocene-age, unconsolidated alluvial fan deposits.

At milepost 15, look north to see Cimarron Mesa, capped with basaltic and andesitic flows of the Red Hill/Quemado volcanic field. This small volcanic field erupted from more than forty vents in two main pulses between 8 and 5 million years ago. The small volcanic vents from which the lava came appear often as small cones on the tops of the lava flows. To the far north of this small field lies Zuni Salt Lake, a shallow ephemeral lake in a volcanic maar. The lake, a sacred site to the Zuni people, is listed on the National Register of Historic Places.

As the highway climbs onto a basalt flow west of milepost 12, look for the red soil at its base, baked to brick by the heat of the molten rock that flowed over the surface. Red Hill (milepost 11) is northwest of Cimarron Mesa. The cinder cone is part of the Red Hill/Quemado volcanic field of volcanic maars, scoria cones, and associated basaltic flows that are less than 7 million years old. The volcanic field is situated along the Jemez lineament—a pervasive line of structural weakness that extends from here to the northeast part of the state, along which many volcanic eruptions have occurred.

As US 60 continues across basalt flows, far to the north from milepost 5 are hills capped with intertonguing Cretaceous Mancos Shale and Dakota Sandstone. These essentially horizontal rocks mark the southern edge of the Colorado Plateau. The mesas capped with horizontal lava blend with those formed entirely of horizontal layers of sedimentary rock and are flanked with volcaniclastic sedimentary units.

Red Hill, a cinder cone north of milepost 11.

Inset: Two volcanic bombs near Red Hill have finer texture on the outside, where they cooled fastest. The interior of the bombs, insulated in their shells, cooled more slowly, and were able to form a larger crystalline structure. The last eruption at Red Hill has been dated at only 71,000 years ago.

To the west, in Arizona, the Springerville volcanic field erupted from more than four hundred vents, spreading across 1,200 square miles near Springerville and Show Low, Arizona. Pliocene to Pleistocene volcanoes dominate this young basalt field. Basalt erupted from cinder cones, maars, and lava domes. This field is composed of lavas more fluid than those that constructed the Mogollon-Datil volcanic field. Basaltic lavas filled stream valleys and flowed in thin sheets across low areas; they now cap many mesas. The lava flows are young enough not to be broken by faults or tilted and folded by Earth's movements. The White Mountains of Arizona, composed of these young volcanics, reach summits of 11,400 feet at Mt. Baldy.

US 180
Deming—Silver City—Arizona
164 miles

US 180 leads northwest from Deming on a broad plain of alluvium derived from the Mimbres River and piedmont deposits including the Gila Group conglomerates of late Oligocene to middle Pleistocene in age. The road between Deming and Silver City follows a buried fault, one of many northwest-trending faults that have sliced and faulted the mountains that surround this long valley. Surrounding ranges, left high while the basins faulted down around them, are elongated northwest to southeast. This Basin and Range topography, which covers much of the Southwest, resulted from the rotation and extension of the Southwest as motion of the Pacific plate caused east-west tension on the North American continent.

The geology of these ranges is complex, with Proterozoic basement rock, Paleozoic and Mesozoic sedimentary layers, and abundant younger, Cenozoic-age volcanic and sedimentary units. Many of these ranges also expose the Great Unconformity, where there is a gap of at least 650 million years of rock record between Proterozoic basement rock and overlying sedimentary units.

About 10 miles north of Deming, the Cookes Range lies to the east. Rising to 8,404 feet above the plain of the Mimbres River, the tall, Matterhorn-like central peak is Cookes Peak, a gray granodiorite intrusion that was emplaced 38.8 million years ago associated with Mogollon-Datil volcanic field volcanism. The Cookes Range is discussed in more detail in I-10: Deming—Las Cruces on page 292.

The tailings pile north of milepost 133 hints at the scale of the mining in the nearby mountains. Chino Mine, also known as the Santa Rita Mine or Santa Rita del Cobre, lies in the mountains northeast of Bayard (milepost 123). This open-pit porphyry copper mine is the third-largest open-pit mine in the United States. It produced nearly 90,000 tons of copper in 2019. A viewpoint of the mine with informational signs is a few miles north of US 180 on NM 152 (milepost 121 in Santa Clara) toward Hanover. The New Mexico Bureau of Geology and Mining Resources website has an excellent analysis of the geology of the mine. The town of Santa Clara sits on Tertiary intrusive rocks, while Cretaceous Mancos Shale and other Pennsylvanian to Permian sedimentary units are in the hills to the north and northeast.

NM 152 passes the Chino Mine and has large pullouts along its south side, allowing views into the mine. The highway connects to NM 35, an alternate route to Gila Cliff Dwellings National Monument. This route is longer but slightly less winding than

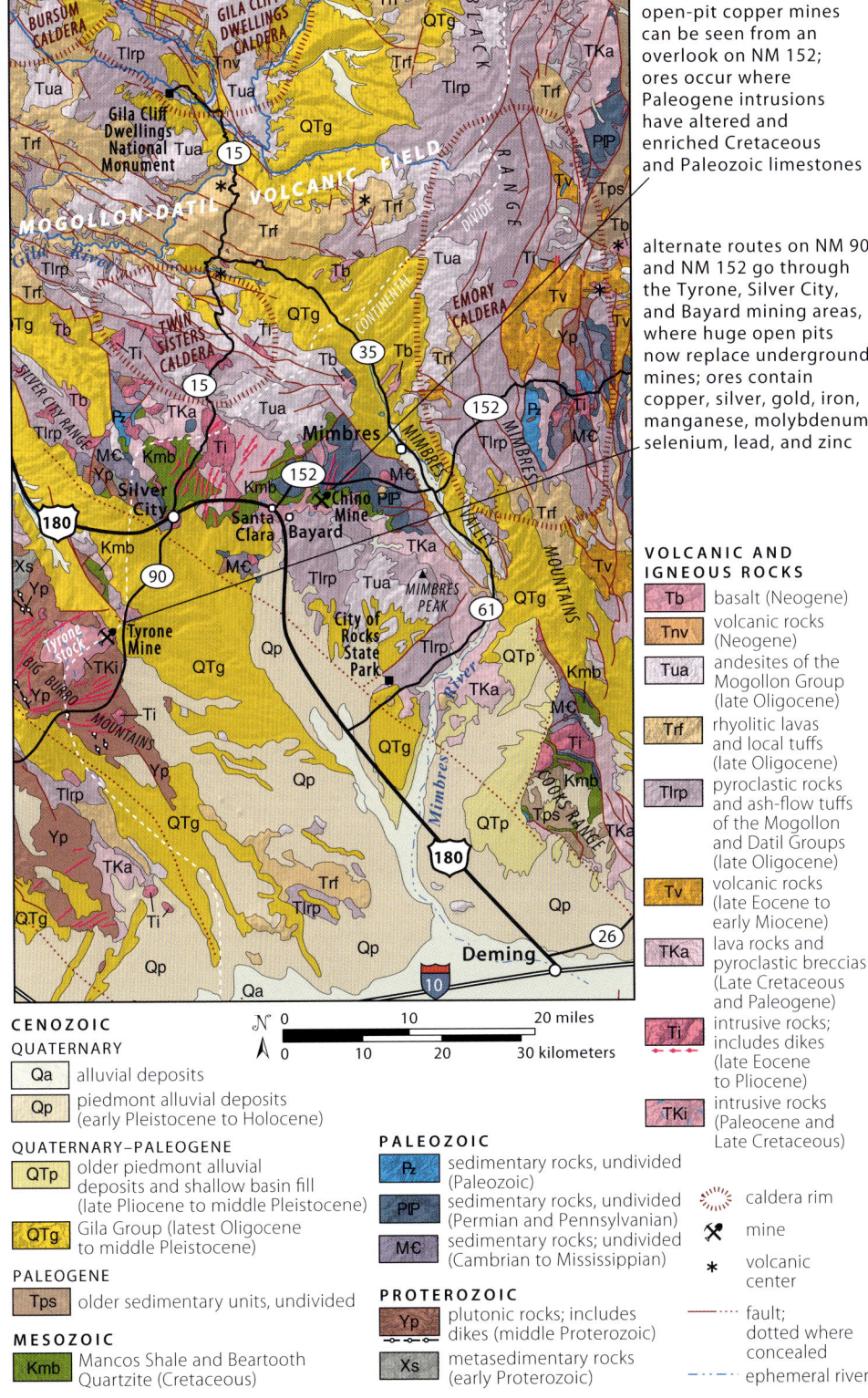

Geology along US 180 between Deming and Silver City.

CITY OF ROCKS STATE PARK

City of Rocks State Park, a collection of large rocks that look like a city from a distance, lies several miles northeast of US 180. From milepost 41, take NM 61 to the east and follow the signs. The Kneeling Nun Tuff, which forms the majority of exposed rocks within the park, erupted 35 million years ago from the Emory caldera. The eruption spewed more than one thousand times the material that was erupted in the 1980 Mount St. Helens eruption. The Emory caldera forms the southern Black Range, visible east of the park. The hot masses of lava that formed these tuffs cooled and cracked, forming columnar joints that allowed moisture to begin the weathering process underground. These columns have emerged from the ground as the surrounding rock has eroded away. Freeze-thaw action and wind erosion continue the weathering process on the now-exposed rock.

The Kneeling Nun Tuff has weathered and eroded into large "buildings" at City of Rocks State Park.

Flakes spall off the Kneeling Nun Tuff at City of Rocks State Park.

The Chino Mine, a large open-pit copper mine, viewed from NM 152.

NM 15 from Silver City. The NM 35 route follows the Mimbres Valley, a fault-bound valley bordered occasionally with basement rock (it forms the base of the hills to the west at milepost 2 on NM 35), but mostly with the Gila Conglomerate and basaltic andesites of the Mogollon Group.

Silver City (milepost 113) is tucked at the base of the Silver City Range to the northwest. The Silver City Museum has exhibits on mining history and the history of flooding through the city. When the first wave of miners settled in the area, they denuded the nearby hills of stabilizing trees and other vegetation. The result was several years of ferocious flooding—right down main street—during the 1890s. Silver City sits largely on alluvium, and the flooding scoured the channel down to bedrock, a depth of 20 to 30 feet below the businesses that fronted the original channel.

In the Big Burro Mountains southwest of Silver City is the Tyrone Mine. The 57- to 53-million-year-old Tyrone stock was emplaced during Laramide compression and now hosts the second-largest porphyry copper deposit in New Mexico. The copper was originally present as low-grade deposits in small plutons—or stocks—of diorite, granodiorite, and monzonite. The copper ore was concentrated when the low-grade deposits interacted with groundwater. Mineralized fluids deposited the high-grade copper in fractures, or breccias. Other minerals deposited in the area include gold, silver, molybdenum, uranium, and pyrite. Native Americans in the area initially excavated turquoise in the area, underground mining was developed in the 1860s, and open-pit stripping began in 1968.

West of Silver City, US 180 climbs sharply, passing through an altered zone of sedimentary rocks of various ages before cresting onto a surface of limestone. Rocks exposed along the highway in roadcuts include the Cretaceous Beartooth Quartzite (milepost 112.5); the Mississippian to Pennsylvanian Lake Valley and Oswaldo Limestone, which is cherty and has dark beds of shale interspersed within it (milepost

When devastating flooding destroyed Silver City's main street, the city responded by moving the street over and walling the channel with rocks. Now known as the Big Ditch, the channel is a park with paved trails along its sides; the over-100-year-old walls still line it.

GILA CLIFF DWELLINGS NATIONAL MONUMENT

The Gila Cliff Dwellings National Monument sits within the Mogollon Mountains, a pile of volcanic outpourings of the Mogollon-Datil volcanic field. Three main geologic units are present within the main canyon of the monument. The oldest (lowermost) unit is the Bloodgood Canyon Tuff, a rhyolite that erupted from the Gila Cliff Dwellings caldera about 28 million years ago. Overlying it is a series of 25-million-year-old basalt and andesite lavas. A mix of conglomerate, sandstone, and siltstone of largely Miocene age known as the Gila Conglomerate caps the lavas. The Gila Conglomerate contains clasts of the underlying volcanics within its matrix, indicating those units were being eroded as the conglomerate was deposited.

The cliff dwellings are in alcoves developed within weakly cemented layers in the Gila Conglomerate. Erosion easily removes the silty, poorly cemented material, forming shallow caves. Research suggests these caves began to form naturally when Cliff Dweller Creek was at this level. Stream power is a very powerful tool for carving alcoves in cliffsides as the river migrates side to side. The stream has since incised more than 200 feet below the cave level. These caves were then worked and built upon by early inhabitants. A geologic trail guide is available from the National Park Service. The monument is surrounded by the Gila Wilderness, which on June 3, 1924, became the first congressionally designated Wilderness Area.

180 THE VOLCANIC SOUTHWEST 311

The Gila Cliff Dwellings occupy large alcoves in the ledges of Gila Conglomerate. The inset shows a large clast of vesicular volcanic rock (lower half) in the reddish conglomerate. The vesicles are filled with white minerals.

Cliff Dweller Creek flows below cliffs of the monument. The lower, darker rock is composed of basalt and andesite lavas; the higher rock is the Gila Conglomerate.

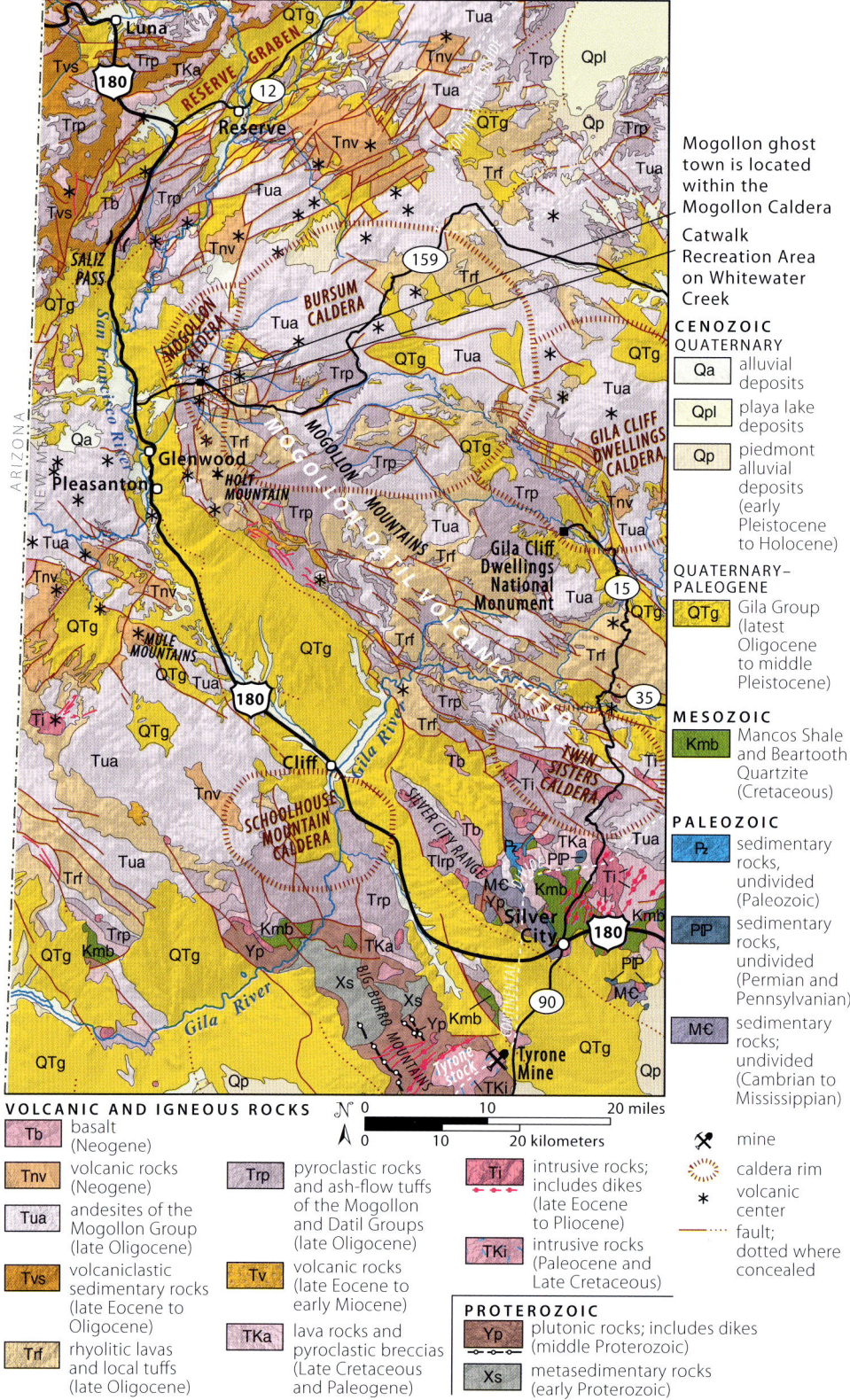

Geology along US 180 between Silver City and Arizona.

112); the Devonian Percha Shale, a dark-green-gray shale (milepost 111); and the Ordovician to Silurian Montoya Formation and Fusselman Dolomite, here grouped because they appear to be a single carbonate unit. By milepost 110, the road is back onto the Gila Group volcaniclastic conglomerate and sandstones with abundant felsic volcanic and rhyolite clasts.

The Continental Divide, crossed between mileposts 109 and 108, traces the mountain peaks to the west. West of the divide, US 180 follows another down-dropped basin in the Basin and Range. The ephemeral Mangas Creek flows north toward the Gila River, whose waters ultimately drain into the Colorado River. The Big Burro Mountains, west at milepost 95, are the Datil Group rhyolite ash-flow tuffs (36 to 31 million years old). US 180 reaches the riparian vegetation along the Gila River at milepost 88 as the road curves around a large roadcut of Mogollon Group volcanics.

The highway crosses the Gila River between mileposts 86 and 85. Pebbly sedimentary layers in roadcuts at the northern edge of the town of Cliff (milepost 84) are the Gila Group. These sedimentary beds filled the basin here as it dropped along faults. More of these sediments are seen farther north. The Gila Group's sediments, tremendously thick in this region, were derived from the immense piles of volcanic rock that are adjacent to the road from here to the New Mexico–Arizona border. The Mogollon-Datil volcanic field, a massive field of volcanoes, domes, and calderas, erupted about 40 to 24 million years ago, expelling material across 15,000 square miles of what is now the Gila Mountains. The resulting volcanic flows, breccias, and tuffs provide ample material for erosion and basin filling over many millions of years.

The Mule Mountains, far to the west at milepost 68, are Cenozoic rhyolite and dacite flows. The massive ash-flow tuffs of the Gila Mountains to the east dominate the view. Pleasanton (milepost 55) sits along the San Francisco River, a perennial source of water in this dry valley.

North of Glenwood, US 180 follows the San Francisco River valley upstream for about 10 miles, crossing the river at milepost 40. Rocky bluffs at the bridge and in impressive roadcuts north of the bridge are in the Gila Group.

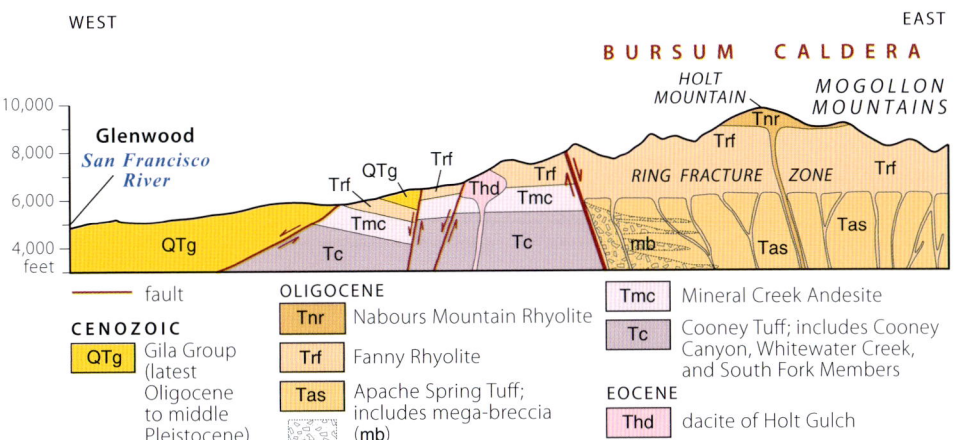

Cross section through the Bursum Caldera between Shelton Canyon on the Gila River south of Glenwood and Holt Mountain in the Mogollon Mountains. —Modified from Ratte, Lynch, and McIntosh, 2006

THE CATWALK

The Catwalk above Whitewater Creek east of Glenwood.

In a canyon to the east of Glenwood (milepost 51), the need for water in gold and silver mining required a pipeline to channel water from Whitewater Canyon to the town of Graham. This operation was short lived, and the Civilian Conservation Corps built the Catwalk, an elevated trail along the route of the water pipeline, in the 1930s. The Catwalk was dedicated as a National Recreation Trail in 1973 and is now an accessible trail with picnic area, restrooms, and further hiking access.

Evident in the rocks of the canyon and the surrounding Mogollon Mountains near Glenwood are the enormous, 28-million-year-old Bursum caldera, the older Mogollon caldera (evidence of which was nearly destroyed by the Bursum), and the more easterly Gila Cliff Dwellings caldera, which also erupted 28 million years ago. As the Catwalk National Recreation Trail begins, the rocks are Cooney Tuff, which is the oldest ignimbrite in the Mogollon Mountains at 34 million years and here is between 2,500 and 3,000 feet thick. It erupted from the Mogollon caldera. Along the canyon, look for thick, dark layers interspersed with lighter layers of rock. These dark, nearly black layers, visible up close at the top of the swimming hole stairs, are interpreted to be an eruption's volcaniclastic fallout that contains pumice lapilli and bubble-wall glass shards. They sit between thicker moderately welded tuff layers. The trail crosses several faults that drop various ignimbrite layers against each other.

In 2012, Whitewater Canyon flooded spectacularly during monsoonal storms a few months after the enormous Whitewater-Baldy wildfire removed moisture-absorbing vegetation and soils. Bare landscape, without any vegetation, is vulnerable to flooding during the extreme monsoon rain events. This flooding destroyed the Catwalk National Recreation Trail, but it has since been rebuilt. Check for seasonal road closures before visiting.

North of milepost 34, the highway begins climbing in earnest toward Saliz Pass (milepost 31), leaving behind the Gila Group sediments and passing numerous roadcuts in faulted rhyolitic pyroclastic rocks of the Mogollon Group. The rhyolite tends to break into parallel fractures that produce rectangular, blocky outcrops. Volcanic breccia is visible in the roadcuts. The route is in the intensely faulted rocks on the southeast side of the down-dropped Reserve graben, an extensional half graben.

The Reserve graben is filled with sediments of the Gila Group. Near and north of milepost 26, US 180 winds through these basin fill units. The road is in the middle of the graben at the junction with NM 12. It then crosses several northeast-trending faults on the northwest side of the graben and passes through the up-faulted San Francisco Mountains back into an area of mostly volcaniclastic rocks. Luna (milepost 8) lies along the upper San Francisco River, much smaller here and closer to its headwaters in Arizona. US 180 crosses the river near milepost 3. More layers of volcaniclastic rocks can be seen between milepost 3 and the Arizona border.

The brilliant gypsum dunes of White Sands National Park, majestic as New Mexico's summer clouds, are swept slowly northeastward by the wind.

GLOSSARY

Some definitions were simplified from the *Glossary of Geology*, published by the American Geological Institute.

aa. The Hawai'ian word for a lava flow that has a rough, jagged surface.

accretion. Process in which material from a subducting plate is added onto the leading edge of an overriding plate at a convergent plate boundary.

agate. An often-banded rock formed with cryptocrystalline quartz.

aggregate. Sediment—usually sand or gravel—or crushed stone used in making concrete, mortar, plaster, and road material. Also used alone as railroad ballast and construction fill.

alluvial fan. A gently sloping, fan-shaped accumulation of sediments deposited by a stream where it flows out of a narrow valley onto a wider, flatter area.

alluvium. A general term for clay, silt, sand, gravel, and other unconsolidated sedimentary material deposited in relatively recent geologic time by running water.

ammonites. A group of cephalopods (squid relatives) of the Paleozoic and Mesozoic eras with coiled, chambered shells similar to that of the modern chambered nautilus.

amphibolite. A type of rock formed by regional metamorphism of basalt under medium to high pressures and temperatures. It consists largely of the minerals amphibole and plagioclase.

andesite. A medium- to dark-colored volcanic rock that is between basalt and rhyolite in silica content. It was named after the Andes Mountains in South America, where it is common.

angular unconformity. An unconformity in which the bedding planes between two groups of rocks are not parallel. The older, underlying strata typically dip more steeply than the overlying strata, and the contact is an erosional surface developed on the tilted older rocks.

anticline. A fold in which layered rocks have been bowed upward, producing an archlike profile.

aquifer. A porous underground body of rock from which water can be obtained. The aquifer is replenished by rain in the recharge zone where permeable rock layers reach the surface.

arkosic. Said of sand made up of mostly quartz with at least 25 percent feldspar minerals. Arkoses form commonly in continental interiors near young granitic mountain chains.

arroyo. A usually dry, steep-sided gully eroded by fast-flowing water in an arid region.

artesian. In an artesian aquifer pressurized groundwater trapped between layers of impermeable rock rises to the surface via a well drilled into the aquifer.

ash. Tiny fragments of volcanic rock blown into the air and distributed across the countryside near an erupting volcano.

ash fall. Accumulation of volcanic ash particles that have fallen out of the atmosphere.

ash-flow tuff. The rock formed from the consolidation and compaction of an ash-flow deposit.

badlands. A type of topography characterized by a great number of intricate streams separated by narrow ridges with steep slopes. Badlands develop in areas with little vegetation or poorly consolidated rocks.

ballast. Gravel, broken stone, or similar material used as a foundation for roads, especially that laid in the roadbed of a railroad.

basalt. A fine-grained, dark-colored volcanic igneous rock made up mostly of the calcium-rich minerals plagioclase feldspar and pyroxene. It is the very abundant extrusive equivalent of gabbro.

basement. The igneous and metamorphic rocks upon which a sequence of sedimentary rocks was deposited.

basin. A low area in the landscape where water may accumulate as a lake. Also, a large-scale downwarping of the Earth's crust in which relatively thick sequences of sediments are deposited.

Basin and Range. A physiographic province in the western United States characterized by long, linear, fault-block mountains separated by intervening fault-block valleys; developed because of crustal extension.

batholith. A mass of coarse, granular granitic igneous rock exposed over an area greater than about 30 square miles and consisting of two or more plutons.

bed. A single layer of sedimentary rock. The layering of beds is called bedding. The boundary between beds is a **bedding plane**.

bedrock. The hard rock at or near the surface that has not been moved by erosion or human activity. It is commonly concealed by younger, uncemented sediments deposited by water, ice, or wind.

biotite. A dark, platy mineral in the mica group.

bioturbation. The reworking of sediment by organisms.

brachiopod. An invertebrate animal with two unequal shells (unlike clam shells), each of which is bilaterally symmetrical. Brachiopods were abundant and diverse until Cenozoic time, with only a few living today.

braided river. A river with a floodplain composed of many interconnected, shifting channels separated by coarse sand or gravel bars.

breccia. A sedimentary or volcanic rock containing angular, pebble-size to larger rock fragments.

bryozoan. A group of invertebrate marine colonial animals that secrete calcareous skeletons with varied forms. Some colonies branch and resemble plants, though they are animals.

burrow. A trace left by an organism, such as a worm, as it ate its way through sediment or dug a home.

calcareous. Said of a rock containing calcium carbonate along with other constituents.

calcite. A light-colored mineral composed of calcium carbonate that is the main constituent of limestone, most marble, and many marine fossils.

calcium carbonate. A chemical compound of calcium, carbon, and oxygen that can crystallize as either calcite or aragonite.

caldera. A volcanic structure, typically a basin or depression, formed by the collapse of the surface due to the eruption of a large volume of lava or volcanic ash.

caliche. A layer of calcium carbonate that forms in soils and sediments in dry regions. Also called calcrete.

caprock. An overlying rock layer that is unusually hard and protects the underlying layers from erosion.

carbonate. Rocks such as limestone or dolostone that are formed with calcium, magnesium carbonate, or both.

cement. The substance, usually composed of silica (quartz), calcium carbonate minerals, or iron oxide, that binds the grains of a sedimentary rock.

chert. A sedimentary rock composed of quartz crystals too small to see with the naked eye. It forms as a chemical deposit, found primarily as nodules precipitated from water percolating through limy sediments after deposition, but it can also be precipitated directly from water on the seafloor.

chlorite. A green, platy mineral characteristic of low-temperature metamorphism of mafic rocks—rocks with iron and magnesium.

cinder cone. A steep-sided, cone-shaped accumulation of cinders that surrounds a basaltic vent.

clast. A grain or fragment of a rock. A sedimentary rock is said to be **clastic** if it is composed of broken fragments, such as sand grains, derived from preexisting rocks.

claystone. A fine-grained sedimentary rock composed mostly of clay minerals but lacking the fine layering typical of shale.

coal. A black or, rarely, dark-brown combustible rock composed mostly of carbonaceous material formed by the compaction and induration of variably decomposed plant material. In the geologic record, coal typically represents a swamp or marsh environment.

coarse-grained. Said of sedimentary rocks that have clasts, or particles, that are relatively large, usually averaging 2 millimeters in diameter or larger. Also said of igneous rocks with relatively large crystals.

coastal plain. An area of very low relief next to the coast.

colluvium. Unconsolidated sediments deposited at the base of a slope by rainwash, sheetwash, or by slowly moving down the slope.

columnar jointing. The fracturing in a lava flow that causes the flow to break into columns.

composite volcano. Synonymous with stratovolcano but used to explicitly describe a volcano that consists of several parts, typically a series of domes and vents.

conglomerate. A coarse sedimentary rock made up of pebbles, cobbles, boulders, or a combiantion of the three eroded from older rocks. The large size implies a vigorous process of deposition, such as that of a mountain stream.

contact. The boundary or surface between two different rock types, formations, or ages of rocks. Contacts may be depositional bedding planes, faults, edges of intrusive bodies, angular unconformities, or metamorphic grades.

continental shelf. The submerged margin of a continent. Modern continental shelves extend from the coastline to water depths of about 600 feet.

corals. Marine, bottom-dwelling, mostly colonial animals that secrete an external skeleton composed of calcium carbonate. Their skeletons form major components of modern and ancient reefs.

country rock. The preexisting bedrock intruded by or surrounded by igneous rocks.

craton. A part of the Earth's crust, usually in the interior of a continent, that has not been pervasively deformed for at least 1 billion years.

crinoid. A group of invertebrate animals, also called "sea lilies," related to sea stars and sea urchins. They resemble a plant, with a stem attached to the sea bottom and arms radiating from the top, and were abundant in the Paleozoic Era.

cross-bedding. Laminations, or cross-beds, in sand or sandstone that are inclined about 30 degrees to the main bedding. These inclined beds represent the lee, or downwind, faces of dunes or ripples, and their downslope direction indicates the direction of the wind or water current that deposited them.

crust. The outermost layer of Earth. **Oceanic crust**, made of basalt and rocks of similar composition, ranges from 3 to 5 miles thick. **Continental crust**, made mainly of lighter-colored, less dense rock, such as granite or gneiss, often with a veneer of sedimentary rock, normally ranges from 20 to 40 miles thick.

crystal. A many-faced solid bounded by smooth planar surfaces that reflect an orderly internal arrangement of atoms.

crystalline. Said of a rock formed of interlocking mineral crystals, usually igneous or metamorphic.

cuesta. A low ridge with a steep face on one side and a gentle slope on the other side.

dacite. A volcanic rock that contains about 65 percent silica, slightly less than that of rhyolite.

debris flow. A mass of soil, mud, and rock that may move slowly or up to 80 miles per hour.

deformation. A general term for the folding, faulting, shearing, extension, or compression of rock.

delta. A body of sediment deposited where a river enters a standing body of water.

deposition. The process of sediment settling out of water or air.

desert varnish. A distinctive, usually shiny, dark-brown surface coating of manganese and iron on rocks. It may be related to the presence of organic acids formed by the decomposition of lichen.

dike. A tabular body of igneous rock that cuts across the structure of older adjacent rock into which it was intruded as fluid magma.

diorite. An intrusive igneous rock that is between gabbro and granite in silica content. It is the intrusive equivalent of andesite.

dip. The downslope direction (the direction water would run) on an inclined (or dipping) bedding surface.

dissolution. The process of dissolving rock, usually by acidic rainwater or groundwater.

dolomite. A sedimentary rock akin to limestone that contains magnesium as well as calcium carbonate. Dolomite typically forms when fluids moving through buried limestone precipitate magnesium.

dolostone. A nonclastic sedimentary rock consisting largely of the mineral dolomite, which is composed of calcium magnesium carbonate.

downsection. To move down through the sequence of layered sedimentary rock units.

drainage basin. The area, or watershed, from which all water drains into a river.

erosion. A general term for several processes that loosen, dissolve, or weather and then transport earthen materials and thus wear away landscapes.

escarpment. The steep face of a ridge or plateau along which the land drops abruptly to a lower level.

evaporite. A nonclastic sedimentary rock typically formed by the partial or total evaporation of brine.

extrusive igneous rocks. Rocks that solidify from magma on the Earth's surface.

fanglomerate. A sedimentary rock consisting of waterworn fragments of various sizes, deposited in an alluvial fan and later cemented into a firm rock.

fault. A fracture along which the rocks on either side have been displaced from inches to miles.

fault block. A body of rock adjacent to a fault. A mountain range that is uplifted along one side by a normal fault and shows a consistent tilt in its bedding away from the fault is called a **tilted fault block**.

fault system. A number of faults spaced miles apart, with about the same orientation, that serve the same geologic function, for example, the edge of a mountain range.

fault zone. A relatively narrow region in which multiple faults, more or less parallel but often interconnected, have developed.

feldspar. The most abundant mineral group in Earth's crust. Feldspars are divided between potassium-bearing feldspars (for example, orthoclase) and calcium- and sodium-bearing plagioclase feldspars. They are especially characteristic of igneous rocks. Feldspar is not as common as quartz in soil or sedimentary rocks because weathering alters feldspar to clay.

felsic. A chemical term applied to igneous rocks that, if crystalline, would be composed mostly of light-colored minerals such as quartz, feldspar, and muscovite.

fine-grained. Said of sedimentary rocks that have clasts, or particles, that are relatively small, usually averaging less than 2 millimeters in diameter. Also said of igneous rocks with relatively small, hard-to-see crystals.

floodplain. An area adjacent to a riverbed that may lie underwater when the river overflows its banks.

fold. Rock layers that have been bent, usually as a result of compressive tectonic forces. A slight fold is called open or gentle. A V-shaped fold is called tight.

foreland basin. A depression that forms where crust adjacent to an orogenic belt receives additional load produced by the overthickened crust.

formation. The basic subdivision of sedimentary rocks that can be mapped from place to place.

fossil. Any remains or traces of animals or plants preserved in sediment or sedimentary rocks. Impressions, tracks, burrows, and other traces that are not actual remains are called **trace fossils**.

fracture. A general term for a break or crack in a rock. If no movement has occurred along the break, the fracture is a joint. If movement has occurred, the fracture is a fault or extension fracture.

gabbro. A dark-colored, coarse-grained intrusive igneous rock composed mostly of calcium plagioclase feldspar and pyroxene. When the same magma erupts at the surface, it forms basalt.

gastropod. Any mollusk belonging to the class Gastropoda, such as snails.

geode. A cavity in rock lined with crystals. If the surrounding rock weathers away, the geode may remain as a round rock that reveals its crystals when broken.

geothermal. Pertaining to heat released from within the Earth. **Geothermal power** is energy derived from the internal heat of the Earth.

glacial outwash. Sediment deposited by the large quantities of meltwater emerging from the terminus of a glacier.

glaciation. The formation and movement of glaciers or large ice sheets.

gneiss. A coarse-grained metamorphic rock that has a striped appearance caused by alternating bands of light-colored minerals (quartz and feldspar) and bands of dark-colored, platy or flaky minerals (biotite and hornblende).

graben. A crustal block that is downdropped between two inwardly dipping normal faults. A **half graben** forms where there is a normal fault only on one side.

granite. A light-colored, coarse-grained intrusive igneous rock made mostly of plagioclase, potassium feldspar, and quartz, but which may also contain mica, hornblende, or both.

granitic gneiss. Gneiss that is dominated by quartz and feldspar, with relatively thin bands of dark minerals. It most likely formed from the metamorphism of granite.

granodiorite. An intrusive igneous rock with a chemical composition between diorite and granite. It is made up of quartz, plagioclase, and a little potassium feldspar, biotite, and hornblende.

greenstone. Volcanic rocks, typically basalt, that have been metamorphosed and have grown green metamorphic minerals.

groundwater. Subsurface water in the zone of saturation, the area below where rainwater percolates through the soil and underlying rock. Groundwater may include ancient salt-water and brine.

group. A formal rock unit containing two or more formations.

gypsum. A mineral or rock composed of calcium sulfate that formed from the evaporation of water (making it an evaporite rock).

high-grade. Said of an ore with a relatively high ore-mineral content.

hogback. A ridge with a summit consisting of both outcrops and steep slopes of resistant rock with roughly the same degree of incline, typically greater than 20 degrees.

horst. A crustal block that is uplifted between two outwardly dipping normal faults.

hummocky. Said of topographic land or ice forms that abound in small hills and depression meters to tens of meters across, such as a hummocky dune.

hydrothermal. Pertaining to geothermally heated water.

ice age. A time of extensive continental-scale glacial activity.

igneous. Rock that cooled from molten material, either from magma within the Earth (intrusive or plutonic) or from lava at the surface (extrusive or volcanic).

ignimbrite. The deposit of a pyroclastic flow.

impermeable. Having a texture that does not permit water to move through. Clay is often considered a relatively impermeable sediment.

interbedded. Said of rock units with alternating layers of differing rock types or characteristics, or both.

intermontane. Situated between or surrounded by mountains.

intrusive. Rocks cooled from magma that penetrated other rocks.

iridium. A metallic white cubic mineral consisting of more than 80 percent platinum, the remainder being osmium, palladium, or related elements.

island arc. An offshore volcanic arc or linear chain of volcanoes formed along a convergent plate margin.

joint. A crack in rock along which no movement has occurred, so it is not a fault. Joints often occur parallel to each other in groups or sets.

karst. A distinctive type of topography that formed from the dissolution of carbonate rocks and is characterized by sinkholes, caves, and underground streams.

laccolith. A mushroom-shaped igneous intrusion formed when rising magma spreads between layers of sedimentary rock.

lava. Melted rock, or magma, that erupts at Earth's surface.

limestone. A sedimentary rock composed of calcium carbonate. It commonly is largely made up of the calcareous skeletons of invertebrate fossils.

lithosphere. The outer rigid shell of Earth that is broken into the tectonic plates. On average, continental lithosphere is about 100 miles thick and old oceanic lithosphere is about 60 miles thick.

lode. A mineral deposit consisting of a zone of veins, veinlets, disseminations, or planar breccias.

low-grade. Said of an ore with a relatively low ore-mineral content. Also pertains to rocks metamorphosed under conditions of low pressure and temperature.

maar volcano. An explosive, crater-forming eruption caused by the interaction of magma with groundwater.

mafic. Rich in iron, magnesium, or both. Applied to silicate minerals (especially amphibole, pyroxene, or olivine), rocks containing them, or the magma that produces such rocks. Mafic minerals are typically green or black. Mafic magma usually arises from the partial melting of rock within the mantle. Such magma is characteristic of rift zones but is also part of the suite of igneous rocks that form above plates that are being subducted into the mantle.

magma. Molten rock. Termed **lava** where it erupts at the surface of the Earth.

mantle. The part of Earth between the interior core and the outer crust.

marine. Pertaining to the sea.

meander. A looplike bend in a river that tends to form in flatlands where there is loose, relatively fine sediment.

member. A formal stratigraphic unit representing some specially developed part of a formation. It is ranked lower than a formation and higher than a bed.

metamorphic. Said of minerals and rocks with compositions and textures that were changed by heat and/or pressure. For example, shale that was metamorphosed to slate and sandstone to quartzite.

metamorphism. The process through which the texture and, often, the mineralogy of rock (by the formation of new crystals, or recrystallization) change due to heat and pressure but without the rock melting.

metasedimentary. A metamorphosed sedimentary rock. Pressure and heat caused sediment grains to partly recrystallize, but original bedding and sedimentary structures remain largely intact.

mica. A family of silicate minerals, including biotite and muscovite, that breaks easily into thin flakes. Micas are common in many kinds of igneous and metamorphic rocks.

mineral. A naturally occurring chemical element or compound with a characteristic crystal form.

mountain building event. An event in which mountains rise. During these events, rocks are typically folded, faulted, and/or metamorphosed. Intrusive and extrusive igneous activity often accompanies it. Also known as **orogeny**.

mud crack. A sedimentary structure forming a roughly polygonal pattern caused by the shrinkage of wet clay or silt, typically due to drying.

mudstone. A sedimentary rock made mainly of clay. Unlike shale, it doesn't tend to split into thin pieces.

muscovite. A common, colorless to light-brown mineral of the mica group. It is present in many igneous and metamorphic rocks, and much less commonly in sedimentary rocks.

nodule. A rounded fist-size or smaller concretion usually found in limestone or dolostone.

normal fault. A fault along which one fault block slides down a sloping fault surface relative to the fault block on the other side of the fault. A normal fault forms as the result of forces that are pulling an area apart.

North American Plate. One of eight huge moving plates that make up the outer, solid part of the Earth.

obsidian. Volcanic glass, typically high in silica and dark gray to black. Impurities may give rise to brown or red colors in the obsidian.

olivine. A glassy green or brown mineral that is common in gabbro and basalt.

ore. A mineral or aggregate of minerals from which one or more valuable substances, especially metals, can be profitably extracted.

orogeny. The process of mountain formation.

outcrop. An exposure of bedrock.

outwash. See **glacial outwash.**

oxide. A compound in which oxygen combines with a positively charged ion. Geologically, oxide minerals include both metallic ores and gems.

pahoehoe. Hawai'ian word for a basaltic lava flow that has a smooth, undulating, or ropy surface. These surface features are due to the movement of fluid lava under a congealing surface crust.

paleosol. A fossilized or ancient soil that formed on a landscape in the distant past.

Pangaea. A supercontinent that assembled through plate convergences about 300 million years ago. It broke into the modern continents through divergences beginning about 200 million years ago.

pediment. A broad, gently sloping, bedrock-floored erosion surface or plain of low relief in an arid or semiarid region at the base of a mountain front. Commonly mantled with a thin veneer of alluvium.

pegmatite. A very coarse-grained, usually granitic, igneous rock with crystals at least 2.5 centimeters long.

physiographic province. A geographic region defined by differences in the landscape.

plagioclase. A mineral in the feldspar group containing sodium, calcium, or both.

plate tectonics. The theory that Earth's lithosphere (its crust plus the rigid uppermost mantle) consists of a mosaic of tectonic plates that move independently of each other and interact at their boundaries. Most large-scale structures of the Earth's surface are formed by the relative movements of these plates. For example, mountains form where two plates collide.

playa. A smooth, fine-grained lakebed in a desert valley, normally dry but sometimes flooded.

playa lake. A shallow intermittent lake that occupies the lowest part of a small, closed basin.

pluton. A solidified body of intrusive igneous rock, such as granite, that crystallizes beneath the surface. The adjective **plutonic** contrasts with volcanic, which refers to igneous rocks that form at the surface.

porphyry. An igneous rock with large visible crystals of feldspar or quartz surrounded by a fine-grained groundmass.

pumice. A pyroclastic rock that consists of volcanic glass with a frothy texture because of an abundance of airholes.

pyrite. A mineral made of iron sulfide. Also called fool's gold.

pyroclastic. Describing fragmental volcanic particles, fragmented during explosive eruptions.

pyroxene. A group of mostly dark, stubby minerals that are rich in iron, calcium, magnesium, and silica and occur in igneous and metamorphic rocks.

quartz. One of the most common minerals in the Earth's crust. The most common variety is colorless and clear like glass. Quartz is composed of silicon and oxygen, is the main constituent of most sand, and is common in sedimentary rocks and light-colored igneous rocks.

quartzite. A sandstone metamorphosed by heat, pressure, or both, so that the rock breaks across individual sand grains rather than around them.

radiocarbon dating. Determining the age of a rock by using the properties of the radioactive carbon-14 isotope.

radiometric dating. The calculation of age based on the rate of time it takes for radioactive elements to decay.

red bed. Sediment of any grain size that is reddish in color, usually due to coatings or cements of hematite or other iron oxide minerals.

regression. The seaward movement of a shoreline as sea level drops or the land rises. See also **transgression**.

reservoir (for hydrocarbon). Any porous and permeable rock that contains oil or gas.

resistant. Said of a rock or rock outcrop that withstands the effects of weathering or erosion.

reverse fault. A fault in which the block overlying a dipping fault appears to have moved up relative to the block underlying the fault. Reverse faults commonly form as a result of compression.

rhyolite. A light-colored, fine-grained volcanic rock (an extrusive igneous rock) containing the same proportions of quartz and potassium feldspar as granite.

rift. A long, narrow, down-dropped zone where Earth's crust is separating. The process of a tectonic plate being pulled apart or splitting is called **rifting**.

ripple mark. An undulatory surface on loose, granular sand or fine gravel sculpted by either wind, water currents, or waves. The marks can be preserved in sedimentary rocks.

sandstone. A clastic sedimentary rock consisting primarily of sand-size grains, usually of the mineral quartz.

scarp. A linear cliff produced by faulting or erosion. The term is an abbreviation for escarpment.

schist. A metamorphic rock that is strongly layered due to an abundance of visible, platy minerals.

sediment. Earth material that has settled out of water or air. It can consist of pieces of preexisting rock, body parts of organisms, or mineral crystals that precipitated out of water.

sedimentary rock. Sediment that has been naturally compacted or cemented to form solid rock.

seismic. Pertaining to the phenomenon of movements, or earthquakes, in the Earth's crust.

selenite. A clear, colorless variety of the mineral gypsum.

sequence. A series of major rock units deposited under the same or closely related environmental conditions within a particular region.

shale. A fine-grained sedimentary rock made mainly of clay that tends to split into thin pieces parallel to its bedding.

shear zone. The zone in which deformation occurs when two bodies of rock slide past each other, resulting in crushed and brecciated rock with parallel fractures due to shear strain.

shield volcano. A gently sloped volcano typically made of basalt. In profile, shield volcanoes resemble shields.

silica. Silicon dioxide, the compound that makes up quartz in all its varieties, including chert. A rock or magma that is rich in silica is **silicic**.

sill. A tabular, intrusive igneous body that is parallel to the bedding or foliation in the layered rocks that it intrudes.

siltstone. A clastic sedimentary rock consisting primarily of silt-size grains, particles of rock larger than clay but finer than sand. It can have texture and composition similar to shale but lacks the thin-bedded, platy appearance and tends to be better cemented.

sinkhole. A surface depression that results from the collapse of an underlying cavity.

strata. Beds within layered sedimentary rocks. A single bed is called a **stratum**.

stratified. Said of a well-layered sedimentary rock.

stratigraphy. The science of rock strata, including the arrangement of rock layers and their relationships.

strike-slip fault. A fault showing sideways movement, or offset, along a near-vertical fault.

strip mine. A surface mine in which the valuable rock is near the surface and is recovered by removing the overburden. Mining stops when the rock is too deep to pay for removal of the overburden. If the overburden and final mine site are not reclaimed, the area typically is unusable.

structure. A fold, fault, or other geologic feature that results from deformational forces in the Earth's crust.

subduction. The process by which the edge of a tectonic plate capped by oceanic crust descends below an adjacent plate into the asthenosphere, where it melts. The molten rock can make its way back to the surface through volcanoes or as intrusions in the Earth's crust. The long, narrow feature produced by subduction is called a **subduction zone**.

sulfide. A compound containing sulfur. Sulfide minerals often contain valuable metals, such as copper and zinc.

supercontinent. A clustering of all or most of Earth's continental masses into one major landmass; this has occurred at least three times in geologic history.

suspended sediment. Sediment in a stream that remains lifted or transported due to the turbulence and energy of the water. When the currents slow, suspended sediment falls to the bottom.

syncline. A fold in which layered rocks have been bowed downward, producing a smile-like profile.

tafone. A granitic or gneissic block or boulder hollowed out by cavernous weathering.

tephra. A general term for the material ejected from a volcano.

terrace. An erosional remnant of a former floodplain or coastline, standing above the present river or coast.

thrust fault. A fault dipping less than 45 degrees that formed by tectonic forces of horizontal compression. Generally the rock above the fault moved upward and over the rock below it.

trace fossil. A fossil, such as a burrow or a fecal pellet, that preserves evidence of past life but is not the actual skeletal remains.

transform fault. A special type of strike-slip fault along a mid-ocean ridge or plate boundary.

transgression. The expansion of the sea over land areas and the consequent geologic evidence, such as deep-water deposits overlying shallow-water deposits.

trap. A subsurface barrier to the upward movement of oil or gas.

travertine. A form of calcite deposited from solution by groundwater or surface water; for example, in cave formations.

trilobite. An extinct arthropod with an external calcareous skeleton and appendages for swimming and walking.

tuff. A volcanic rock made mostly of consolidated pyroclastic material, chiefly ash and pumice, derived from ash falls or pyroclastic flows. A **welded tuff** is distinctly harder because the enormous mass and great heat of the flow causes the material to fuse, or weld, together. An **ash-flow tuff** is derived from an ash flow.

unconformity. A break or gap in any sequence of strata that implies an interval of time during which no strata were deposited, or strata were eroded before deposition began again.

upsection. To move up through the sequence of layered sedimentary rock units.

varve. Layers of alternately finer and coarser silt or clay believed to be evidence of annual cycles of deposition in a body of still water.

vein. A mineral deposit that formed along a crack in rock by the precipitation of minerals from water. Quartz and calcite are the most common vein materials.

vent (silicic and mafic). The actual place where volcanic materials erupt. Vents are either eruptive localities on large volcanoes or mark much smaller volcanoes.

vesicle. A cavity of variable size and shape in lava, formed by a trapped gas bubble during solidification of the lava. A rock with vesicles is said to be **vesicular**.

volcanic arc. A chain of island volcanoes that formed above an ocean-floor subduction zone.

volcanic ash. Fine rock fragments from an explosive volcanic eruption.

volcaniclastic. Pertaining to a clastic rock containing volcanic rock fragments in any proportion and without regard to origin or environment.

weathering. The process by which rocks break down near Earth's surface due to exposure to air, water, and the action of organisms. An unweathered rock surface is called "fresh."

welded tuff. A glassy pyroclastic volcanic rock that has been hardened by the fusing together of its glass shards under the combined action of heat, weight of overlying rocks, and hot gasses.

REFERENCES

Aby, S. 2020. *Geologic Map of the Echo Amphitheater 7.5-minute Quadrangle, Rio Arriba County, New Mexico.* New Mexico Bureau of Geology and Mineral Resources Open-File Geologic Map GM-280.

Aby, S., A. Gellis, and M. Pavich. 1997. The Rio Puerco arroyo cycle and the history of landscape changes. In *Impacts of Climate Change and Land Use in the Southwestern United States.* An interactive workshop of the US Global Change Research Program. Available online at https://geochange.er.usgs.gov/sw/

Aby, S., K. Karlstrom, D. Koing, K. Kempter, and P. Davis. 2010. *Geologic Map of the Las Tablas 7.5-minute Quadrangle, Rio Arriba County, New Mexico.* New Mexico Bureau of Geology and Mineral Resources Open-File Geologic Map GM-200.

Allen, B. D. 2000. *Geologic Map of the Edgewood Quadrangle, Torrance and Santa Fe Counties, New Mexico.* New Mexico Bureau of Geology and Mineral Resources Map Series GM-35.

Allen, B. D. 2005. Ice age lakes in New Mexico. In *New Mexico's Ice Ages.* New Mexico Museum of Natural History and Science Bulletin 28, eds. S. G. Lucas, G. S. Morgan, and K. E. Zeigler, 107–14.

Anderson, O. J. 1994. *Geology of the Zuni Salt Lake 7.5 minute Quadrangle.* New Mexico Bureau of Geology and Mineral Resources Open-File Report 405.

Anderson, O. J., C. H. Maxwell, and S. G. Lucas. 2003. *Geology of Fort Wingate Quadrangle, McKinley County, New Mexico.* New Mexico Bureau of Geology and Mineral Resources Open-File Report 473.

Armstrong, A. K., and L. D. Holcomb. 1989. Stratigraphy, Facies, and Paleotectonic History of Mississippian Rocks in the San Juan Basin of Northwestern New Mexico and Adjacent Areas. In *Evolution of Sedimentary Basins, San Juan Basin.* USGS Bulletin 1808-B-D.

Ashcroft, B. 1988. Miner and merchant in Socorro's boom town economy, 1880–1893. *New Mexico Historical Review* 63 (2): 103–17.

Aubele, J. C., and L. S. Crumpler. 2001. Raton-Clayton and Ocate volcanic fields. In *Geology of the Llano Estacado*, New Mexico Geological Society 52nd Annual Fall Field Conference Guidebook, eds. S. G. Lucas and D. S. Ulmer-Scholle, 69–76.

Axen, G., J. van Wijk, F. Phillips, and others. 2019. The Socorro Magma Body. *New Mexico Earth Matters* 19 (1): 1–5.

Bachhuber, F. W., and W. A. McClellan. 1977. Paleoecology of marine Foraminifera in the pluvial Estancia Valley, Central New Mexico. *Quaternary Research* 7 (2): 254–67.

Baer, S., K. E. Karlstrom, P. Bauer, and S. D. Connell. 2004. *Geologic Map of the Manzano Peak 7.5 Minute Quadrangle.* New Mexico Bureau of Geology and Mineral Resources Geologic Map GM-61.

Bahadori, A., W. E. Holt, R. Feng, and others. 2022. Coupled influence of tectonics, climate, and surface processes on landscape evolution in southwestern North America. *Nature Communications* 13 (4437). Published online at nature.com

Balk, R. 1962. *Geologic Map and Sections of Tres Hermanas Mountains.* New Mexico Bureau of Geology and Mineral Resources Geologic Map GM-16.

Baltz, E. H., and D. A. Myers. 1999. *Stratigraphic Framework of Upper Paleozoic Rocks, Southeastern Sangre de Cristo Mountains, New Mexico.* New Mexico Bureau of Geology and Mineral Resources Memoir 48.

Bauer, P. W. 2011. *The Rio Grande: A River Guide to the Geology and Landscapes of Northern New Mexico.* New Mexico Bureau of Geology and Mineral Resources.

Bauer, P. W., K. E. Karlstrom, S. A. Bowring, A. G. Smith, and L. B. Goodwin. 1993. Proterozoic plutonism and regional deformation: New constraints from the southern Manzano Mountains, central New Mexico. *New Mexico Geology* 15 (3): 49–77.

Bauer, P. W., and K. I. Kelson. 2001. *Geologic Map of the Taos 7.5-Minute Quadrangle, Taos County, New Mexico.* New Mexico Bureau of Geology and Mineral Resources Geologic Map GM-43.

Bauer, P. W., K. I. Kelson, and S. B. Aby. 2005. *Geologic Map of the Peñasco 7.5-Minute Quadrangle, Taos County, New Mexico.* New Mexico Bureau of Geology and Mineral Resources Geologic Map GM-62.

Bauer, P. W., K. I. Kelson, V. J. S. Grauch, and others. 2016. *Geologic Map and Cross Sections of the Embudo Fault Zone in the Southern Taos Valley, Taos County, New Mexico.* New Mexico Bureau of Geology and Mineral Resources Open-File Report 584.

Bauer, P. W., K. I. Kelson, J. Lyman, M. R. Heynekamp, and D. McCraw. 2005. *Geologic Map of the Ranchos de Taos Quadrangle, Taos County, New Mexico.* New Mexico Bureau of Geology and Mineral Resources Geologic Map GM-33.

Beane, R. E., C. Jaramillo, and L. E. Bloom. 1975. Geology and base metal mineralization of the southern Jarilla Mountains, Otero County, New Mexico. In *Las Cruces Country*, New Mexico Geological Society 26th Annual Fall Field Conference Guidebook, eds. W. R. Seager, R. E. Clemons, and J. F. Callender, 151–56.

Bennett, M. R., D. Bustos, J. S. Pigati, and others. 2021. Evidence of humans in North America during the last Glacial Maximum. *Science* 373 (6562): 1528–31.

Brand, B. D., A. B. Clark, and S. Semken. 2009. Eruptive dynamics and depositional processes of Narbona Pass maar volcano, Navajo volcanic field, Navajo Nation, New Mexico (USA). *Bulletin of Volcanology* 71: 49–77.

Brandes, N. 2021. *New Mexico Rocks! A Guide to Geologic Sites in the Land of Enchantment.* Missoula, MT: Mountain Press Publishing Company.

Brister, B. S., and R. R. Gries. 1994. *Tertiary stratigraphy and tectonic development of the Alamosa basin (northern San Luis Basin), Rio Grande rift, south-central Colorado.* GSA Special Paper 291.

Broadhead. R. F. 1997. *Subsurface Geology and Oil and Gas Potential of Estancia Basin, New Mexico.* New Mexico Bureau of Geology and Mineral Resources Bulletin 157.

Carlson, M. P. 1999. Transcontinental Arch: A pattern formed by rejuvenation of local features across central North America. *Tectonophysics* 305 (1-3): 225–33.

Carpenter, K., and J. M. Parrish. 1985. Late Triassic vertebrates from Revuelto Creek, Quay County, New Mexico. In *Santa Rosa, Tucumcari Region*, New Mexico Geological Society 36th Annual Fall Field Conference Guidebook, eds. S. G. Lucas and J. Zidek, 197–98.

Cather, S. 2004. Laramide orogeny in central and northern New Mexico and southern Colorado. In *The Geology of New Mexico: A Geologic History*. New Mexico Geological Society, 203–48.

Cather, S. M., R. M. Chamberlin, and J. C. Ratte. 1994. Tertiary stratigraphy and nomenclature for western New Mexico and eastern Arizona. New Mexico Geological Society Field Conference Series 45: 259–66.

Cather, S., S. D. Connell, R. M Chamberlin, and others. 2008. The Chuska erg: Paleogeomorphic and paleoclimatic implications of an Oligocene sand sea on the Colorado Plateau. *GSA Bulletin* 120 (1/2): 13–33.

Cather, S. M., W. C. McIntosh, and S. A Kelley. 2004. *Tectonics, Geochronology, and Volcanism in the Southern Rocky Mountains and Rio Grande Rift.* New Mexico Bureau of Geology and Mineral Resources Bulletin 160.

CD-ROM Working Group. 2002. Structure and Evolution of the Lithosphere beneath the Rocky Mountains: Initial Results from the CD-ROM Experiment. *GSA Today* March: 4–10.

Chamberlin, R. M. 1999. *Geologic Map of the Socorro Quadrangle, Socorro County, New Mexico.* New Mexico Bureau of Geology and Mineral Resources Geologic Map GM-34.

Chamberlin, R. M., and J. S. Harris, 1994. Upper Eocene and Oligocene volcaniclastic sedimentary stratigraphy of the Quemado-Escondido Mountain area, Catron County, New Mexico. *Mogollon Slope*, New Mexico Geological Society 45th Annual Fall Field Conference Guidebook, eds. R. M. Chamberlin and others, 269–75.

Chamberlin, R. M., W. C. McIntosh, and T. L. Eggleston. 2004. $^{40}Ar/^{39}Ar$ geochronology and the eruptive history of the eastern sector of the Oligocene Socorro caldera, central Rio Grande rift, New Mexico. In *Tectonics, Geochronology, and Volcanism in the Southern Rocky Mountains and Rio Grande Rift.* New Mexico Bureau of Geology and Mineral Resources Bulletin 160: 251–79.

Chapin, C. E. 2012. Origin of the Colorado Mineral Belt. *Geosphere* 8 (1): 28–43.

Chapin, C. E., S. A. Kelley, and S. M. Cather. 2014. The Rocky Mountain Front, southwestern USA. *Geosphere* 10 (5): 1043–60.

Clow, T., W. M. Behr, and M. A. Helper. 2019. Pleistocene to recent geomorphic and incision history of the northern Rio Grande gorge, New Mexico: Constraints from field mapping and cosmogenic ^{3}He surface exposure dating. *Geosphere* 15 (3): 820–38.

Connell, S. D. 2008. *Geologic Map of the Albuquerque–Rio Rancho Metropolitan Area and Vicinity, Bernalillo and Sandoval Counties, New Mexico.* New Mexico Bureau of Geology and Mineral Resources Geologic Map GM-78.

Connell, S. D. 2011. *Preliminary Study of the Geologic Framework of the Colorado Plateau–Middle Rio Grande Basin Transition, New Mexico.* New Mexico Bureau of Geology and Mineral Resources Open-File Report 539.

Connell, S. D., D. W. Love, and N. W. Dunbar. 2007. Geomorphology and stratigraphy of inset fluvial deposits along the Rio Grande valley in the central Albuquerque Basin, New Mexico. *New Mexico Geology* 29 (1): 13–31.

Contaldo, G. J., and J. E. Mueller. 1991. Earth fissures in the Mimbres Basin, southwestern New Mexico. *New Mexico Geology* 13 (4): 69–74.

Crews, S. G. 1994. Tectonic control of synrift sedimentation patterns, Reserve graben, southwestern New Mexico. *Mogollon Slope*, New Mexico Geological Society 45th Annual Fall Field Conference Guidebook, eds. R. M. Chamberlin and others, 125–34.

Crouse, D. L., M. C. Hultgren, and L. A. Woodward. 1992. *Geology of French Mesa Quadrangle, Rio Arriba County, New Mexico.* New Mexico Bureau of Geology and Mineral Resources Geologic Map GM-67.

Drakos, P. G., J. Riesterer, and K. Bemis. 2013. Recharge sources and characteristics of springs on the Zuni Reservation, New Mexico. In *Geology of Route 66 Region: Flagstaff to Grants*, New Mexico Geological Society 64th Annual Field Trip Guidebook, eds. K. Zeigler, J. M. Timmons, S. Timmons, and S. Semken, 205–13.

Drenth, B. J., V. J. S. Grauch, K. J. Turner, and others. 2019. A shallow rift basin segmented in space and time: The southern San Luis Basin, Rio Grande rift, northern New Mexico, USA. *Rocky Mountain Geology* 54 (2): 97–131.

Dunbar, N., S. Lucas, M. Zimmerer, and A. Jochems. 2019. Organ Mountains–Desert Peaks and Prehistoric Trackways National Monuments. *Lite Geology* 44: 3–5.

Dunbar, N. W., and F. M. Phillips. 2004. Cosmogenic ^{36}Cl ages of lava flows in the Zuni-Bandera volcanic field, north-central New Mexico, USA. In *Tectonics, Geochronology, and Volcanism in the Southern Rocky Mountains and Rio Grande Rift*. New Mexico Bureau of Geology and Mineral Resources Bulletin 160: 309–17.

Elliott, L. A., and J. K. Warren. 1989. Stratigraphy and depositional environment of lower San Andres formation in subsurface and equivalent outcrops: Chaves, Lincoln, and Roosevelt Counties, New Mexico. *American Association of Petroleum Geologists Bulletin* 73 (11): 1307–25.

Elston, W. E. 2008. When batholiths exploded: The Mogollon-Datil volcanic field, southwestern New Mexico. In *Geology of the Gila Wilderness-Silver City area*, New Mexico Geological Society 59th Annual Fall Field Conference Guidebook, eds. G. Mack, J. Witcher, and V. W. Lueth, 117–28.

Elston, W. E., E. G. Deal, and M. J. Logsdon. 1983. *Geology and Geothermal Waters of Lightning Dock region, Animas Valley and Pyramid Mountains, Hidalgo County, New Mexico*. New Mexico Bureau of Geology and Mineral Resources Circular 177.

Fackelman, S. P., J. R. Morrow, C. Koeberl, and T. H. McElvain. 2008. Shatter cone and microscopic shock-alteration evidence for a post-Paleoproterozoic terrestrial impact structure near Santa Fe, New Mexico, USA. *Earth and Planetary Science Letters* 270 (3-4): 290–99.

Fawcett, P. J., J. Heikoop, F. Goff, and others. 2007. Two middle Pleistocene glacial-interglacial cycles from the Valle Grande, Jemez Mountains, northern New Mexico. In *Geology of the Jemez Region II*, New Mexico Geological Society 58th Annual Fall Field Conference Guidebook, eds. B. S. Kues, S. A. Kelley, and V. W. Lueth, 409–17.

Fleischhauer, H. L. 1978. Summary of the late Quaternary geology of Lake Animas, Hidalgo County, New Mexico. In *Land of Cochise (Southeastern Arizona)*, New Mexico Geological Society 29th Annual Fall Field Conference Guidebook, eds. J. F. Callender, J. Wilt, R. E. Clemons, and H. L. James, 283–84.

Fleischhauer, H. L., and W. J. Stone. 1982. *Quaternary Geology of Lake Animas, Hidalgo County, New Mexico*. New Mexico Bureau of Geology and Mineral Resources Circular 174.

Gardner, J. N., F. Goff, S. Kelly, and E. Jacobs. 2010. Rhyolites and associated deposits of the Valles-Toledo caldera complex. *New Mexico Geology* 32: 1–16.

Gile, L. H. 2002. Lake Jornada, an early-middle Pleistocene lake in the Jornada del Muerto Basin, southern New Mexico. *New Mexico Geology* 24 (1): 3–14.

Goff, F., S. A. Kelley, C. J. Goff, and others. 2015. *Geologic Map of Mount Taylor, Cibola, and McKinley Counties, New Mexico*. New Mexico Bureau of Geology and Mineral Resources Open-File Report 571.

Goff, F., S. A. Kelley, C. J. Goff, and others. 2019. *Geologic Map of Mount Taylor Volcano Area, New Mexico*. New Mexico Bureau of Geology and Mineral Resources Geologic Map GM-80.

Goff, F., L. Shevenell, J. N. Gardner, F. D. Vuataz, and C. O. Grigsby. 1988. The hydrothermal outflow plume of Valles Caldera, New Mexico, and a comparison with other outflow plumes. *Journal of Geophysical Research* 93 (B6): 6041–58.

Gonzales, D. A. 2010. The enigmatic Late Cretaceous McDermott Formation. In *Geology of the Four Corners Country*, New Mexico Geological Society 61st Annual Fall Field Conference Guidebook, eds. J. E. Fassett, K. E. Zeigler, and V. W. Lueth, 157–62.

Gonzalez, G. R. 2019. *Fault Kinematics of the southern Rio Grande rift: Exploring the possibility of fault reactivation*. Master's Thesis, University of Texas, El Paso, TX.

Grambling, J. A. 1979. Precambrian geology of the Truchas Peaks region, north-central New Mexico and some regional implications. In *Santa Fe Country*, New Mexico Geological Society 30th Annual Fall Field Conference Guidebook, eds. R. V. Ingersoll, L. A. Woodward, and H. L. James, 135–43.

Grambling, T. A., K. E. Karlstrom, M. E. Holland, and N. L. Grambling. 2016. Proterozoic magmatism and regional contact metamorphism in the Sandia-Manzano Mountains, New Mexico, USA. In *Geology of the Belen Area*, New Mexico Geological Society 67th Annual Fall Field Conference Guidebook, eds. B. A. Frey and others, 169–75.

Grauch, V. J. S., P. W. Bauer, B. J. Drenth, and K. I. Kelson. 2017. A shifting rift: Geophysical insights into the evolution of Rio Grande rift margins and the Embudo transfer zone near Taos, New Mexico. *Geosphere* 13 (3): 870–910.

Grauch, V. J. S., and S. D. Connell. 2013. New perspectives on the geometry of the Albuquerque Basin, Rio Grande rift, New Mexico: Insights from geophysical models of rift thickness. In *New Perspectives on Rio Grande Rift Basins: From Tectonics to Groundwater*, GSA Special Paper 494, eds. M. R. Hudson and V. J. S. Grauch, 427–62.

Griswold, G. B. 1961. *Mineral Deposits of Luna County, New Mexico*. New Mexico Bureau of Geology and Mineral Resources Bulletin 72.

Hawley, J. W., B. J. Hibbs, J. F. Kennedy, and others. 2000. *Trans-international Boundary Aquifers in Southwestern New Mexico*. New Mexico Water Resources Research Institute, New Mexico State University, Las Cruces, NM.

Hedlund, D. C., D. E. Hendzel, and R. F. Kness. 1984. *Mineral Resource Potential of the Sandia Mountain Wilderness, Bernalillo and Sandoval Counties, New Mexico*. USGS Miscellaneous Field Studies Map MF-1631-A.

Hildebrand, R. S., C. A. Ferguson, and S. Skotnicki. 2008. *Geologic Map of the Silver City Quadrangle, Grant County, New Mexico*. New Mexico Bureau of Geology and Mineral Resources Geologic Map GM-164.

Hill, C. A. 1987. *Geology of Carlsbad Cavern and Other Caves in the Guadalupe Mountains, New Mexico and Texas*. New Mexico Bureau of Geology and Mineral Resources Bulletin 117.

Hobbs, K. M. 2016. *Sedimentation, pedogenesis, and paleoclimate conditions in the Paleocene San Juan Basin*, New Mexico, USA. Dissertation, University of New Mexico.

Hobbs, K. M., and P. J. Fawcett. 2021. A physical and chemical sedimentary record of Laramide tectonic shifts in the Cretaceous-Paleogene San Juan Basin, New Mexico, USA. *Geosphere* 17 (3): 854–75.

Holland, M. E., T. A. Grambling, K. E. Karlstrom, and others. 2020. Geochronologic and Hf-isotope framework of Proterozoic rocks from central New Mexico, USA: Formation of the Mazatzal crustal province in an extended continental margin arc. *Precambrian Research* 347: 105820.

Horn, M., and J. M. Timmons. 2006. *Geologic Map of the Ojitos Frios Quadrangle, San Miguel County, New Mexico*. New Mexico Bureau of Geology and Mineral Resources Geologic Map GM-130.

Huff, G. F. 2004. *Simulation of Groundwater Flow in the Basin-fill Aquifer of the Tularosa Basin, south-central New Mexico: Predevelopment through 2040*. USGS Scientific Investigations Report 5197.

Hunt, A. P. 2010. Clayton Lake State Park. In *The Geology of Northern New Mexico's Parks, Monuments, and Public Lands*, eds. Price and L. Greer, 319–24.

Hunt, A. P., and S. G. Lucas. 1998. Vertebrate tracks and the myth of the belly-dragging, tail-dragging tetrapods of the Late Paleozoic in Permian stratigraphy and paleontology of the

Robledo Mountains, New Mexico. El Paso Geological Society Field Guide, *New Mexico Museum of Natural History and Science Bulletin* 12: 67–69.

Julian, B., and J. Zidek, eds. 1991. *Field Guide to Geologic Excursions in New Mexico and Adjacent Areas of Texas and Colorado.* Prepared for the Geological Society of America Rocky Mountain and South-Central Sections Annual Meeting, Albuquerque, New Mexico, April 21–24, 1991. New Mexico Bureau of Geology and Mineral Resources Bulletin 137.

Kan, H. 2012. *Kinematics analysis of Laramide deformation in north-central New Mexico.* Master's Thesis, University of Houston.

Karlstrom, K. E., S. D. Connell, C. A. Ferguson, and others. 1994. *Geology of the Tijeras Quadrangle, Bernalillo County, New Mexico.* New Mexico Bureau of Geology and Mineral Resources Open-File map series GM-4.

Karlstrom, K. E., and G. R. Keller, eds. 2005. *The Rocky Mountain Region: An Evolving Lithosphere.* Geophysical Monograph Series 154.

Kelley, S. A. No Date. *Tyrone Mine.* New Mexico Bureau of Geology and Mineral Resources, Landmark. Available online.

Kelley, S., P. Barkmann, R. Benson, J. Lovekin, and L. Dunn. 2021. *Geologic Road Log: Cumbres & Toltec Scenic Railroad.* New Mexico Bureau of Geology and Mineral Resources.

Kelley, S., and R. Chamberlin. 2012. Our growing understanding of the Rio Grande rift. *New Mexico Earth Matters*, Summer, 1–4.

Kelley, S., D. J. Koning, F. Goff, and others. 2014. Stratigraphy of the northwestern Sierra Blanco volcanic field. In *Geology of the Sacramento Mountains Region*, New Mexico Geological Society 65th Annual Fall Field Conference, eds. G. Rawling, V. T. McLemore, S. Timmons, and N. Dunbar, 197–208.

Kelley, V. C. 1967. Tectonics of the Zuni-Defiance Region, New Mexico and Arizona. In *Defiance, Zuni, Mt. Taylor Region (Arizona and New Mexico)*, New Mexico Geological Society 18th Annual Fall Field Conference Guidebook, ed. F. D. Trauger, 28–31.

Kelley, V. C. 1972. Geology of the Santa Rosa area. In *East-Central New Mexico.* New Mexico Geological Society 58th Annual Fall Field Conference Guidebook, eds. V. C. Kelley and F. D. Trauger, 218–220.

Kelley, V. C. 1978. *Geology of the Española Basin, New Mexico.* New Mexico Bureau of Geology and Mineral Resources Geologic Map GM-48.

Kelley, V. C., and A. M. Kudo. 1978. *Volcanoes and Related Basalts of the Albuquerque Basin, New Mexico.* New Mexico Bureau of Geology and Mineral Resources Circular 156.

Kelson, K. I., and others. 2001. *Geologic Map of the Taos Quadrangle, Taos County, New Mexico.* New Mexico Bureau of Geology and Mineral Resources Geologic Map GM-43.

Kelson, K. I., P. A. Bauer, and G. Rawling. 2010. *Geologic Map of the Arroyo Seco Quadrangle, Taos County, New Mexico.* New Mexico Bureau of Geology and Mineral Resources Geologic Map GM-170.

Kempter, K., and F. Goff. Unpublished. Diablo Maar Volcano. Available online at www.santafenewmexican.com

Kirtland, D. W., R. E. Denison, and R. Evans. 1995. *Middle Jurassic Todilto Formation of Northern New Mexico and Southwestern Colorado: Marine or Nonmarine?* New Mexico Bureau of Geology and Mineral Resources Bulletin 147.

Kluth, C. F., and P. J. Coney. 1981. Plate tectonics of the Ancestral Rocky Mountains. *Geology* 9 (1): 10–15.

Koning, D. J. 2002. *Geologic Map of the Española Quadrangle, Rio Arriba and Santa Fe Counties, New Mexico.* New Mexico Bureau of Geology and Mineral Resources Geologic Map GM-56.

Koning, D. J. 2002. *Geologic Map of the Cundiyo Quadrangle, Santa Fe County, New Mexico.* New Mexico Bureau of Geology and Mineral Resources Geologic Map GM-54.

Koning, D. J. 2004. *Geologic Map of the Lyden Quadrangle, Rio Arriba County, New Mexico.* New Mexico Bureau of Geology and Mineral Resources Geologic Map GM-83.

Koning, D. J. 2010. *Preliminary Geologic Map of the Carrizozo East Quadrangle, Lincoln County, New Mexico.* New Mexico Bureau of Geology and Mineral Resources Geologic Map GM-211.

Koning, D. J., B. Hallett, and C. Shaw. 2007. *Geologic Map of the Alamogordo North 7.5-minute Quadrangle, Otero County, New Mexico.* New Mexico Bureau of Geology and Mineral Resources Geologic Map GM-153.

Koning, D. J., A. P Jochems, K. M. Hobbs, K. S. Pearthree, and D. W. Love. 2020. *Geologic Map of the Paraje Well 7.5-minute Quadrangle, Socorro County, New Mexico.* New Mexico Bureau of Geology and Mineral Resources Geologic Map GM-286.

Koning, D. J., W. McIntosh, and N. Dunbar. 2011. Geology of southern Black Mesa, Española Basin, New Mexico: New stratigraphic age control and interpretation of the southern Embudo fault system of the Rio Grande rift. In *Geology of the Tusas Mountains and Ojo Caliente*, New Mexico Geological Society 62nd Annual Fall Field Conference Guidebook, eds. D. J. Koning and others, 191–214.

Land, L. A. 2003. Evaporite Karst and Regional Ground-Water Circulation in the Lower Pecos Valley of Southeastern New Mexico. Oklahoma Geological Survey Circular 109: 227–32.

Land, L. 2016. *Overview of Fresh and Brackish Water Quality in New Mexico.* New Mexico Bureau of Geology Mineral Resources Open-File Report 583.

Land, L., V. W. Lueth, W. Raatz, P. Boston, and D. L. Love, eds. 2006. *Caves and Karst of Southeastern New Mexico.* New Mexico Geological Society 57th Annual Fall Field Conference Guidebook.

Lepper, K., and F. Goff. 2007. Yet another attempt to date the Banco Bonito rhyolite, the youngest volcanic flow in the Valles caldera, New Mexico. *New Mexico Geology* 29 (4): 117–20.

Lipman, P. W., and M. J. Zimmerer. 2019. Magmato-tectonic links: Ignimbrite calderas, regional dike swarms, and the transition from arc to rift in the Southern Rocky Mountains. *Geosphere* 15 (6): 1893–1926.

Lisenbee, A. L. 1999. *Preliminary Geologic Map of the Galisteo Quadrangle, Santa Fe County, New Mexico.* New Mexico Bureau of Geology and Mineral Resources Geologic Map GM-30.

Lisenbee, A. L. 2003. *Geologic Map of the Las Vegas NW Quadrangle, San Miguel and Taos Counties, New Mexico.* New Mexico Bureau of Geology and Mineral Resources Geologic Map GM-78.

Lisenbee, A. L., L. A. Woodward, and J. R. Connolly. 1979. Tijeras-Cañoncito fault system: A Major zone of recurrent movement in north-central New Mexico. In *Santa Fe Country*, New Mexico Geological Society 30th Annual Fall Field Conference Guidebook, eds. R. V. Ingersoll, L. A. Woodward, and H. L. James, 89–99.

Lucas, S. G. 1983. The Baca Formation and the Eocene-Oligocene boundary in New Mexico. In Socorro Region II, New Mexico Geological Society 34th Annual Fall Field Conference Guidebook, eds. C. E. Chapin and J. F. Callender, 187–92.

Lucas, S. G. 1995. Triassic stratigraphy and chronology in New Mexico. *New Mexico Geology* 17 (1): 8–13.

Lucas, S. G., and S. N. Hayden. 1991. Type section of the Permian Bernal Formation and the Permian-Triassic boundary in north-central New Mexico. *New Mexico Geology* 13 (1): 9–15.

Lucas, S. G., G. S. Morgan, and K. E. Zeigler, eds. 2005. *New Mexico's Ice Ages*. New Mexico Museum of Natural History and Science Bulletin 28.

Lucas, S. G., W. J. Nelson, W. Dimichele, and J. A. Spielman. 2013. *The Carboniferous-Permian Transition in Central New Mexico*. New Mexico Museum of Natural History and Science Bulletin 59.

Lucas, S. G., and D. S. Ulmer-Scholle, eds. 2001. *Geology of the Llano Estacado*. New Mexico Geological Society 52nd Annual Fall Field Conference Guidebook.

Lucas, S. G., S. Voigt, A. J. Lerner, J. P. MacDonald, J. A. Spielmann, and M. D. Celeskey. 2011. The Prehistoric Trackways National Monument, Permian of Southern New Mexico, USA. *Ichnology Newsletter* 28: 10–14.

Lucas, S. G., K. E. Zeigler, V. W. Lueth, and D. E. Owen, eds. 2005. *Geology of the Chama Basin*. New Mexico Geological Society 56th Annual Fall Field Conference Guidebook.

Lui, Y., and M. Murphy. 2014. Preexisting Structures and Initiation of an Oblique Rift: The Cañones Fault Zone in North-central New Mexico. Poster at AAPG Annual Convention.

Machette, M. N., and R. G. McGimsey. 1983. *Quaternary and Pliocene faults in the Socorro and western part of the Fort Sumner 1°x2° quadrangles, central New Mexico*. US Geological Survey Miscellaneous Field Studies Map MF-1465-A.

Machette, M. N., R. A. Thompson, D. W. Marchetti, D.W., and R. S. U. Smith. 2013. Evolution of Lake Alamosa and integration of the Rio Grande during the Pliocene and Pleistocene. In *Perspectives on Rio Grande Rift Basins: from Tectonics to Groundwater*. Geological Society of America Special Paper 494, eds. M. R. Hudson and V. J. S. Grauch, 120.

Mack, G., J. Witcher, and V. W. Lueth. 2008. *Geology of the Gila Wilderness–Silver City Area*. New Mexico Geological Society Fall Field Conference Guidebook 59.

Maldonado, F. 2008. *Geologic Map of the Abiquiu Quadrangle, Rio Arriba County, New Mexico*. USGS Scientific Investigations Map 2998.

Maldonado, F., and S. A. Kelley. 2009. Revisions to the stratigraphic nomenclature of the Abiquiu Formation, Abiquiu and contiguous areas, north-central New Mexico. *New Mexico Geology* 31: 3–8.

Maldonado, F., and D. P. Miggins. 2007. Geologic summary of the Abiquiu quadrangle, north-central New Mexico. In *Geology of the Jemez Region II*, New Mexico Geological Society 58th Annual Fall Field Conference Guidebook, eds. B. S. Kues, S. A. Kelley, and V. W. Lueth, 182–87.

Maldonado, F., J. L. Slate, D. W. Love, and others. 2007. *Geologic Map of the Pueblo of Isleta Tribal Lands and Vicinity, Bernalillo, Torrance, and Valencia Counties, Central New Mexico*. USGS Scientific Investigations Map 2913.

Maynard, S. R. 2002. *Geologic Map of the Golden 7.5-minute Quadrangle*. New Mexico Bureau of Geology and Mineral Resources Open-File Geologic Map GM-36.

Maynard, S. R. 2005. Laccoliths of the Ortiz porphyry belt, Santa Fe County, New Mexico. *New Mexico Geology* 27 (1): 3–21.

Maynard, S. R. 2014. *Geology and Mineral Resources of the Ortiz Mine Grant, Santa Fe County, New Mexico*. New Mexico Bureau of Geology and Mineral Resources Open-File Report 560.

Maynard, S. R., A. L Lisenbee, and J. Rogers. 2002. *Geologic Map of the Picture Rock 7.5-minute Quadrangle, Santa Fe County, central New Mexico*. New Mexico Bureau of Geology and Mineral Resources Open-File Report DM-49.

McIntosh, W. C., and S. M. Cather. 1994. $^{40}Ar/^{39}Ar$ geochronology of basaltic rocks and constrains on late Cenozoic stratigraphy and landscape development in the Red Hill–Quemado area,

New Mexico. In *Mogollon Slope*, New Mexico Geological Society 45th Annual Fall Field Conference Guidebook, eds. R. M. Chamberlin and others, 209–24.

McLemore, V. T. 2018. A Geology Guide to Red Rock Park. *Lite Geology.* 43: 4.

McLemore, V. T. 2017. *Mining Districts and Prospect Areas in New Mexico.* New Mexico Bureau of Geology and Mineral Resources Resource Map 24.

McLemore, V. T. 2010. *Geology, Mineral Resources, and Geoarchaeology of the Montoya Butte Quadrangle, including the Ojo Caliente No. 2 Mining District, Socorro County, New Mexico.* New Mexico Bureau of Geology and Mineral Resources Open-File Report 535.

McLemore, V. T. 2001. Brantley Lake State Park. *New Mexico Geology* 23 (4): 123–28.

McLemore, V. T. 2001. Living Desert Zoo and Gardens State Park. *New Mexico Geology* 23: 42–50.

McLemore, V. T. 1996. Copper in New Mexico. *New Mexico Geology* 18 (2): 25–36.

McLemore, V. T. 1991. Geology of the Valley of Fires recreation area. In *Geology of the Sierra Blanca, Sacramento, and Capitan Ranges, New Mexico*, New Mexico Geological Society 42nd Annual Fall Field Conference Guidebook, eds. J. M. Barker, B. S. Kues, G. S. Austin, and S. G. Lucas, 67–70.

McLemore, V. T., and N. Dunbar. 2000. Rockhound State Park and Spring Canyon Recreation Area. *New Mexico Geology* 22 (3): 66–71.

McLemore, V. T., Sutphin, D. M., D. R. Hack, and T. C. Pease. 1996. *Mining History and Mineral Resources of the Mimbres Resource Area, Doña Ana, Luna, Hidalgo, and Grant Counties, New Mexico.* New Mexico Bureau of Geology and Mineral Resources Open-File Report 424.

McMahon, P. B., J. K. Böhlke, and C. P. Carney. 2007. *Vertical Gradients in Water Chemistry and Age in the Northern High Plains Aquifer, Nebraska, 2003.* USGS Scientific Investigations Reports 2006-5294.

McMillan, N., and V. McLemore. 2004. Cambrian-Ordovician magmatism and extension in New Mexico and Colorado. In *Tectonics, Geochronology, and Volcanism in the Southern Rocky Mountains and Rio Grande Rift,* New Mexico Bureau of Geology and Mineral Resources Bulletin 160: 1–11.

Murray, K. D., M. H. Murray, and A. F. Sheehan. 2019. Active deformation near the Rio Grande rift and Colorado Plateau as inferred from continuous global positioning system measurements. *Journal of Geophysical Research: Solid Earth* 124 (2): 1139–2236.

National Park Service. *Geology of Cliff Dweller Canyon.* Available online at www.nps.gov/gicl

Nereson, A., J. Stroud, K. Karlstrom, M. Heizler, and W. McIntosh. 2013. Dynamic topography of the western Great Plains: Geomorphic and ^{40}Ar/^{39}Ar evidence for mantle-driven uplift associated with the Jemez lineament of northeastern New Mexico and southeastern Colorado. *Geosphere* 9 (3): 521–45.

New Mexico Bureau of Geology and Mineral Resources. 2003. *Geologic Map of New Mexico.* Published in cooperation with the USGS.

Newton, B. T., and L. Land. 2016. *Brackish Water Assessment in the Eastern Tularosa Basin, New Mexico.* New Mexico Bureau of Geology and Mineral Resources Open-File Report 582.

Osburn, G. R., and C. E. Chapin. 1983. *Nomenclature for Cenozoic Rocks of Northeast Mogollon-Datil Volcanic Field, New Mexico.* New Mexico Bureau of Geology and Mineral Resources Stratigraphic Chart 1.

Partey, F. K., L. Land, B. Frey, E. Premo, and L. Crossey. 2011. *Final Report on Geochemistry of Bitter Lake National Wildlife Refuge, Roswell, New Mexico.* New Mexico Bureau of Geology and Mineral Resources Open-File Report 526.

Perry, F. V., W. S. Baldridge, J. DePaolo, and M. Shafiqullah. 1990. Evolution of a magmatic system during continental extension: the Mount Taylor volcanic field, New Mexico: *Journal of Geophysical Research* 95: 19327–48.

Price, L. G., ed. 2010. *The Geology of Northern New Mexico's Parks, Monuments, and Public Lands.* Socorro, NM, New Mexico Bureau of Geology and Mineral Resources.

Ratte, J. C. 2008. The early Oligocene Copperas Creek volcano and geology along New Mexico Highway 15 between Sapillo Creek and the Gila Cliff Dwellings National Monument, Grant and Catron counties, New Mexico. *Geology of the Gila Wilderness–Silver City Area*, New Mexico Geological Society 59th Annual Fall Field Conference Guidebook 59, eds. G. J. Mack, J. Witcher, and V. W. Lueth, 129–40.

Ratte, J., S. Lynch, and B. McIntosh. 2006. *Geologic Map of the Holt Mountain Quadrangle, Catron County, New Mexico.* New Mexico Bureau of Geology and Mineral Resources Geologic Map GM-120.

Rawling, G. 2012. *Generalized Geologic Map of the Southern Sacramento Mountains, Otero and Chaves Counties, New Mexico.* New Mexico Bureau of Geology and Mineral Resources Open-File Report 537.

Rawling, G., V. T. McLemore, S. Timmons, and N. Dunbar, eds. 2014. *Geology of the Sacramento Mountains Region.* New Mexico Geological Society 65th Annual Fall Field Conference.

Rawling, G. C., M. W. Morris, and C. T. Pierson. 2015. *Geologic Map of the Thoreau NE 7.5-minute Quadrangle, McKinley County, New Mexico.* New Mexico Bureau of Geology and Mineral Resources Geologic Map GM-148.

Rauzi, S. L. 2009. *Implications of Live Oil Shows in an Eastern Arizona Geothermal Test.* Arizona Geological Survey Open-File Report 94-1.

Read, A. S., K. E Karlstrom, S. Connell, and others. 1995. *Geology of the Sandia Crest Quadrangle, Bernalillo and Sandoval Counties, New Mexico.* New Mexico Bureau of Geology and Mineral Resources Geologic Map GM-6.

Read, A. S., D. J. Koning, G. A. Smith, and others. 2000. *Geologic Map of Santa Fe Quadrangle, Santa Fe County, New Mexico.* New Mexico Bureau of Geology and Mineral Resources Geologic Map GM-32.

Reneau, S. L., P. G. Drakos, and D. Katzman. 2007. Post-resurgence lakes in the Valles Caldera, New Mexico. In *Geology of the Jemez Region II*, New Mexico Geological Society 58th Annual Fall Field Conference Guidebook, eds. B. S. Kues, S. A. Kelley, and V. W. Lueth, 398–408.

Rentz, S., G. Michelfelder, M. A. Coble, and E. Salings. 2018. U-Pb zircon geochronology of calc-alkaline ash-flow tuff units in the Mogollon-Datil volcanic field, southern New Mexico. In *Field Volcanology: A Tribute to the Distinguished Career of Don Swanson.* GSA Special Paper 538, 409–34.

Repasch, M., K. Karlstrom, M. Heizler, and M. Pecha. 2017. Birth and evolution of the Rio Grande fluvial system in the past 8 Ma: Progressive downward integration and the influence of tectonics, volcanism, and climate. *Earth Science Reviews* 168: 113–64.

Ricketts, J. W., K. E. Karlstrom, and M. T. Heizler. 2016. Tectonic evolution of the Lucero uplift. In *Geology of the Belen Area*, New Mexico Geological Society 67th Annual Fall Field Conference Guidebook, eds. B. A. Frey and others, 185–94.

Ricketts, J. W., K. E. Karlstrom, A. Priewisch, L. J. Crossey, V. J. Polyak, and Y. Asmerom. 2014. Quaternary extension in the Rio Grande rift at elevated strain rates recorded in travertine deposits, central New Mexico. *Lithosphere* 6 (1): 3–16.

Ricketts, J. W., S. A. Kelley, K. E. Karlstrom, B. Schmandt, M. S. Donahue, and J. van Wijk. 2016. Synchronous opening of the Rio Grande rift along its entire length at 20-10 Ma supported by apatite (U-Th)/He and fission-track thermochronology, and evaluation of possible driving mechanisms. *GSA Bulletin* 128 (3-4): 397–424.

Robinson, G. D., A. A. Wanek, W. H. Hays, and M. E. McCallum. 1964. *Philmont Country: The Rocks and Landscapes of a Famous New Mexico Ranch.* USGS Professional Paper 505.

Ruleman, C., and M. Machette. 2007. An Overview of the Sangre de Cristo fault system and new insights to interactions between Quaternary faults in the northern Rio Grande rift. In *Rocky Mountain Section Friends of the Pleistocene Field Trip: Quaternary Geology of the San Luis Basin in Colorado and New Mexico,* USGS Open-File Report 2007–1193, eds. M. N. Machette, M-M Coates, and M. L. Johnson, 187–97.

Ruleman, C. A., A. M. Hudson, R. A. Thompson, and others. 2019. Middle Pleistocene formation of the Rio Grande Gorge, San Luis Valley, south-central Colorado and north-central New Mexico, USA: Process, timing, and downstream implications. *Quaternary Science Reviews* 223: 105846.

Sawyer, D. A., and S. A. Minor. 2006. Geologic setting of the La Bajada constriction and Cochiti Pueblo area, New Mexico. In *The Cerrillos Uplift, the La Bajada Constriction, and Hydrogeologic Framework of the Santo Domingo Basin, Rio Grande Rift, New Mexico,* USGS Professional Paper 1720-A, ed. S. A. Minor, 1–21.

Sawyer, D. A., R. R. Shroba, S. A. Minor, and R. A. Thompson. 2002. *Geologic Map of the Tetilla Peak Quadrangle, Santa Fe and Sandoval Counties, New Mexico.* USGS Miscellaneous Field Studies Map MF-2352.

Schilling, J. H. 1956. *Geology of the Questa Molybdenum Mine Area, Taos County, New Mexico.* Socorro: New Mexico Institute of Mining and Technology.

Scholle, P. 2020. *Introduction to the Geology of the Permian Reef Complex, Guadalupe and Delaware Mountains, New Mexico and West Texas.* New Mexico Bureau of Geology and Mineral Resources Online Geologic Tour.

Schroder, J., A. Ellwein, G. Engelmann, and C. Englemann. 2021. *Geology Underfoot in Colorado's Western Slope.* Missoula, MT: Mountain Press Publishing Company.

Seager, W. R. 2010. *Geologic Map of the Hatch Quadrangle, Dona Ana County, New Mexico.* New Mexico Bureau of Geology and Mineral Resources Geologic Map GM-213.

Seager, W. R., J. W. Hawley, F. E. Kottlowski, and S. A. Kelley. 1987. *Geology of East Half of Las Cruces and Northeast El Paso 1x2 degree sheets, New Mexico.* New Mexico Bureau of Geology and Mineral Resources Geologic Map GM-57.

Seager, W. R., F. E. Kottlowski, and J. W. Hawley. 2008. *Geologic Map of the Robledo Mountains and Vicinity, Dona Ana County, New Mexico.* New Mexico Bureau of Geology and Mineral Resources Open-File Report 509.

Seager, W. R., J. O. Thacker, and S. A. Kelley. 2021. *Geologic Map of the Selden Canyon 7.5-Minute Quadrangle, Dona Ana County, New Mexico.* New Mexico Bureau of Geology and Mineral Resources Geologic Map GM-290.

Self, S., G. Heiken, M. L. Sykes, and others. 1996. *Field Excursions to the Jemez Mountains, New Mexico.* New Mexico Bureau of Geology and Mineral Resources Bulletin 134.

Semken, S. 2003. Black rocks protruding up: The Navajo volcanic field. In *Geology of the Zuni Plateau.* New Mexico Geological Society 54th Annual Fall Field Conference Guidebook, eds. S. G. Lucas and others, 133–38.

Sion, B. D., F. M Phillips, G. J. Axen, and others. 2020. Chronology of terraces in the Rio Grande rift, Socorro basin, New Mexico: Implications for terrace formation. *Geosphere* 16 (6): 1457–78.

Smith, R. L., and R. A. Bailey. 1968. *Resurgent Cauldrons*. Studies in Volcanology, Geological Society of America Memoirs 116, Geological Society of America, 613–62.

Stark, J. T., and E. C. Dapples. 1946. Geology of the Los Pinos Mountains, New Mexico. *GSA Bulletin* 57 (12): 1121–1172.

Tappa, M. J., D. S. Coleman, R. D. Mills, and K. M. Samperton. 2011. The plutonic record of a silicic ignimbrite from the Latir volcanic field, New Mexico. *Geochemistry, Geophysics, Geosystems* 12 (10). Available online.

Thacker, J. O., S. A. Kelley, and K. E. Karlstrom. 2021. Late Cretaceous–Recent low-temperature cooling history and tectonic analysis of the Zuni Mountains, west-central New Mexico. *Tectonics* 40 (1).

Thompson, R. A., K. J. Turner, P. W. Lipman, J. A. Wolff, and M. A. Dungan. 2022. *Field-trip guide to continental arc to rift volcanism of the southern Rocky Mountains—Southern Rocky Mountain, Taos Plateau, and Jemez Mountains volcanic fields of southern Colorado and northern New Mexico*. USGS Scientific Investigations Report 2017-5022-R.

Thorman, C. H., and H. Drewes. 1978. Cretaceous–early Tertiary history of the northern Pyramid Mountains, southwestern New Mexico. In *Land of Cochise (Southeastern Arizona)*, New Mexico Geological Society 29th Annual Fall Field Conference Guidebook, eds. J. F. Callender, J. Wilt, R. E. Clemons, and H. L. James, 215–18.

Titley, S. R. 1959. Geological summary of the Magdalena mining district, Socorro County, New Mexico. In *West-Central New Mexico*, New Mexico Geological Society 10th Annual Fall Field Conference Guidebook, eds. J. E. Weir Jr. and E. H. Baltz, 144–48.

Tomlinson, D. W. 2013. Oligocene shortening in the Little Burro Mountains of southwest New Mexico. *Rocky Mountain Geology* 48 (2): 169–83.

Valentine-Darby, 2016. *El Morro National Monument: Natural Resource Condition Assessment*. Natural Resource Report NPS/ELMO/NRR—2016/1192.

van Wijk, J., D. Koning, G. Axen, D. Coblentz, E. Gragg, and B. Sion. 2018. Tectonic subsidence, geoid analysis, and the Miocene-Pliocene unconformity in the Rio Grande rift, southwestern United States: Implications for mantle upwelling as a driving force for rift opening. *Geosphere* 14 (2): 684–709.

Wanek, A. A., C. B. Read, G. D. Robinson, and others. 1964. *Geologic Map and Sections of the Philmont Ranch Region, New Mexico*. USGS Miscellaneous Geological Investigations Map I-425.

Warren, J. K. 1989. Stratigraphy and depositional environment of lower San Andres Formation in subsurface and equivalent outcrops: Chaves, Lincoln, and Roosevelt Counties, New Mexico. *AAPG Bulletin* 73 (11): 1307–25.

Waters, M. R., T. W Stafford Jr., and D. L. Carlson. 2020. The age of Clovis: 13,050 to 12,750 cal yr B. P. *Science Advances* 6 (43).

Wolff, J. A., K. A. Brunstad, and J. N. Gardner. 2011. Reconstruction of the most recent volcanic eruptions from the Valles caldera, New Mexico. *Fuel and Energy Abstracts* 199 (1): 53–68.

Woodward, L. A. 1973. Structural framework and tectonic evolution of the Four Corners region of the Colorado Plateau. In *Monument Valley (Arizona, Utah, and New Mexico)*, New Mexico Geological Society 24th Annual Fall Field Conference Guidebook, ed. H. L. James, 94–98.

Woodward, L. A., M. C. Hultgren, D. L. Crouse, and M. A. Merrick. 1992. Geometry of Nacimiento-Gallina fault system, northern New Mexico. In *San Juan Basin IV*, New Mexico Geological Society 43rd Annual Fall Field Conference Guidebook, eds. S. G. Lucas, B. S. Kues, T. E. Williamson, and A. P. Hunt, 103–8.

Woodward, L. A., W. H. Kaufman, and J. B. Anderson. 1972. Nacimiento Fault and related structures, northern New Mexico. *Geological Society of America Bulletin* 83: 2382–96.

Woodward, L. A., D. McLelland, J. B. Anderson, and W. H. Kaufman. 1972. *Geologic Map and Section of Cuba Quadrangle, New Mexico.* New Mexico Bureau of Geology and Mineral Resources Geologic Map GM-25.

Wynne, J. J. 2021. *Chaco Culture National Historical Park: Natural Resource Condition Assessment.* Natural Resource Report NPS/CHCU/NRR—2021/2304.

Zeller Jr., R. A. 1975. *Structural Geology of Big Hatchet Peak Quadrangle, Hidalgo County, New Mexico.* New Mexico Bureau of Geology and Mineral Resources Circular 146.

Zhang, H., F. Zhang, J. Chen, and others. 2021. Felsic volcanism as a factor driving the end-Permian mass extinction. *Science Advances* 7 (47).

Zimmerer, M. J., J. Lafferty, and M. A. Coble. 2015. The eruptive and magmatic history of the youngest pulse of volcanism at the Valles caldera: Implications for successfully dating late Quaternary eruptions. *Journal of Volcanology and Geothermal Research* 310 (January): 50–57.

INDEX

Page numbers in bold face include photographs.

aa lava, 69, 71, 180
Abiquiu, 140, 141, 143
Abiquiu Formation, 140, 141, 143
Abiquiu Reservoir, 141
Abo Canyon, 196
Abo Formation, 14, 34, 134, **137**, 157, 175, 187, 188, **196**, **197**, 198, 199, **212**, 255
Abo Ruins, 197
accretion, 9, 38
Acoma, 31
Aden Crater, 294
Aden Lava Flow Wilderness, 294, **296**
Afton Basalt, 294
agate, 294, 295
aggregate, 103, 278, 300
Agua Fria Peak, 121
Aguirre Spring Recreation Area, 204
Ah-Shi-Sle-Pah Wilderness Area, 43
alabaster, 140
Alamitos Formation, 113, 217
Alamogordo, 28, 72, 171, 192, **205**, 251, 253
Albuquerque, 131, 151–152, 154, **155**, 160, 178, 183
Albuquerque Basin, 22, 72, 73, 150, 151, 153, 162, 163, 164, 181, 194
Albuquerque Volcanoes, 23, 74, 150, 152, 155, 160, 182, 183
alluvial fans, 29, 75, 76, **167**, **194**, **204**, 281, 284. See also bajadas
alluvium, **112**, 291
Amalia Tuff, **6**, 116, 141
ammonites, 77, 282
amphibolite, 81, 85, 87, 108, 145
Ancestral Puebloans, 130, 182
Ancestral Rocky Mountains, 13, 14, 16, 79, 95, 134, 140, 164, 188, 199, 202, 217, 229, 233, 241, 263; orogeny of, xii, 19
Ancha Formation, 145
Ancho Canyon, 128, 129
andesite, 4, 93, **158**, **311**
Angel Fire, 87
Angel Peak Scenic Area, **57**
anhydrite, 207, 255, 268
Animas, Lake, 27, 289, 290
Animas Formation, 33, 34, 35, 42, 44
Animas Glacier, 54

Animas Mountains, 289
Animas Peak, 173
Animas Peak caldera, 285
Animas River, 45, 54, 56, 57
Animas Valley, 289, 291
Animas Valley fault, 289
anticlines, 32, 52, **63**, 266. See also Zuni Mountain anticline
Anton Chico Formation, 15, 233
Apache Mountains, 269
Apache Point Laboratory, 256
aquifers, 23, 28, 74, 145, 150, 200, 218, 238, 251, 260, 269. See also artesian; groundwater
arroyo, 29, **37**, 60, 64, 103, 164, 167, 229, 256, 267, 274, 278
Arroyo Hondo, 92, 116
Artesia, 192, 257, 258, 260
Artesia Group, v, 77, 79, 217, 231, 263, 265, 266, 267, 268, 274, 275, **276**, 281
artesian, 251, 258, 259, 260, 265, 275, 276
ash. See volcanic ash
ash-flow tuff, 103, 118, 123, 155, 167, 175, 176, 180, 182, 278, 286, 290, 292, 295, **302**, 313. See also ignimbrites; tuff
Atlantic Ocean, 3, 16, 28, 38
Aztec, 54, 56, 57, 85, 156
Aztec Ruins National Monument, 54, **56**, 57

Baca Formation, 20, 300, 304
Baca Basin, 20, 300
Baca Ranch, 124
badlands, 29, **45**, **57**, **58**, 100, 101, **108**, 110, 124, 138, **173**, 181
Baldy Mountain, 84, 85, 87
ballast, 287
Banco Bonito Flow, 132, **133**
Bandelier National Monument, 129–130, 131, 149
Bandelier Tuff, 123, 126, **128**, 129, **130**, 131, 132, 135, **136**, **137**. See also Tshirege Member
Bandera Crater, 67, **70**, 71
Bandera lava flow (basalt), 67, **68**, 69, 70, 71
barranca, 100, 124, 152
Bartlett Mesa, 220, **222**

341

basalt, 4, **6**, 22, **71**, **96**, **118**, 121, 122, **126**, **127**, **182**, **226**, **282**, **311**. *See also* aa lava; Bandera lava flow; Black Mountain–Santo Tomas Basalt; columnar jointing; Cuerbio Basalt; El Alto Basalt; pahoehoe lava; West Potrillo Basalt; Servilleta Basalt
basement rocks, 8, 84, 118, 135, 217
basins, structural, 14, 19, 27, 28, 29, 72, 73, 150, 215, 222, 259, 284. *See also* Albuquerque Basin; Baca Basin; Bisbee Basin; Chama Basin; Delaware Basin; Española Basin; Estancia Basin; Galisteo Basin; Hueco Basin; Midland Basin; Mimbres Basin; Permian Basin; Raton Basin; San Juan Basin; San Luis Basin; Santo Domingo Basin; Socorro Basin; Tularosa Basin; Val Verde Basin
Basin and Range, xii, 21, 23, 24, 27, 67, 73, 76, 180, 229, 284, 289, 290, 291, 292, 296, 306, 313
batholiths, 10, 22, 176, 286, 290
Battleship Ignimbrite, 132
Battleship Rock, 134
Baylor Peak, 176
Bearhead Rhyolite, 149
Beartooth Quartzite, v, 309
beds, **41**, **63**, 121, **133**, **253**
Belen, 61
Bennett Peak, 52
Benton Shale, 222
bentonite, 16, 178, 240, 260
Bents Fort, 220
beryl, 102, 105
Bernal Formation, **80**
Bernalillo, 76, 183
Bidahochi, Lake, 64
Bidahochi Formation, 34, 64
Big Burro Mountains, 290, 309, 313
Big Ditch, **310**
biotite, **7**, 80, 85
bioturbation, 43, 278
Bisbee Basin, 290
Bishop Cap, 176
Bisti/De-Na-Zin Wilderness, 34, 43, **58**
Bitter Lake National Wildlife Refuge, 251, 265
Black Butte, 163, **164**
Black Lake, 87
Black Mesa, 100, 101, 103, **105**, 126, 143, **169**, 170, 228

Black Mountain–Santo Tomas Basalt, 294
Black Range, 170, 174, 292, 308
Blackwater Draw, 248, 250
Blackwater Draw Formation, 248, 250
Bliss Sandstone, 11, 170, 189, 295
Bloodgood Canyon Tuff, 284, 310
Bloomfield, 43, 44, 57
Blue Hole, 231, 232
Bluewater Lake State Park, 38
Bobcat Pass, 120
Bone Spring Formation, 268
Bonita Canyon, 67
Bonney-Miser's Chest mine, 290
Bootheel volcanic field, 285
Border Hills structural zone, 278
Borrego Canyon, 111
Borrego Mesas, 123
Bosque del Apache National Wildlife Refuge, 151, 167
Bottomless Lakes, 265, **275**, 276
brachiopods, 13, 15, **88**, 135, 154, 186, 260, 269
braided streams, 164, 165, 167, 282, 300
Brantley Lake State Park, 267
Brazos, 32, 33, 73, 96
Brazos Box, **95**, **97**
Brazos Canyon, 95
Brazos Mountains, 73, **95**
Brazos Summit, 95
breccia, 4, **5**, 52, 156, 286, 291, **295**, 297, 313, 315
Broken Back Crater, 211
Brushy Basin Member, 64
bryozoans, 154, 186, 216, 260, 269
Bull Canyon Formation, 238
Burro Mountain batholith, 290
Burro uplift, 290
burrows, 52, 59, 245, 278
Bursum caldera, 285, 287, 313, 314
Bursum Formation, 164, 196, 255, 287, 313, 314
Butterfield Trail, 289

Caballo Lake, 173
Caballo Lake State Park, **173**
Caballo Mountains, 170, 171, **173**, 174
Caballo Reservoir, 173, 174
Cabezon Peak, **25**, 61, **62**, 157
calcareous, 121
calcite, 4, 269, 289, 291, 292

calcium carbonate, 90, 100, 124, 135, 188, 212, 233, **235**, 245, 248, 265, **271**, 272, 278, 291
calderas, 22, 119, 121, 285, 286. *See also* Bursum caldera; Emory caldera; Organ caldera; Platoro caldera; Questa caldera; Socorro caldera; Toledo caldera; Valles Caldera
caliche, 90, 100, **108**, 124, 212, **235**, **236**, 245, 265, **274**
Camel Rock, **98**
Cañada Mariana terrace, 169
Canadian Breaks, 237
Canadian River, 28, 82, 215, 219, 220, 224, 225, 228, 235, 237, 238, 248; Canyon, **226**
Cañon Colorado, 199
Cañones fault zone, 123, 140
Capilla Peak, 199
Capitan, 282
Capitan Formation, 15, 269, 270
Capitan Mountains, 276, 278, 281
Capitan pluton, 211, 282
Capitan Reef, xii, 14, 268, 269, 270
caprock, 29, 98, 105, 108, 136, 170, 197, 223, 237, 238, 266
Capulin Mountain, 246, **247**, 248
Capulin Volcano National Monument, ix, 22, 247, 248
carbonates. *See* calcium carbonate; dolomite; limestone; reefs
carbonatites, 12, 165, 166
Carlsbad, 15, 27, 28, 79, 215, 259, 267
Carlsbad Caverns National Park, 268, 270–272
Carrizo Mountain, 238
Carrizozo, 192, 193, 209, 238, 240, 283
Carrizozo Malpais, 192
Catwalk National Scenic Trail, 314
caves, 215, 216, 220, 232, 240, 241, 242, 254, 255, 257, 264, 268–272, 275, 281. *See also* lava tubes; sinkholes
Ceja Formation, 182
cement, 47, 135, 138, 141, 146, 187, 251. *See also* calcium carbonate; silica
cenotes, 216, 264, **275**
Central Basin Platform, 258
Cerrillos, 146, 156, 157, 158, 159, 160
Cerrillos Hill laccolith, **158**
Cerrillos Hills, 146, 156, 157, **158**, 159, 160
Cerrillos Hills igneous complex, 159
Cerrillos Hills State Park, 158–159

Cerro Chafo, 61
Cerro Colorado, 260, 262
Cerro Cuate, 61
Cerro de la Olla, 90
Cerro de los Taoses, 93
Cerros del Rio volcanic field, 25, 126, 127, 145, 146
Cerro Grande Fire, 131
Cerro Guadalupe, 61
Cerro la Jara, **131**, 132
Cerro Pedernal, 123, 140
Cerro Santa Clara, 61
Cerro Toledo, 251
Chaco Canyon, 56, 59
Chaco Culture National Historic Park, 59
chalcedony, 294
chalcopyrite, 291, 300
Chama, 28, 41, 75, 97, 100, 143, 151
Chama Basin, 39, 95, 96, 138
Chama Water Project, 150
Chamita Formation, 100, 106
Chamizal terrace, 169
Chalchihuitl, Mt., 158, 159
chert, 140, 186, 281, 309
Chevron Questa molybdenum mine, 118
Chicxulub Crater, 18, 224
Chihuahuan Desert, 29, 267
Chimayosos Peak, 109
Chinle Group, 15–16, **21**, 34, 38, **63**, 77, 137, 138, 140, **142**, **143**, **147**, 157, 217, **226**, 240, 248, 251, 260, 263; color of, **235**; **262**; fossils in, 141, 231, 238; sandstone, **226**; uranium in, 178. *See also* Bull Canyon Formation; Garita Creek Formation; Santa Rosa Formation
Chino Mine, 306, **309**
chlorite, 81
Chupadera Mesa, 197
Chupadera Mountains, 12, 165, 167, 299
Chupadera Peak, 167
Chuska Mountains, 31, **49**, 50
Chuska Sandstone, 34, 49, 51
Cibola Gneiss, 153, 186
Cibola Granite, 186
Cimarron, 82, 84
Cimarron Canyon, 85, 86, 87
Cimarron Canyon State Park, 85, 86; Palisades at, 85–**86**
Cimarron Cutoff, 228, 242
Cimarron Mesa, 305
Cimarron Range, 82, 84

Cimarron River, 16, 85, 87
cinder cones, 22, 61, 62, 67, 69, **70**, 71, 126, 130, 180, 219, **247**, 248, 294, **305**, 306; quarries in, 102, 103, 146
cirque, **26**, 87
City of Rocks State Park, 308
Civilian Conservation Corps, 314
claystone, 108, 186
Clayton, 25, 84, 217, 219, 223, 225, 242, 246, 248
Clayton Lake State Park, 244–245
Clear Fork, 258
Cliff Dweller Creek, 310, **311**
Cliff House Sandstone, 34, **46**, 52, 53, **59**, 60
Clines Corners, 188, 229, 263
Cloudcroft, 192, 255, 256
Clovis, 28, 248, 250
coal, 4, 34, 37, **41**, 52, 217; mining of, 45, 52, 84, 157, 222, 223, 282
coastal plain, 15, 79, 198, 205
Cochiti Dam, 151
Cochiti Formation, 149
Cochiti Pueblo, 146
Coconino Formation, 31
Colorado Plateau, 20, 21, 28, 30–35, 47, 178
Colorado River, 29, 31, 37, 38, 44, 52, 67, 95, 150, 284, 313
columnar jointing, **50**, 90, **126**, **127**, 130, 143, 180, 308
Conchas Lake, 220; State Park, 235
conglomerate, 4, 34, **112**, **141**, **167**, 217, 284, **311**. *See also* Gila Group
Continental Divide, 28, 38, 41, 67, 284, 291, 304, 313; National Scenic Trail, 94
continental shelf, 163, 224, 265
Cookes Peak, 292, 306
Cookes Range, 175, 292, 306
Cooney Tuff, 313, 314
copper, 85, 119, 121, 191, 198, 202, 286, 290, 292, 295, 297, 300, 306, 309. *See also* turquoise
coral, 154, 186, 216, 260, 269
Corona, 241
Cotton City, 289
Cottonwood Lake, 275
Coyote, 140
Coyote Creek State Park, 87
cratons, 2, 9
Crevasse Canyon Formation, 17, 34, 47, 64, 68
crinoids, 13, 135, 154, 186, 216, 260, 269

cross-bedding, 37, 39, 49, 64, 66, 79, 137, 149, 187, 198, **208**, 233, 236, 251
crust (Earth's), 2–3, 12, 20, 21, 23. *See also* subduction; faults; Rio Grande Rift; grabens
crystals, 4, **6**, **7**, **63**, 185, **204**. *See also* minerals
Cuba, 25, 33, 58, 60, 138
Cub Mountain Formation, 282, 283
Cuerbio Basalt, 126
cuestas, **46**, 138, 197
cutbanks, 256, 257
Cutler Formation, 34, 140

dacite, 4, 22, 93, 102, 117, 118, 123, 141, 286, 294, 295, 313
Dakota Sandstone, 65, **66**, **77**, 178, 225, 244–245, 281–282
Datil, 163, 213
Datil Group, 290, 302, 313
debris flows, 113, 192, 302
Defiance uplift, 14, 31, 33, 49, 51, 300
Delaware Basin, 14, 15, 215, 258, 267, 268, 269, 275
deltas, 187, 196, 248
Deming, 27, 28, 175, 284, 285, 291, 292, 294, 306
Desert Peaks National Monument, 176, 297
desert varnish, 182, 192, 281
Des Moines, 246, 248
Devils Throne, **158**
Diablo Canyon Recreation Area, 126
Diamondtail Formation, 159
diatremes, 51, 52
dikes, 10, **21**, 22, 35, 52, **53**, **111**, 119, 152, 186, 251, 253, 289; feeder, 181, **225**; swarms, **42**, **43**, 166, **282**. *See also* Dulce dike swarm; Galisteo Dike
dinosaurs, 16, 17, 18, 64, 141, 143, **152**, 224, 237, 244–45
diorite, 4, 118, 297, 309
dissolution, 15, 216, 257, **279**, 280. *See also* caves; cenotes; sinkholes
Dog Canyon National Recreation Trail, 191
Dog Lake Formation, 188
Dog Springs Formation, 302, 304
dolomite, 14, 176, **191**, **192**, 216, 267, 269, 270, **279**, 292. *See also* Fusselman Dolomite, Goat Seep Dolomite
Dolores Gulch, 156
dolostone, 207, 267

Domingo, 146
Doña Ana, 177
Doña Ana Mountains, 175, 202
Dripping Springs Natural Area, 176
Dulce dike swarm, 42, 43
dunes, 29, 34, 45, 66, 101, 165, 178, 183, 194, 212, 217, 229, 238, 256, **274**; gypsum, 188, 191, 203, 206–209. *See also* Entrada Sandstone; Glorieta Sandstone; White Sands National Park; Mescalera Sands

Eagle Draw, 257
Eagle Nest, 87, 114, 121
Eagle Nest Dam, 87
Eagle Nest Lake, 82, 85; State Park, 87
Eagle Tail Mountain, 84, **222**, 225
Earth, 2
East Defiance monocline, 31, 51
East Potrillo Mountains, 296
East Robledo fault, 175
Edgewood, 188
El Alto Basalt, 141
El Cajete Pumice, **133**
El Camino Real, 203
El Cerro de los Lunas, 163
Elephant Butte, **171**; Dam, 170; Reservoir, 44, 170, **171**
El Malpais National Monument, 22, 64, 67, 69–71, 180
El Morro National Monument, **x**, 5, **66**
El Paso, 27, 28, 189, 191, 203, 209, 215, 259, 268, 269
El Rito Conglomerate, **141**
El Rito Sport Climbing Area, **141**
Embudo fault zone, 24, 101, 106, 114
Embudo Stream Gauging Station, 105
Emory caldera, 285, 308
Enchanted Circle Scenic Byway, 114
Encino, 27, 199, 263
Entrada Sandstone, **16**, 33, 34, 37, 38, **39**, 63, 65, 66, 138, **142**, 145, **147**, 217, 236
erosion, 10, 16, 19, 29, 31, 39, 51, **54**, 75, **136**, 219, 257, 286; differential, 84, 191, 253; surfaces, 95, **110**, 145, 178, 216; by wind, 175. *See also* badlands; caprocks; hoodoos; pediments; peneplains; terraces
escarpments, 49, 79, 80, 109, 146, 215, 228, 238, 251, 260, 263, 270
Escondido Mountain, 304
Española, 100, 101, 103, 123, 126, 131

Española Basin, 22, 24, 72, 73, 98, 100, 101, 103, 106, 114, 123, 126, 131, 145, 146
Espinaso Formation, 160
Estancia, Lake, 27, 198, 218
Estancia Basin, 188, 197, **198**
evaporites, 4, 23, 63, 146, 169, 191, 207, 213, 216, 231, 253, 255, 265, 274, 289. *See also* gypsum; salt

fanglomerate, 126, 295
Farmington, 14, 27, 33, 35, 45, 46, 72
faults, 19, 23, 24, 29, 97, 101, 137, **253**, 266, 284; normal, 22, **74**, 114, **147**, 171; reverse, 156, 196; strike-slip, 21, 60; thrust, 25, 259; transform, 106. *See also* Animas Valley fault; Cañoncito fault; Cañones fault; East Robledo fault; Embudo fault; Fowler Pass fault; Jemez fault; Jicarilla fault; La Bajada fault; Lost Cabin fault; Manzano fault; Montosa fault; Nacimiento fault; Picuris-Pecos fault; Pajarito fault; San Andres fault; Santa Barbara fault; Tijeras fault
fault blocks, 32, 150, 153, 155, 194, 272, 290
feldspar, 4, **7**, 185, **204**
Fence Lake Formation, 304, 305
Fenton Lake State Park, 134
flash floods, 44, 299
Flechado Formation, 13, 107, **114**
floodplains, 34, 45, 76, 217, 259
Florida Mountains, 294, 295
Folsom, 152, 246, 248
Ford Butte, 52
foreland basin, 258
Fort Seldon National Monument, 175
Fort Stanton, 281
Fort Sumner, 212, **214**
Fort Union National Monument, **228**
fossils, 4, 12–13, 57, **88**, 141, 177, 233; plants, 177, 186, 187, 222; trace, **59**, 198, 212; trackways, 176, **244**, 245. *See also* brachiopods; bryozoans; burrows; corals; crinoids; dinosaurs; gastropods; petrified wood; reptiles; trilobites
Four Corners, 45
Four Corners Generating Station, 45, 52
Four Hills, 185
Fowler Pass fault, 85, 87
Fra Cristobal Mountains, 170, **171**
Franklin Mountains, 178, 189, 202
Fresnal Canyon, 254

Frijoles Canyon, 123, 124, 129, 130, 131
Fruitland Formation, 17, 35, 45, 46, 138
Fusselman Dolomite, 171, 189, 291, 292, 313

gabbro, 4, 6, 60, 85, 118, 251, 297
Galisteo Basin, 18
Galisteo Creek, 157
Galisteo Dike, **262**, 263
Galisteo Formation, **159**, 160
Gallina-Archuleta arch, 41, 95, 138
Gallinas Mountains, 240
Gallup, 16, 19, 27, 28, 31, 32, 33, 38, 47, 178
Gallup Sag, 31, 37, 64
Gallup Sandstone, 17, 34, 37, 65, 68
Garden of the Gods, 159
Garita Creek Formation, 233
garnets, 81, 102, 291, 296
gastropods, 198, 260, 278
geodes, 294, 295
Ghost Ranch, 15, 16, 141, 142, 143
Gila Cliff Dwellings, 306, 310, **311**, 314
Gila Cliff Dwellings caldera, 285, 310, 314
Gila Group, 284, 290, 291, 306, 313, 315; conglomerate of, 284, 309, 310, **311**
Gila Mountains, 313
Gila River, 27, 29, 289, 313
Gila Wilderness, 310
glacial outwash, 54, 94, 102
glaciation, 26, 45, 85, 87, 94, 188
Glorieta Mesa, 79, **80**
Glorieta Pass, 80
Glorieta Sandstone, 33, 34, 67, 77, 79, **80**, 217, 229, 278, 279, 280
gneiss, 4, **8**, 10, 31, 60, 64, 80, 81, 87, 106, 135, 202, 295; granitic, 67, 119, 260. *See also* Cibola Gneiss
Goat Seep Dolomite, 269
Gobbler Formation, **253**, 255
Gobernador Canyon, 44
gold, 21, 87, 115, 118, 145, 191, 193, 238, 240, 290, 291, 292, 309; placer, 85, 121, 156, 157, 160, 193; veins, 186, 202, 294, 297
grabens, 73, 75, 169, 181, 189, 205, 212, 251, 284, 289, 300. *See also* Reserve graben
Gran Quivira, 197
Graneros Shale, 77, 84, 217, 228, 245
granite, 4, 6, **7**, 31, 38, 60, 64, 67, **68**, **93**, 94, 102, 116, 119, **154**, **173**, **176**, 194, **200**, **203**, **204**, 205, **241**, 263. *See also* Cibola Granite; grus; Sandia Granite; Southern Granite-Rhyolite Province; Tres Piedras Granite; Tusas Mountain Granite
granodiorite, 60, 81, 118, 189, 292, 297, 306, 309
Grants, 22, 33, 38, 39, 65, 68, 69, 164, 180
Grants Mineral Belt, 178
Grayburg Formation, 217, 251, 265, 268, 276
Great Plains, 20, 28, 215–283
Great Unconformity, 10, 31, 135, 153, 163, 290, 292, 306
Greenhorn Formation, 77, 84
greenstone, 85, 196
Grenville orogeny, xii, 10
groundwater, 74, 150, 178, 258–259, 275, 278, 292. *See also* aquifers; artesian; maar volcanoes
grus, 94, 154, 185
Guadalupe Mountain, 117
Guadalupe Mountains, 15, 268, 269, 270
Guadalupe Mountains National Park, 268, 269, 272
Guadalupe Peak, 269
Guadalupe Ridge, 268, 269, 270
Guadalupita Canyon, 87
Guaje Pumice, 129, 251
gypsum, 16, **61**, **63**, 80, 140, **270**, 272, 275; dunes of, 188, 191, 203, **206–209**; mining of, 138, 146; stringers of, **212**, 229, 274, **276**. *See also* Artesia Group; dissolution; Dog Lake Formation; evaporites; San Andres Formation; selenite; Todilto Formation; Yeso Formation

Harding Pegmatite Mine, 105
Hatch, 174, 175
Hells Mesa Tuff, 299
Heron Lake, 150
High Road to Taos, 108–114
Historic Civil Engineering Landmark, 105, 170
Hogback, the, 38, 46, 52, 53, 65
hogbacks, 20, 31, 32, 37, **38**, **46**, 65, **77**, 85, 138, 211, 227, 228
Hondo junction, 280
hoodoos, 42, 101, 108, 130, 134, **136**, **147**
Hopewell Lake, 94
hornblende, 4, 80, 85
horsts, 284, 290, 292, 299
Hot Springs, 171. *See also* Truth or Consequences

hot springs, 4, 101, 134, 137, 175
Hueco Basin, 84, 75
Hueco Formation, 175, 177, 189, 202, 291
Huerfano Mountain, 58
hydrothermal, 119, 134, 240, 290, 295, 300

ice age, 26, 27, 75, 103, 169, 182, 198, 203, 207, 209, 220, 228, 238, 265, 301
Ice Cave, **70**, 71
igneous rock, 1, 4, 6, **7**
ignimbrite, 6, 20, 26, 123, 124, **128**, 132, 286, 314. *See also* tuff
ignimbrite flare-up, xii, 20, 21, 73, 74, 116, 156, 285
Imperial Valley, 73
Indian Divide, 283
Inscription Rock, **66**
inverted topography, 84, 219, 227, 245
iridium, 18, 224
Iron Hill, 12
island arc, 7, 9, 118, 153. *See also* Mazatzal province; Yavapai province

Jarilla Mountains, 189, 191
Jemez Canyon, 132
Jemez Falls, 132
Jemez fault zone, 135
Jemez Historic Site, 135, **136**
Jemez lineament, xii, 9, 23, 25, 26, 67, 106, 122, 146, 180, 219, 226, 227, 305
Jemez Mountains, 1, 26, 28, 121–124, **155**. *See also* Valles Caldera
Jemez Pueblo, 137
Jemez River, 132, 133, 134, **135**
Jemez Springs, 121, 123, 134, 135, 137
Jemez volcanic field, 23, 123, 126
Jicarilla fault, 109
Jicarilla Mountains, 238, 240
joints, 39, 90, 105, 130, **142**, **154**, **176**, **204**, **287**. *See also* columnar jointing
Jornada, Lake, 169, 174, 213
Jornada del Muerto, 72, 73, 169, 174, 175, 200, 212

karst, 13, 15, 216, 231, 254. *See also* caves; dissolution; sinkholes; weathering
Kasha-Katuwe Tent Rocks National Monument, 74, **147**–**149**
Kelly Formation, 300
Kilbourne Hole, **296**
Kiowa National Grassland, 242

Kirtland, 33, 45, 46, 138, 185
Kirtland Formation, 18
Kneeling Nun Tuff, **308**
K-Pg (K-T) boundary, 18, 224

La Bajada constriction, 146
La Bajada fault zone, **147**
La Bajada Hill, 146
Laborcita Formation, 255
laccolith, 22, 54, 156, 157, 158, 160, 227. *See also* Cerrillos Hill laccolith
La Cienega, 146
La Cueva Member, 123
Ladron Peak, 163
Laguna, 178, 181
Laguna del Perro, 198
La Jara Canyon, 44
La Jara Creek, 44
La Junta, 117, 118, 220
Lake Valley, 309
Lake Valley Formation, 171, 191, 254
Lake Valley Limestone, **191**, 309
La Mesa, 87, 131, 178
Lamy, 260, 262
landslides, 49, 80, 82, 85, 96, 101, 119, **137**, 154, 157; deposits of, **49**, 68, 105, 109, 134, 178, 223
Laramide orogeny, xii, 17–18, 31, 35, 51, 138, 222
Las Cañas terrace, 169
Las Conchas Fire, 131
Las Conchas Trailhead, 132
Las Cruces, 73, 169, 174, 175, 177, 192, 200, 202, 203, 212, 297, 306
Las Huertas Canyon, 152
Las Vegas, 18, 77, 228
lava, 6, 22, 50, 69, 92, 121, **147**, **162**, **166**, **180**, 181, 219, **247**, 286, 299; aa lava, 69, 71, 180; pahoehoe, 69, 71, 180, **211**, 238. *See also* Bandera lava low; basalt; McCartys lava flow; rhyolite; volcanic fields
lava domes, 121, 131, 181, 219, 286, 297
lava tubes (tunnels), 69, **70**, 71, 180, 182, **211**, 294
Lea Lake, 275
Leasburg Dam State Park, 175
Lemitar Mountains, 12, 72, 165, 299
Lewis Shale, 35, 42, 46, 97, 138
limestone, 4, 34, **56**, **82**, **88**, **192**, 216, 217, 241, **254**, 266, **270**, **279**. *See also* Artesia

Group; Bone Spring Formation; Capitan Formation; fossils; Greenhorn Formation; karst; Lake Valley Limestone; Oswaldo Limestone; Madera Group; San Andres Formation; Todilto Limestone
Lincoln Folds, 280
Lincoln National Forest, 283
lithium, 102, 104
lithosphere, 2, 3, 9
Little Black Peak, 192, 209, 211, 238
Little Florida Mountains, **294**
Little San Pascual Mountains, 167
Living Desert Zoo and Gardens State Park, 267
Llano de Albuquerque, 152, 182
Llano Estacado, 215, 218, **236**, 237, 238, 248, 272, **274**
Lobato Formation, 123, 141
Lobo Hill, 12
lode, 85, 145. *See also* mining; ore; *specific metal*
Loma Parda terrace, 169
Lone Mountain, 240
Long Valley, 121
Lordsburg, 27, 28, 285
Lordsburg Draw, 290
Lordsburg mining district, 290
Los Alamos, 74, 123, 130
Los Alamos National Laboratory, 124, 131
Los Barrancos, 100, 124
Los Cerrillos, 158, 159
Los Cordovas, 106, 114
Los Duranes Formation, 76
Los Gigantes, 65, **67**
Los Lunas, 163
Los Pinos Formation, 93
Los Pinos Mountains, 163, 196
Los Pinos River, 44
Lost Cabin fault, 87
Lost Padre Gold Mine, 203
Lucero, Lake, 203, 207, 208, 255
Lucero uplift, 19, 300
Lucero volcanic field, 25
Lumberton, 42, 43
Luna, 28, 163, 315
Lybrook, 58

maar volcanoes, 49, 50, 51, 123, 127, 130, 146, 294, 296, 305, 306
Madera Group, **5**, **88**, 153, 185, 186, 188, 199, 217; limestone, **10**, 13, 134, 135, **136**, 150, 152, 154, **155**, 157, **187**, 229, 260
Madrid, 157
mafic, 52, 69, 119, 123, 124, 145, 165, 180, 181, 186, 286
Magdalena Group, 254
Magdalena mining district, 300
Magdalena Mountains, 21, 166, 297, 300
Magdalena Ridge Observatory, 166, 300
magma, 1, 6, 22, 25, 71, **86**, 286, 296. *See also* dikes; plutons; volcanoes
mammals, 20, 57, 100, 126, 152
Mancos Shale, 17, 34, **46**, 52, **54**, 96, 157, 282
Mangas Mountains, 304
Manhattan Project, 131, 213; National Historic Park, 131
mantle, 2, 3, 15, 20, 22, 23, 73, 122, 130, 165, 166, 296
Manzano fault zone, 163, 194
Manzano Mountains, 7, 9, 22, 160, 162, 163, 185, 186, 194, 196, 197, **198**, **199**
Manzano Peak, 160, 199
Manzano Ruins, 199
Marathon orogeny, 229
Maxson Crater, 228
Mayhill, 255
Mazatzal orogeny, xii, 9, 153, 186, 196
Mazatzal Province, 8, 9, 25, 122, 160, 219
McCartys lava flow, 69, **180**
McKittrick Canyon, 272
meanders, 60, 132, 143, 198, 256, 257, 265
Menefee Formation, 34, 41, **46**, **47**, 52, 157
Mesa Negra, 68
Mesa Prieta, 61
Mesa Rica Sandstone, 237
Mesa Verde, 53, 56, 187
Mesa Verde National Park, 54
Mesaverde Group, 17, 34, 35, **41**, **54**, 59, 96, 97, 157, **170**, 282, 283. *See also* Cliff House Sandstone; Menefee Formation; Point Lookout Sandstone
Mescalero Sands Recreation Area, **274**
Mesilla Valley, 175, 176, 177, 297
Mesilla Valley Bosque State Park, 177
metamorphic rocks, 4, **8**, 38, 106, 116, 153, 163, 170, 175, 185–186, 188, 260. *See also* amphibolite; gneiss; quartzite; schist
metamorphism, 94, 196
metasedimentary, 93, 94, 111, **112**, 119, **120**, 165, 186, 199, 290, 300
meteor impacts, 18, 81, 224

mica, 4, 102, 154, 186
Midland Basin, 258
Mimbres Basin, 72
Mimbres Mountains, 167, 174, 292
Mimbres River, 291, 292, 306
Mimbres Valley, 291, 309
Mineral Hill, 203
minerals, 86, 102, 105, 121, 291, 294, 296, 300; clay, 182, 237, 240; in vesicles, 295, 311. *See also specific mineral names*
minette, 52
mining, 21. See also Chevron-Questa; Chino Mine; coal, mining; copper; gold; Harding Mine; Lordsburg mining district; Magdalena mining district; Ortiz mining district; silver; Tyrone Mine; uranium; White Oaks mining district; zinc
Miranda granite, 116
Moenkopi Formation, 15, 274
Mogollon-Datil volcanic field, 21, 166, 167, 175, 285, 287, 290, 292, 296, 297, 300, 302, 304, 306, 310, 313
Mogollon Group, 118, 174, 287, 300, 309, 313, 315
Mogollon Mountains, 310, 313, 314
molybdenum, 21, 118, 119, 121, 309
Monero, 41
Monte Largo, 12
monoclines, 31, 32, 38, 49, 51, 52, 126
Montoya Formation, 313
Monument Rock, 302, 304
monzonite, 4, 12, 159, 176, 189, 204, 291, 300, 309
moraines, 26
Mora River, 226, 228
Moreno Valley, 87, 120, 121
Morrison Formation, 16, 33, 34, 47, 60, 63, 64, **77**, 84, 138, 146, 178, 217, 222, 226, 237, 263
Mountainair, 197, 198, 199
Mountain Branch of the Santa Fe Trail, 228
Mt. Taylor volcanic field, 23, 25, 61, 62, 68. *See also* Taylor, Mt.
mud cracks, 198, 212, 244, **245**, 289
mudstone, 26, 34, 80, **88**, **137**, 217, **235**, 266
Mule Mountains, 313
Mystic Lode, 85

Nacimiento fault, 17, 42, 60, 61
Nacimiento Formation, 33, 34, 35, 44, 54, **56**, 57, 58, 138

Nacimiento uplift, 31, 32, 95, 137, 138
Narbona Pass, 49, 50, 51
Native Americans, 124, 135, 140, 158, 197, 198, 233, 240, 281, 309
Navajo Lake State Park, 34, 44
Navajo Refinery, 257, **258**
Navajo volcanic field, 35, 49, 52
Needles, 204
Newcomb, 52
Newkirk, 233, 235
New Mexico Museum of Natural History and Science, 64, **152**
Niobrara Formation, 84, 217, 222
nodules, 159, 295, 300
Nogal Canyon, 213
normal faults. *See* faults, normal
North American Plate, 1, 3, 21, 42, 229
North Cebollita Mesa, 68
Nutria monocline (and hogback), 38, 65

Oak Ridge, 131
Oasis State Park, 250
obsidian, 122, 123, 124, 132, 134, 140, 149
Ocaté volcanic field, 227, 228
Ogallala aquifer, 218, 238, 272
Ogallala Formation, 216–218, 226, 235–**236**, 237, 248, 263, 266, 272, **274**
oil and gas, 15, 20, 263, 265–66, 269, 272
Ojo Alamo Formation, 18, 34, 35, 44, **45**, 138
Ojo Caliente, **101**, 102
Ojo Caliente Member, 100, 101
Ojo Huelos Member, 16
Ojo Sarco, 111
O'Keefe, Georgia, 107, 123, 143
Oliver Lee Memorial State Park, 191
olivine, 90, 93, 130, 146, 180, 192, 211, 219, 220, 296
Oñate Formation, 191, 254
Ophiomorpha, **59**
ore, 118, 119, 121, 166, 178, 290, 292, 297, 300, 309
Organ, 200, 202
Organ caldera, 285
Organ Mountains, 169, **176**, 177, **200**, 202, 203, 204, 205, 287; Organ Needle, 176
Organ Mountains–Desert Peaks National Monument, 176, 297
orogenies, xii, 160. *See also* Grenville orogeny; Laramide orogeny; Marathon orogeny; Mazatzal orogeny; Picuris Orogeny; Yavapai orogeny

Orogrande, 189, 191
Ortega Quartzite, 9, 87, 94, 95, **97**, 116, 263
Ortiz Arroyo, 157
Ortiz mining district, 145
Ortiz Mountains, 145, 146, 155, 156, 160
Ortiz porphyry belt, 20, 156, 157, 160
Ortiz surface, 145, 146, 160
Oscura Mountains, 193, 212, 213
Oswaldo Limestone, 309
Otero, Lake, 207, 209
Otowi Bridge, 126
Otowi Member, 123, 124, **128**, 134
oxides, 47, 80, 159, 178, 187, 198, 289, 295

Pacific Plate, 3, 20, 21, 23, 284, 306
pahoehoe, 69, 71, 180, **211**, 238
Pajarito fault, 60, 131
Pajarito Plateau, 123, 126, 129, 131, **148**
paleosol, 174
Palisades, 85, **86**
Palo Duro Canyon, 248
Palo Flechado Pass, 87, 88
Palomas Formation, 174
Pangaea, xii, 13, 15, 16, 212, 229
Pecos Baldy, 109
Pecos Canyon, 81, 232
Pecos Canyon State Park, 81–82
Pecos diamonds, 265
Pecos National Historic Park, 80
Pecos River, 27, 28, 79, 81–82, 220, 231–233, 265, 266, 276, 278
Pecos Slope, 251, 278
Pecos Valley, 236, 274, 276
Pedernal arch, 215
Pedernal Hills, 188, 198, 229, 263
Pedernal uplift, 14
pediments, 82, 109, **110**, 194
pegmatite, 102, 105, 185
Peloncillo Mountains, **287**, 289
Peñasco, 111
peneplains, 10, 95
Peralta Tuff, **74**, **149**
Percha Dam State Park, 173
Percha Shale, 171, 313
perlite, 102, 149, 166, 297
Permian Basin, 15, 258, 259, 265, 267, 278. *See also* Delaware Basin; Midland Basin
petrified wood, 16, 160, 176, 187, 231, 237
Petroglyph National Monument, 182
petroglyphs, 129, **182**, 192, **193**
Pictured Cliffs Sandstone, 35, 42, 46

Picuris Formation, 109, 111, **112**, **113**
Picuris Mountains, 9, 109
Picuris orogeny, xii, 9
Pierre Shale, 82, 84, 85, 121, 217, 222, **225**
Pie Town, 304
Pine Creek, 176, 204
plagioclase, 90, 180, 182
Plains of San Agustin, 27, 300, 301, 302
plate tectonics, 2, 3
Platoro caldera, 42, 285
playas, 26, 29, 162, 189, **207**, 289, 301; lakes, 169, 174, 188, 203, 207, 213, 218, 251, 253
Plaza Blanca, 143
Pleistocene, 26–27; volcanism of, 98, 123, 150. *See also* glaciation; ice ages
plutons, 7, 22, 85, 116, 196, 278, 287, 295, 309. *See also* batholiths, Capitan pluton; Cibola Gneiss
Point Lookout Sandstone, 17, 34, 41, **46**, 157
Pojoaque, 108, 124
Pojoaque River, 124
porphyry, 20, 111, 156, 157, 158, 159, 160, 289, 306, 309
Portales, 248, 250
Portales Valley, 250
Potrillo volcanic field, 177, 292, 293, 294, **296**
Prehistoric Trackways National Monument, **177**, 297
Prewitt, 37, 38, 39
Pueblo Bonito, **59**
Pueblo of Isleta, 163
Puerco River, 35, **37**
pumice, 6, 122, 123, 126, 132, **133**, 141, 251, 314
Pyramid Mountains, 289, 290
pyrite, 121, 156, 159, 186, 203, 292, 300, 309
pyroxene, 180, 296

Quarai Ruins, 199
quartz, 4, **7**, 95, 122, 123, 134, 154, 206, 265, 292, 300; chalcedony, 295; clasts of, 81; sand, 206, 213, 229, 281; veins of, 119, 121, 185, 186, 202. *See also* chert
quartzite, 4, **8**, **94**, **113**, **120**, 160, 196. *See also* Beartooth Quartzite; Ortega Quartzite; Sais Quartzite; White Ridge Quartzite
quartz latite, 118, 289
quartz monzonite, 291

Quemado, 304, 305
Questa, 21, 116–118
Questa caldera, 116, 285

Rabbit Ear Mountain, 242, **246**
radiocarbon dating, 152, 250
radiometric dating, 1, 12, 126
Radium Springs, 175
Railroad Mountain dike, 250
Rancheria Formation, 191, 254
Raton, 19, 82, 121, 219, 220–225, 228, 242
Raton Basin, 82, 84, 222, 225, 228, 246
Raton-Clayton volcanic field, 25, 84, 219, 223, 225, 242, 246, 248
Raton Formation, 18, 82, 84, 85, 217, 222, 223, **224**, 248
Raton Mesa, 82, 84
Raton Pass, 220, 222
red beds, 14, 16, 63, 77, 134, **137**, **140**, 177, 202, 227, 255, 281, 300
Red Hill, **305**
Red Mesa, **137**
Redonda Formation, 237
Redondo Peak, 124, 131, 132, 134
Red River, 21, 90, 93, 117, 118, 119, 120
Red Rock State Park, 34, 39
reefs, 259, 267, 268–69. *See also* Capitan Reef
regression, 11, 35
reptiles, 14, 152, 177, 187, 225, 231, 233, 238
Reserve graben, 315
Revuelto Creek, 238
rhyolite, 4, 22, 117, 118, 119, 121, 122, 132, 134, 175, 286, **294**–295, 310, 313, 315; dikes of, 289. *See also* Bearhead Rhyolite; lava domes; perlite
rifting, 21, 73, 116, 155, 189, 212. *See also* Rio Grande rift
Rincon Ridge, 153, **155**
ring fractures, 121, 122, 124, 132, 286, 299, 313
Rio Bonito, 280
Rio Brazos, 95, 97
Rio Cebolla, 134
Rio Chama, 28, 75, 97, 143
Rio Fernando, 87, 90, 114
Rio Grande, 21, 27, 28, 75–76, **92**, 103, **117**, 127, 150–151, **153**, 162, 183, 194. *See also* aquifers; terraces
Rio Grande del Norte National Monument, 90, 117–118

Rio Grande Gorge, 90, 95, 106, **107**, 118; Bridge, 90
Rio Grande rift, xii, 1, 3, 21, 25, 34, 72–76, 88–89, 90, 103, 150, 163, 177, 181, 213. *See also* Española Basin; San Luis Basin; Taos embayment
Rio Grande valley, 73, 75, 146, 155, 156, 160, **164**, 167, 169, 173, 175, 183, 213, 297
Rio Hondo, 92, 93, 116, 278, 280
Rio Medio, 109
Rio Nutria, 65
Rio Ojo Caliente, 101
Rio Peñasco, 255, 257
Rio Pueblo de Taos, 90, 114
Rio Puerco, 25, 28, 60, 61, 62, 140, 157, 164, 181, 183
Rio Rancho, 76
Rio Salado, 165
Rio Salado Sand Dunes Historical Marker, 165
Rio San Jose, 38, 180, 181
Rio Santa Barbara, 111
Rio Tesuque, 98
Rio Tusas, 93
ripple marks, 198, 228, 245
Robledo Mountains, 175, **177**, 297
rock glaciers, 85
Rockhound State Park, 294–295
Rocky Mountains, 4, 18, 27, 31, 39, 42, 73, 77, 94, 96, 155, 162, 216, 218, 222, 225, 228, 236, 248, 250, 272. *See also* Ancestral Rocky Mountains
Rodinia, xii, 10, 160
Roosevelt Wind Project, 251
Roswell, 14, 19, 27, 215, 251, 259, 263, 265, 272, 275, 276, 278
Rough and Ready Hills, 296
Rubio Peak Formation, **295**
Ruth Hall Museum, 143

Sabina Wilderness, 226
Sacramento Mountains, 171, 189, 191, 192, **205**, 206, 251, **253**, 254, 255, 265, 276
Sais Quartzite, 199
Salado coal fields, 282
Salado Formation, 268
Salinas Pueblo Missions National Monument, 197, 199
Saliz Pass, 315
salt, 61, 197, 205, 211, 216, 251, 253, 255, 269, 272. 279, 289. *See also* evaporites

San Acacia, 165
San Agustin, 27, 300, 301, 302
San Andreas fault, 21, 23
San Andres fault, 189
San Andres Formation, 15, 34, 77, 79, 171, 197, 207, 211, 217, 229, 253, 255, **279**, 280; caverns in, 216, 276, 281; quarries in, 67; as reservoir, 251, 257, 259, 265, 278; sinkholes in, **231**, 240, 241, 263, **275**
San Andres Mountains, 7, **164**, 174, 175, 189, 191, 202, **204**, 205, 207, 251, 255
San Antonio, 134, 192, 238
San Antonio Mountain, **102**, 103
San Augustin Pass, 203
Sandia Cave, 152
Sandia Crest, 10, 153, 154, 157
Sandia Formation, v, **10**, 87, 89, 114, 135, 155, 185, 186, 196, 217
Sandia Granite, **7**, **10**, 153, 154, **155**, **185**, 186
Sandia Mountains, 5, 8, 10, 31, 73, 150–152, **153**–155, 183, 185, 187. *See also* Great Unconformity; Madera Group
Sandia Peak, 153, 154, **155**; Ski Area, 157
Sandia Peak Tramway, 153
Sandia Point, 152
Sandia Pueblo, 153
Sandia Tram, 154
Sandoval County, 146
San Diego Canyon, 123, 124, 132, 134, 135, **136**
sand dunes. *See* dunes
sandstones, 4, 34, **37**, 39, *52*, **56**, **58**, **61**, **88**, **89**, **112**, **114**, **137**, **140**, **170**, **197**, **235**, **254**, **279**. *See also* Bliss Sandstone; Chuska Sandstone; Cliff House Sandstone; Dakota Sandstone; Entrada Sandstone; Gallup Sandstone; Glorieta Sandstone; Mesa Rica Sandstone; Mesaverde Group; Ojo Alamo Sandstone; Ojo Caliente Sandstone; Pictured Cliffs Sandstone; Point Lookout Sandstone; San Jose Formation; Trinidad Sandstone; Zuni Sandstone
San Francisco Mountains, 315
San Francisco River, 284, 313, 315
Sangre de Cristo Formation, 77, 79, **80**, 81, 217
Sangre de Cristo Mountains, 26, 79, 82, 84, 103, 106, 107, 116. *See also* Pecos Canyon; Pecos River; Rocky Mountains; Taos Mountains; Wheeler Peak

San Gregorio de Abo, Mission of, **197**
San Jon, 238
San Jose, 33, 38, 180, 181
San Jose Formation, 18, 34, 35, **42**, **43**, 44, 57, 58, 138
San Juan Basin, 20, 28, 32–35, 37, 41, 54; resources in, 32, 44, 52, 57, 178. *See also* Fruitland Formation; Kirtland Formation; San Jose Formation
San Juan Mountains, 32, 41, 44, 54, 94, 102
San Juan River, 41, 44, 45, 46, 52, 54, 57, 150
San Juan volcanic field, 20, 35, 42, 75, 95
San Lorenzo Canyon, **165**
San Luis, 73, 114
San Luis Basin, 22, 24, 72, 101, 103, 106
San Mateo, 167
San Mateo Mountains, 170, 297
San Pedro, 155, 157
San Pedro Arroyo Formation, 16
San Pedro Mountains, 150, 156
San Pedro-Ortiz porphyry belt, 157
San Rafael, 68, 69
San Rafael Group, 34, 47, 156, 236
Santa Barbara fault, 109,
Santa Clara, 61, 100, 101, 306
Santa Clara Canyon, 123
Santa Clara Lake, 109
Santa Cruz River, 109
Santa Fe, 77, 98, 145, 155, 260
Santa Fe Group, 23, 34, 74–75, **98**, **101**, 103, **105**, **108**, 145, 150, 162, **165**, **167**, **173**, 175; aquifers in, 23, 74, 150, 200; badlands in, 100, 103, 181; faults in, 100, 106, 169; lava in, 163; terraces of, 76, 152, 182, 194. *See also* Ancha Formation; Chamita Formation; Los Pinos Formation; Palomas Formation; Tesuque Formation
Santa Fe Railroad, 37
Santa Fe River, 146
Santa Fe Trail, 77, 80, 82, 84, 220, 227, 228, 242
Santa Rita Mine, 306
Santa Rosa Formation, 77, 79, 217, 226, 229, 231, **232**, 233, 251, 281
Santa Rosa Lake, 220, 231
Santo Domingo Basin, 146
Santo Tomas Basalt, 294
San Ysidro, 16, 63, 64, 137, 138
Sauk Sea, 11
Sawtooth Mountains, **304**

scarps, 96, 131, 163, 207, 212, 255. *See also* escarpments
schists, 4, 81, **120**, 145, 160, 196
Sedgwick, Mt., 65
sedimentary rocks, 4, 5, 12–13, 89, **137**, **147**, **205**, 213, **226**. *See also* claystones; conglomerates; limestones; mudstones; sandstones; shales; siltstones
selenite, 79, 169, 207, 208, 213
Servilleta Basalt, 89, 90, 100, 101, 103, **105**, 106, 143
Seven Rivers Formation, 259, 265, 266, 267, 268, 274
Seven Springs shear zone, 134
Sevier deformation, xii, 16
Sevilleta National Wildlife Refuge, 164, 165
shales, 4, 34, 54, **56**, 64, **88**, **89**, **114**, **170**, 217, 224, 237, **254**, 260, 266. *See also* Benton Shale; Colorado Shale; Graneros Shale; Lewis Shale; Mancos Shale; Menefee Formation; Morrison Formation; Percha Shale; Pierre Shale; Tucumcari Shale; Wolfcamp Shale
shatter cones, 81
shear zone, 186
Sheep Springs, 49
shield volcanoes, 69, 93, **102**, 163, 180, 225, 246, 294
Shinarump Formation, 34, 281
Ship Rock, **21**, 35, 47, 51, 52, **53**
Shiprock (town), 44, 46, 52
Sierra Blanca, 17, 26, 192, 193, 276, 278
Sierra Blanca igneous complex, 280, 282
Sierra Blanca volcanic field, 74
Sierra Grande arch, 222, 225, 246, 248
Sierra Madre Occidental, 285
Sierra Nacimiento, 60, 137
Sierra Negra, 143
silica, 3, 22, 51, 101, 132, 206, 281, 287, 300
silicates, 2, 300
sills, 22, 85, 86, 105, 127, 156, 158, 241
siltstones, 4, 34, 39, **56**, **114**, **147**, 217, **279**
silver, 21, 119, 156, 202, 283, 287, 290, 291, 292, 294, 296, 300, 309
Silver City, 21, 28, 306, 309, 310
sinkholes (sinks), 13, 216, 220, 229, **231**, 240, 241, **242**, 251, 256, 257, 263, 265, 267, 272, 275
Sleeping Lady Hills, 296
Sleeping Ute Mountain, 54
Sly Gap Formation, 191, 254

Socorro, 160, 167, 297
Socorro Basin, 72, 164, 167
Socorro caldera, 165, 166, 285, 297, 299
Socorro magma body, 165
Socorro Mountains, 297
Socorro Peak, **166**, 297
Soda Dam, 134, **135**
Soledad Canyon Day Use Area, 176
South Baldy, 300
Southern Granite-Rhyolite Province, 8, 10, 12, 25
Southern Pacific Railway, 287
southern Rocky Mountains. *See* Rocky Mountains
South Mountain, 131, 132, 156, 157
Spanish missionaries, 80, 197, 198
Spears Group, 290
Spring Canyon Recreation Area, 294–**295**
Springer, 226
Staked Plains, 215, 218, 236, 248
stratigraphic columns, 34, 217
stratigraphy, 41, 75, 134, 237, 250, 291, 292
stratovolcanoes, 22, 117, 155, 181, 192, 246, 295
subduction, 2, 3, 9, 16, 18, 21, 23, 25, 94, 219, 260
Sugarite Canyon State Park, 223
sulfides, 159, 203, 269, 270, 300
Sulfur Springs, 134
Summerville Formation, 34, 37, 236, 237
Sumner Lake State Park, 233
Sunshine Mesa, 233
Sunspot Scenic Byway, 253, 256
supercontinents, xii, 10, 13, 160, 229
synclines, 32, 33, 37, 39, 41, 64, 81, **137**, 157, 187, 255, 266

tafoni, **5**, **128**, 129, **128**, 130
Tansill Formation, 265, 268, 269, 270
Taos, 82, 90, 103, 108, 114
Taos embayment, 114
Taos Mountains, 73, 82, 87, 88, **89**, 90, 93
Taos Plateau, 89, **92**, 93, 116, 117;
Taos Plateau volcanic field, 90, 100, 101, **102**, 103, 106, **107**, 118, 143
Taos Pueblo, **89**, 107
Taos Ski Valley, 116
Taylor, Mt., 22, 23, 25, 34, 38, 61, 62, 68, 155, 180, 181
tectonic plates, 3, 20
tent rocks, **136**, 148

tephra, 251
terraces, 45, 75, 76, 97, 103, 152, 162, 163, 167, 174, 183, 194
Tesuque Formation, 89, 100, 101, 106, 108, 109, **110**, 124, 126, 127, 145
Tetilla Peak, 146
Thoreau, 33, 37
Three River stock, 192
Three Rivers Campground, 192
Three Rivers Petroglyphs Site, 192, 193
thrust faults. *See* faults, thrust
thundereggs, 295
Tierra Amarilla, 96, 97
Tijeras, 155
Tijeras Arroyo, 160
Tijeras Canyon, 160, 185, 186
Tijeras fault zone, 160, 186, 187; with Cañoncito fault system, 156
Tijeras Greenstone, 186
Tio Bartolo terrace, 169
titanotheras, 160
Tocito dome, 52
Todilto Formation, 16, 34, 60, **61**, **63**, 138, 146, 180, 236, 237
Tohatchi, 49
Toledo caldera, 26, 123, 285
Tonque Arroya Member, 63
trachydacite, 85
trackways, 176, **244**, 245
Trampas, 111
Trampas Group, 116
Transcontinental Arch, xii, 11, 12
transgression, 11, 35
Trans-Pecos volcanic field, 285
travertine, 4, 31, 134, **135**, 163, 171, **256**, 289
Tres Piedras, 93, 102
Tres Piedras Granite, 93, 94
trilobite, 186
Trinidad Sandstone, 84, 217, 222
Trinity Site, 213
Truchas Peak, 102, 103, 107, 108, 109, 143
Truchas surface, **110**,
Truth or Consequences, 28, 167, 171, 173, 202, 285, 297
Tsankawi Prehistoric Site, 129
Tsankawi Pumice Bed, 124
Tschicoma Formation, 123, 141
Tshirege Member, 123, 124, **128**, 129, **136**, 134
Tucumcari, 233, 237–238
Tucumcari Basin, 14, 215

Tucumcari Mountain, **237**
Tucumcari Shale, 237
Tuerto Gravel, 145, 157, 160
tuffs, 20, 94, **129**, 130, **147**, **166**, 285, 286; welded, 6, 121–124, 130, 131, 134, 192, 278, 286, 294, 302, 314. *See also* Amalia Tuff; Bandelier Tuff; Bloodgood Canyon Tuff; Cooney Tuff; Hells Mesa Tuff; ignimbrites; Kneeling Nun Tuff; Peralta Tuff
Tularosa, 192
Tularosa Basin, 28, 72, 189
Tularosa Valley, 73, 175, 189, 191, 203, **204**, 205, 206, 207, 209, 212, 251, 255, 283
tungsten, 156
Turkey Mountains, 227
turquoise, **156**, 158–159, 309
Turquoise Hill, 159
Turquoise Trail National Scenic Byway, 146, 155, 156, 187
Tusas Mountain Granite, 94
Tusas Mountains, 9, 73, 89, 93, 94, 96, 102
Tyrone mining district, 290, 309

Uncompahgre uplift, 14, 16
unconformities, 10, 16, 34, 79, 159, **165**, 196, 202, 217, 266, 282; angular, **16**, 159, 165. *See also* Great Unconformity
uranium, 1, 12, 166, 178, 180, 237, 290, 309
Ute Lake State Park, 238
Ute Mountain, 103, 117
Ute Park, 82, 85

Vadito Group, 9, 87, 94, 105, 116, 120
Val Verde Basin, 258
Valle de La Parida terrace, 169
Valle de Oro National Wildlife Refuge, 151
Valle Grande, 121, 131
Valles Caldera, 21, 22, 25, 26, 27, 121, 122, 123, 124, **131**, 132, 140, 155
Valles Caldera National Preserve, 124
Valles Caldera Preservation Act, 124
Valley of Fires National Recreation Area, 211
Valmont, 191
varves, 236, 269
Vaughn, 241, 242, 263
veins, **8**, 79, 80, 118, 156, 159, 240, 281, 289, 290, 291, 294, 300; of gypsum, **212**, 229, 274, **276**; of quartz, 119, 121, 185, 186, 202
vents, 51, 71, 122, **164**, 247

Magdalena Sandoval Donahue grew up in northern New Mexico, and was fascinated by the mountains and valleys of the high desert at the southern end of the Rocky Mountains. She received a BS in Geological Sciences and a BS in Fine Arts from the University of Oregon. She obtained her MS and PhD from the University of New Mexico, and much of her research focused on the evolution of topography and mountain ranges in Colorado. She is the coauthor of *Colorado Rocks! A Guide to Geologic Sites in the Centennial State*. She lives in Albuquerque with her husband, John, and three children.

Growing up and traveling around the world with two parents who had PhDs in geology, **Lucy Chronic** participated in discussions and adventures focusing on her parents' amazingly varied interests. When it came to her educational decisions, the family's love of geology won out, and she earned a BA in geology from Carleton College in Minnesota and an MS in paleontology from the University of Wyoming. Lucy's professional life has been eclectic, with writing and science the threads that tie her various pursuits together. She has worked as an archaeologist, educator, scientific writer, fire lookout, and interpreter in state and national parks. She and her mom, Halka Chronic, coauthored the second editions of *Pages of Stone: Geology of the Grand Canyon and Plateau Country National Parks and Monuments*. Lucy, Halka, and her sister, Felicie, coauthored the second edition of *Roadside Geology of Utah*. Lucy resides in the mountains of central Idaho with her husband, Chris Hinze, and enjoys her daughters, Betsy and Haley, during their all-too-infrequent visits home.

Vermejo Formation, 84, 217, 222
Vermejo River, 84, 222
Very Large Array, 301
vesicles, **6**, 71, 90, 180, 182, 211, 295, 311
Villanueva State Park, 79
volcanic arcs, 60, 81
volcanic ash, 4, 22, 34, 61, 74, 124, **133**, **147**, 150, 213, 217, 237, **247**. *See also* bentonite; pumice; tuff
volcanic bombs, 71, 130, 132, 296, **305**
volcanic necks, **61**
volcanic fields. *See* Cerros del Rio; Jemez; Mogollon-Datil; Navajo; Ocaté; Potrillo; Raton-Clayton; San Juan; Sierra Blanca; Taos Plateau; Zuni-Bandera
volcanoes, 22, 35, 51, 52, 103, 122, 152, 181, 192, 211, 219, 286; composite, 278. *See also* calderas; cinder cones; maar volcanoes; shield volcanoes; stratovolcanoes; volcanic fields

Wagon Mound, 226, 227
Waldo Gulch, 159
Walker Group, 192
Walking Sands Rest Area, 165
Watrous, 228
weathering, 5, **66**, 85, **88**, **93**, 94, **120**, **133**, **154**, **159**, **185**; along joints, 102, 130, 186, 241, 308; of pyrite, 203; rind, **193**. *See also* grus; karst; tafoni
welded tuff. *See* tuff, welded
West Potrillo Basalt, 294
West Potrillo Mountains, 292
Western Interior Seaway, xii, 13, 17, 35, 37, 42, 47, 60, 66, 222, 228, 281, 282

Wheeler Peak, 28, 87, 102, 107, 113, 114, **116**, 120
White Mesa, 16, 63, 138
White Mountains, 219, 306
White Oaks, 193, 238, 240
White Oaks mining district, 283
White Ridge Quartzite, 199
White Rock, 122, 124, 129
White Rock Canyon, 129, 146
White Sands Missile Range, 213
White Sands National Park, **iii**, 7, 26, 189, 191, 203, **206**–**209**, 211, 255
Whitewater-Baldy wildfire, 314
Whitewater Canyon, 314
Window Rock, 31

xenoliths, 296

Yates Formation, 258, 267, 268
Yavapai orogeny, xii, 7
Yavapai Province, 8, 9, 12, 25, 122, 219,
Yeso Formation, 34, 79, **80**, 187, 199, 216, 229, 240, 241, 255, 256, **279**, 280
York Canyon Mine, 222

zinc, 202, 290, 291, 292, 297, 300
Zuni-Bandera volcanic field, 25, 67, 69, 180
Zuni Mountain anticline, 38
Zuni Mountains, 7, 31, 38, 64, 65, 67, 68, 69
Zuni Salt Lake, 305
Zuni Sandstone, **5**, 65, **66**, **67**, **69**
Zuni uplift, 14, 19, 38, 180, 300